The International Library of Bioethics

Volume 111

The *International Library of Bioethics* comprises international and interdisciplinary volumes on foundational and applied issues in bioethics. The series focuses on perennial topics of aging, mental health, preventive medicine, medical research issues, end of life, biolaw, and other areas of bioethics, whilst expanding into other current and future questions involving the living world. The *International Library of Bioethics* examines the role of bioethics against the background of increasing globalization and interdependency of the world's cultures and government.

The series explores meaningful ways to answer the questions of how to direct biotechnology to human and other living things' ends, how to deal with changed values in the areas of religion, society, and culture, and how to formulate a new way of thinking. That is, to create a new, practical bioethics.

The international Library of Bioethics welcomes book proposals representing the broad interest of the series' focus. We especially encourage proposals addressing aspects of changes in biological and medical research and clinical health care, health policy, medical and biotechnology, and other applied ethical areas involving living things, with an emphasis on those interventions and alterations that force us to re-examine foundational issues.

David B. Resnik

The Ethics of Research with Human Subjects

Protecting People, Advancing Science, Promoting Trust

Second Edition

Springer

David B. Resnik
National Institute of Environmental Health Sciences
National Institutes of Health
Durham, NC, USA

ISSN 2662-9186 ISSN 2662-9194 (electronic)
The International Library of Bioethics
ISBN 978-3-031-82756-3 ISBN 978-3-031-82757-0 (eBook)
https://doi.org/10.1007/978-3-031-82757-0

This Springer imprint is published by the registered company Springer Nature Switzerland AG
The registered company address is: Gewerbestrasse 11, 6330 Cham, Switzerland

Acknowledgments

I would like to thank the following people for reading drafts of the manuscript and providing helpful comments: Bruce Androphy, Jessica Berg, Linda Birnbaum, Kyle Brunner, Michael Fessler, Symma Finn, Vanessa Flores, Stavros Garantziotis, Mohammad Hosseini, Brandon Konecny, Amy McGuire, Mark Miller, Liam O'Fallon, Henry Richardson, Adil Shamoo, Elise Smith, Devin Sullivan, Paul Wade, Joshua Warmack, and David Wendler. This research was supported by the Intramural program of the National Institute of Environmental Health Sciences (NIEHS), National Institutes of Health (NIH). It does not represent the views of the NIEHS, NIH, or US government.

Contents

Abbreviations

AAMC	Association of American Medical Colleges
AAHRPP	Association for the Accreditation of Human Research Protection Programs
AI	Artificial intelligence
ARD	Advance research directive
ART	Anti-retroviral therapy
CDC	Centers for Disease Control and Prevention
CHEERS	Children's Environmental Exposure Research Study
CIA	Central Intelligence Agency
CIOMS	Council for International Organizations of Medical Sciences
COI	Conflict of interest
COVID-19	Coronavirus of 2019
DHHS	Department of Health and Human Services
DMC	Decision-making capacity
DNA	Deoxyribonucleic acid
DOD	Department of Defense
DSMB	Data and Safety Monitoring Board
EU	European Union
EPA	Environmental Protection Agency
FDA	Food and Drug Administration
GFPR	General Data Protection Regulation
HIPAA	Health Insurance Portability and Accessibility Act
HIV	Human immunodeficiency virus
HRPP	Human Research Protection Program
IRB	Institutional Review Board
LAR	Legally authorized representative
NASA	National Aeronautics and Space Administration
NBAC	National Bioethics Advisory Commission
NIH	National Institutes of Health
NPRM	Notice of proposed rulemaking

NRC	National Research Council
NSF	National Science Foundation
OHRP	Office of Human Research Protections
ORI	Office of Research Integrity
PHS	Public Health Service
QA/QI	Quality assurance/quality improvement
RCT	Randomized controlled trial
REB	Research Ethics Board
SAE	Serious adverse event
SACHRP	Secretary's Advisory Committee on Human Research Protections
SOPs	Standard operating procedures
SUPPORT	Surfactant, Positive Pressure, and Oxygenation Randomized Trial
UP	Unanticipated problem
WMA	World Medical Association

Chapter 1
Introduction

Abstract Research with human subjects exemplifies the perennial conflict between the good of the individual and the good of society. Policies and procedures that protect the rights and welfare of human subjects may hinder scientific research that benefits society. Due to this conflict, research with human subjects continues to be one of the most controversial topics in bioethics, despite ample government regulation, institutional oversight, and professional guidance. This chapter introduces the reader to a philosophical framework for thinking about these conflicts and issues. The framework, which will be described and defended in greater depth in subsequent chapters in the book, is grounded on widely accepted moral principles and the ethos of trust. The chapter also discusses seven controversial cases that serve to illustrate the importance of developing a philosophical framework for thinking about the ethics of research with human subjects. Cases discussed in the chapter include perinatal HIV prevention trials, the SUPPORT study, the Facebook deception study, xenotransplantation experiments, COVID-19 challenge studies, and the Henrietta Lacks case.

Keywords Ethics · Human subjects · Regulation · Trust · Oversight · Philosophy · Ethical dilemmas

Research with human subjects exemplifies the perennial conflict between the good of the individual and the good of society.[1] Policies and procedures that protect the rights and welfare of human subjects may hinder scientific research that benefits

---

[1] Some authors use the term 'participant' instead of 'subject' because they regard the term 'subject' as demeaning (Chalmers 1999). Though I will sometimes use the term 'participant' in this book, I will, for the most part, stick with the term 'subject.' My reasons for this word choice are threefold. First, federal regulations and other guidance documents use the term 'subject.' Second, the word 'participant' is somewhat misleading because it implies a degree of active participation and collaboration that is not always present in research. Sometimes people are involved in research passively, e.g., when an investigator analyzes existing biological samples to discover relationships between genetics and disease. Third, 'participant' is a feel-good term that can obscure the very real potential for exploitation or mistreatment that can occur when investigators study people. Using the term 'subject' reminds us that people are being studied.

D. B. Resnik, *The Ethics of Research with Human Subjects*, The International Library of Bioethics 111, https://doi.org/10.1007/978-3-031-82757-0_1

society. Due to this conflict, research with human subjects continues to be one of the most controversial topics in bioethics, despite ample government regulation, institutional oversight, and professional guidance. Every few months, a new issue, problem, or scandal enters the media spotlight and elicits predictable responses from concerned citizens, compliance officials, attorneys, and investigators. Concerned citizens, compliance officials, and lawyers often react by calling for additional regulation, oversight, or liability, while investigators usually respond to this familiar refrain that they are already inundated with red tape and that new rules will impede important scientific research without yielding significant benefits for human subjects or society (Klitzman 2015). The goal of this book it to shed some light on these recurring dilemmas and help society move toward practical solutions by developing a framework for thinking about research with human subjects grounded on widely-accepted moral principles and the ethos of trust. To better understand the value of such a framework, let's consider the following cases that illustrate some of the types of ethical problems and issues that arise in research involving human subjects.

1.1 Perinatal HIV Prevention Trials

In the 1990s, many developing[2] nations faced an epidemic of perinatal transmission of the human immunodeficiency virus (HIV), with hundreds of thousands of babies becoming infected from their mothers each year. In response to this public health crisis, the National Institutes of Health (NIH) and Centers for Disease Control and Prevention (CDC) sponsored 18 randomized, controlled trials (RCTs) involving 17,000 women to determine whether a short course of the drug zidovudine would be effective at preventing this type of infection. Sixteen of the trials were conducted in developing nations, including Burkina Faso, Côte d'Ivoire, the Dominican Republic Ethiopia, Kenya, Malawi, South Africa, Tanzania, Thailand, Uganda, and Zimbabwe (Lurie and Wolfe 1997). A drug regimen known as the 076 protocol had been proven to be effective at reducing perinatal HIV transmission, but it used about $800 worth of ziduvudine and was difficult to implement, as it required the administration of the drug throughout pregnancy. As a result, the 076 protocol was not affordable or practical for most people in developing countries. The short course was more affordable and practical, because it used $80 worth of ziduvudine administered to the mother before, during, and after labor and delivery, and to the child at birth (Varmus and

[2] In this book I will use the term 'developing nation' to refer to nations classified as 'least developed' by the United Nations (2023). I realize that word-choice matters here and that there are different ways of referring to groups of nations, such as "low- and middle-income countries," "poor countries," or "global south countries" (Kahn et al. 2022). I find the idea of economic and industrial development to be more useful than these other terms for discussing research ethics issues, because degree of ethical/regulatory oversight tends to be positively associated with development status. By referring to a nation as 'developing' or 'developed' I do not mean to imply any moral judgement about that nation.

Satcher 1997). The purpose of the trials was to determine whether the short course would be more effective than the treatment that patients had available to them in the host nations, which was no antiretroviral therapy. The trials were reviewed and approved by institutional review boards (IRBs) and ethics committees at the NIH and CDC, and in the host nations. They also had the support of the Joint United Nations Programme on HIV/AIDS (Varmus and Satcher 1997).[3]

While the trials were underway, two scientists from the consumer interest group Public Citizen, Peter Lurie and Sydney Wolfe, published an article in the *New England Journal of Medicine* in which they argued that 15 of the trials with placebo control groups were unethical[4] because ziduvudine been proven effective at reducing perinatal HIV transmission. They argued that these trials should have used an active control design, i.e., a design comparing different ziduvudine regimens. They also noted that three of the trials, including two conducted in the US, used an active control design. It was unethical to use placebo control groups, according to Lurie and Wolfe, because this involved withholding an effective treatment from seriously ill mothers and their babies (Lurie and Wolfe 1997). *New England Journal of Medicine* editor Marcia Angell (1997) wrote an editorial supporting Lurie and Wolfe's critique. Angell compared the trials to the infamous Tuskegee Syphilis study[5] and accused the researchers of subscribing to a double-standard: one for developed nations and a different one for developing nations (Angell 1997). NIH Director Harold Varmus and CDC Director David Satcher responded to these critiques by arguing that placebo control groups were necessary to determine whether the short course would work better than the prevailing standard of care in the host countries and to ensure that the trials would be scientifically rigorous and could be completed in a timely fashion. They said that committees reviewing the studies had considered the active control design but decided to include placebo groups because an active control design would require a larger sample than a placebo control design since it would involve detecting smaller differences between treatment groups.[6] Thus, RCTs with active control groups would require more research subjects and take longer to complete than those with placebo control groups. They might also yield inconclusive results (Varmus and Satcher 1997).

---

[3] In the U.S. committees that oversee research with human subjects are called IRBs. In other countries, they may be called research ethics boards (RECs) or research ethics committees (REBs). The points I make concerning IRBs also apply to RECs and REBs.

[4] Some philosophers hold that 'ethics' refers to the standards of conduct for a particular group or profession, e.g., medical ethics, whereas 'morality' refers to more general standards. I will use the terms 'ethics' and 'morality' more or less interchangeably in this book. I do not find the distinction between 'ethics' and 'morality' to be very useful because laypeople usually do not make this distinction. Using a philosophical distinction which does not reflect common practice may be confusing to readers not schooled in this particular way of speaking.

[5] See discussion of this study in Chap. 2.

[6] The sample size needed to yield statistically significant results from a study is inversely proportional to the size of the effect one is attempting to detect: the smaller the effect, the larger the required sample size (Blair and Taylor 2007).

The controversial HIV prevention studies involved a conflict between the welfare of individuals and the good of society (Resnik 1998). The active control design provides more benefits to subjects in the study than the placebo control design but may provide fewer benefits to society because it may take longer to conduct. The placebo design denies benefits to subjects in the control groups but may benefit more people overall because it will take less time to conduct (Varmus and Satcher 1997). As it turned out, the trials proved that a short course of zidovudine is more effective than a placebo in reducing perinatal HIV transmission (Dabs et al. 1999). While the HIV prevention trials did not lead to new government regulations, they did increase awareness among researchers and bioethicists about the ethical issues in conducting research in developing nations (Wendler et al. 2004). The World Medical Association (WMA) also decided to revise its Declaration of Helsinki to clarify its stance on the use of placebos in medical research (Lie et al. 2004). The latest revision of the WMA guidelines states that placebos may be used when:

> [N]o proven intervention exists, the use of placebo, or no intervention, is acceptable; or [w] here for compelling and scientifically sound methodological reasons the use of any intervention less effective than the best proven one, the use of placebo, or no intervention is necessary to determine the efficacy or safety of an intervention and the patients who receive any intervention less effective than the best proven one, placebo, or no intervention will not be subject to additional risks of serious or irreversible harm as a result of not receiving the best proven intervention. Extreme care must be taken to avoid abuse of this option (World Medical Association 2013).

The 2013 revision of the Helsinki Declaration would not have permitted the controversial HIV prevention trials because subjects receiving a placebo might experience serious or irreversible harm. Other international groups have rejected the WMA's restrictive stance on the use of placebos in medical research, and the Food and Drug Administration (FDA) no longer requires that international research submitted to the agency follow the Helsinki Declaration (Lie et al. 2004).

1.2 The SUPPORT Study

Another example of the conflict between the rights and welfare of individuals and the common good occurred in an NIH-sponsored study titled "Surfactant, Positive Pressure, and Oxygenation Randomized Trial (SUPPORT)". The aims of the study were to: (1) understand more about the benefits and risks of using continuous airway pressure (CAP) in helping the lungs of neonates in the intensive care unit (ICU) remain inflated; and (2) to determine the optimal level of blood oxygen saturation for neonates receiving CAP (SUPPORT Study Group 2010). Investigators at 22 institutions enrolled 1316 infants born between 24 and 28 weeks of gestation in the study. The University of Alabama at Birmingham (UAB) was the lead institution for the study's second aim. IRBs at each site approved the study (Resnik 2013). The neonates were randomly assigned to receive blood oxygenation maintained at a range of 85–89% or 91–95%. The standard of care (i.e., accepted medical practice) at that time was to maintain oxygenation at 85–95%, so the range in the study fell within the standard of care. Previous studies had shown that higher levels of oxygenation are associated with an increased risk of retinopathy of prematurity (ROP) and that lower levels are associated with an increased risk of brain damage or death (Resnik 2013). The consent document used in the study informed the parents of the risks of monitoring blood oxygenation levels, but it did not inform the parents of the risks of random assignment to different treatment groups (Resnik 2013).

On March 7, 2013, the Office of Human Research Protections (OHRP), which oversees Department of Health and Human Services (DHHS)-funded research, sent a determination letter to UAB alleging non-compliance with DHHS regulations, otherwise known as the Common Rule (Office of Human Research Protections 2013).[7] The Common Rule,[8] which has been adopted by 19 federal agencies, requires that research subjects or their representatives be informed about reasonably foreseeable risks (Department of Health and Human Services 2017 at 45 CFR 46.116 b2). OHRP claimed that the study violated this requirement because the consent document did not inform the parents about the risks of randomization. According to OHRP, neonates assigned to the higher oxygenation group faced an increased risk of ROP, while those in the lower oxygenation group faced an increased risk of brain damage or death (Office of Human Research Protections 2013). Public Citizen learned about OHRP's determination letter and asked the DHHS to apologize to the parents of the neonates and conduct an independent investigation of the ethics of the study (Resnik 2013).

Numerous scientists and bioethicists, including SUPPORT investigators and NIH Director Francis Collins and Deputy Director for Science, Outreach and Policy, Kathy Hudson, published commentaries in the *New England Journal of Medicine* and other journals defending the study (Carlo et al. 2013; Drazen et al. 2013; Hudson et al. 2013; Magnus and Caplan 2013; Wilfond et al. 2013). SUPPORT defenders argued that the parents did not need to be informed about the risks of randomization because these risks were not reasonably foreseeable because they were speculative and not based on empirical data (Resnik 2013). Indeed, one of the goals of the study was to determine the risks of blood oxygenation at different ranges (Hudson et al. 2013). Since all neonates were maintained at an oxygen range within the standard of care, the study did not expose them to risks that they would have otherwise been exposed to. SUPPORT defenders also criticized OHRP for its response to the study, arguing that the agency was inappropriately interfering with important research (Drazen et al. 2013; Magnus and Caplan 2013).

SUPPORT critics wrote commentaries and letters defending OHRP's actions (Macklin et al. 2013). Critics argued that the risks of the study had not been appropriately disclosed because the research-related risks were different from the risks of the standard of care, which allows physicians to individualize treatment based on

---

[7] NIH is an agency within DHHS.

[8] Unless noted otherwise, references to the Common Rule used in this book will be to the 2017 revision. I will discuss the federal regulations in more depth in Chap. 2.

the patient's medical condition. ICU neonates not participating in the study might be maintained at blood oxygen levels determined to be best for them by their physician. For example, a physician might determine that 90% oxygen is the best level for his or her patient. Neonates in study, however, would be maintained at blood oxygen levels in the two treatment groups, as determined by the study protocol not by their individual medical needs (Macklin and Shepherd 2013; Resnik 2013). The risks of randomization might be speculative, but they were still possible and should have been disclosed to the parents (Macklin and Shepherd 2013).

Although commentators on the SUPPORT study framed the issues in terms of the risks of randomization and the obligation to disclose those risks, the controversy also involved, at a deeper level, a conflict between the rights and welfare of research subjects and the advancement of scientific knowledge. It is conceivable that many parents would not have agreed to enroll their children in this study if they had been told about the risks of the randomization. They might have decided that it would be better for their children to receive individualized treatment provided by a physician instead of treatment based on the study's design. Disclosing the risks of randomization would have provided parents with some information pertinent to their decision to enroll their children in the study but it could have hampered the research by making it more difficult to recruit patients.

One of the key features of a clinical trial is that physicians have very little leeway when it comes to deviating from the study protocol. Physicians treating patients in a clinical trial are supposed to follow rules described by the protocol instead of individualizing treatment. They can deviate from the protocol only to protect the health of the patient, and when they do this, they must inform the IRB. Following the dictates of the study protocol is an important part of rigorous research design since this helps to control the conditions of the experiment (Miller and Brody 2002). If physicians conducting a clinical trial individualized their treatment, it might be impossible to interpret the results of the study, due to uncontrolled variability related to study interventions and procedures. Patients in a clinical trial need to understand how their treatment may differ from what they would receive outside of the study (Macklin and Shepherd 2013).

The controversy concerning the SUPPORT study did not lead to new research regulations, but it increased awareness among investigators and bioethicists about the ethical issues involved in clinical trials that attempt to understand differences between treatment modalities within the standard of care, also known as comparative effectiveness research. Additionally, OHRP held a public meeting on issues related to the study and released some guidance on disclosing reasonably foreseeable risks when conducting research that evaluates different treatments falling within the standard of care. According to the guidance:

> [I]f a research study examining standards of care includes as a purpose evaluating identified risks associated with those standards of care, the identified risks associated with the standards of care being evaluated that are different from the risks of standards of care at least some of the subjects would be exposed to outside of the research study are generally considered by OHRP to be reasonably foreseeable risks of research. Reasonably foreseeable risks must be described to prospective subjects when seeking their informed consent (Office of Human Research Protections 2014).

1.3 Hospital Quality Improvement Research

OHRP oversight also generated considerable controversy in 2007, when the agency investigated a hospital quality improvement research project coordinated by researchers at Johns Hopkins University (JHU) and funded by the Agency for Healthcare Research and Quality (Miller and Emanuel 2008). The project, which began in October 2003, sought to determine whether following an infection control protocol and using a checklist can reduce intravenous (IV) catheter infectious in the intensive care unit (ICU). The protocol included handwashing, cleaning the catheter site with chlorhexidine before insertion, removal of unnecessary catheters, and other procedures (Pronovost et al. 2006). The study was a prospective cohort design, meaning that all participating institutions would follow the protocol and checklist, with data collection before implementation and at 3-month intervals up to 18 months after implementation. One hundred and three ICUs from 67 hospitals in Michigan provided data for the study. Data collection took place from March 2004 to September 2005. The study found that the following the protocol and checklist reduced IV catheter infections significantly: the infection rate fell from 7.7 per 1000 patient days at baseline to 1.4 after 16–18 months (Pronovost et al. 2006). OHRP began investigating the study when the investigators published their results. The JHU IRB had decided that informed consent from the patients was not necessary because the study was exempt from IRB review. In its July 19, 2007, determination letter, OHRP stated that the project was not exempt from IRB review and that JHU had failed to comply with DHHS regulations concerning informed consent (Miller and Emanuel 2008). After receiving this letter, the JHU IRB suspended the project.

While there is little question that the project required IRB review because it did not fit one of the categories for exempt research,[9] the need for informed consent from the patients is debatable (Miller and Emanuel 2008). Hospitals routinely implement quality improvement projects designed to protect health and safety without patient consent. As Miller and Emanuel (2008) note, the hospitals in the JHU study could have implemented the infection control protocol and checklist without participating in the study or obtaining patient consent. Miller and Emanuel (2008) argue that the IRB could have waived consent for the study because the federal regulations allow IRBs to waive consent for studies when: "(1) The research involves no more than minimal risk to the subjects; (2) The waiver or alteration will not adversely affect the rights and welfare of the subjects; (3) The research could not practicably be carried out without the waiver or alteration; and (4) Whenever appropriate, the subjects will be provided with additional pertinent information after participation (Department of Health and Human Services 2017 45 CFR 46.116f)." According to Miller and Emanuel (2008), an IRB could have approved a waiver of consent because the study posed minimal risks to subjects; the research could not have been conducted without a consent waiver because it would have been nearly impossible to obtain consent from all ICU patients or their representatives; and

[9] See Chap. 2 for discussion of exempt research.

waiving consent would not have adversely affected the rights or welfare of the patients.

Since requiring consent for this quality improvement research would probably have prevented it from occurring, the JHU/OHRP controversy also illustrates the dilemma between protecting the rights and welfare of human subjects and promoting scientific research. Although this controversy did not lead to any immediate changes in research regulations, it raised awareness about ethical and oversight issues in hospital quality improvement research (Kass et al. 2013).

1.4 Henrietta Lacks

A widely used biomedical research tool known as HeLa, the first human cell line successfully grown in culture, has been used by millions of scientists and students across the world since the mid-1950s. Journalist Rebecca Skloot (2010) became curious about this anonymous cell line and wanted to learn whom it came from. She discovered that Henrietta Lacks, an African American woman who underwent treatment for cervical cancer at Johns Hopkins Hospital in 1951, provided the cell line. Skloot interviewed Lacks' family and found out that the tissues used to produce the cell line were extracted from Lacks' left over tumor tissue without her consent, which was at that time a common practice. For many decades, biomedical scientists have been using biological samples left over from surgical procedures or clinical tests for research projects. Though consent must be obtained to enroll someone in a research study involving the collection of biological samples, it need not be obtained to use de-identified[10] biological samples which are left over from medical procedures or tests.

Skloot also learned that Lacks' family received no financial compensation for the tissues, which was also not unusual. Skloot's book became a popular best-seller, *The Immortal Life of Henrietta Lacks*, which was later made into a film. Skloot decided to donate some of the profits from the book and film to a non-profit foundation she established, the Henrietta Lacks Foundation (2017), which provides support to individuals and families adversely impacted by research without consent. In 2013, the NIH reached an agreement with Lacks' family concerning access to genomic data from the cell line. The agreement gives the family control over access to the data and recognition on scientific papers (National Institutes of Health 2013). In 2023, the Lacks family reached a legal settlement with Thermo Fisher, a company that profited from selling Hela Cells to researchers and research institutions (Wadman 2023). Another lawsuit against Ultragenyx Pharmaceuticals is still being litigated (Dye and Menikoff 2024).

[10] 'De-identified' means that personal identifiers, such as name and address, have been removed.

The Lacks case focused national attention using biological specimens in research, which also can involve a conflict between the rights of individuals and the good of society. Proposed revisions to the Common Rule published in 2015 defined human biological specimens as research subjects, which would have mandated consent for most research involving human biological samples (Department of Homeland Security et al. 2015). While this provision would have enhanced respect for individual autonomy, it was dropped in response to concerns that it would interfere with important biomedical research.[11]

It almost impossible to think about the Lacks case without considering its racial implications (Merelli 2023). As will be discussed at greater length in Chap. 2, African Americans have often been the victims of unethical and exploitative experiments, such as the Tuskegee Syphilis Study, and the Lacks case belongs to this historical pattern. However, while although Lacks was exploited by researchers and private companies, it is not clear that she was exploited because of her race, since it was common practice at that time to use left over tissue for research without obtaining consent. Indeed, in Chap. 2, we will a very similar case involving a white man, John Moore.

1.5 The Facebook Behavioral Manipulation Experiment

Informed consent was also a major concern in a social psychology experiment conducted using the social network Facebook in January 2012. The investigators manipulated the newsfeeds of 689,003 randomly selected Facebook users to determine whether emotional states can be transferred by social interactions in the absence of verbal cues. They were able to show that users who receive positive feeds are more likely to post positive feeds, while those who receive negative feeds are more likely to post negative ones. The effect was small but significant (Kramer et al. 2014). The authors of the study said that Facebook's user agreement constituted informed consent for the research. All Facebook users must sign a user agreement to participate in this social media forum. The agreement gives the company permission to collect data about the user and conduct research (Facebook 2017). The research generated headlines—and controversy—when it was published in *Proceedings of the National Academy of Sciences* 2014. The journal published an editorial expressing concern about the study's informed consent process, which was below the standards set by the Common Rule. However, the editorial noted that Facebook, as a private company, was not obligated to follow the Common Rule (Verma 2014). While the lead author, Alan Kramer, was a Facebook employee, the two other authors, James Guillory and Jeffrey Hancock, were associated with Cornell University, which follows the Common Rule. However, Cornell's IRB

[11] We will discuss consent for the use of biological specimens in more depth in Chap. 5.

determined that the federal regulations did not apply to the study, because Guillory and Hancock were not engaged in research with human subjects as they had no interactions with the participants or their private information (Klitzman and Applebaum 2014). While some defended the study because it provided useful information with virtually no risk to the participants, others said the study was unethical because consent via the user agreement did not provide Facebook users with enough information concerning their potential participation in research (Goel 2014; Klitzman and Applebaum 2014).

The Facebook case is an example of significant gap in the U.S. federal research regulations. The human research regulations apply to research funded by federal agencies that follow the Common Rule, or research submitted to the Food and Drug Administration (FDA) or Environmental Protection Agency (EPA) in support of product approvals. Additionally, most institutions that receive federal funding have agreed to apply the Common Rule to all human subjects research they oversee, even research that is not specifically covered by federal regulations (Shamoo and Resnik 2022). Some have argued that the U.S. should close this gap in the federal regulations to ensure uniform protections for all human subjects (Shamoo and Schwartz 2008). Others, however, have pointed out extending the scope of the federal regulations could impose significant burdens on private companies who are engaged in low-risk research, such as public opinion surveys and marketing studies (Resnik 2008).

1.6 COVID-19 Challenge Studies

In the fall of 2019, a virus named SARS-CoV-2 (sever acquired acute respiratory infection, coronavirus 2) spread quickly from Wuhan, China to the rest of the world and caused a global pandemic. Over seven million people died from COVID-19 (coronavirus disease of 2019) and hundreds of millions of people became seriously ill or were hospitalized. COVID-19 caused a public health crisis on a scale not seen since the 1918 Spanish flu pandemic. (World Health Organization 2024).

When the pandemic began, there were no effective treatments or vaccines for COVID-19, so most countries implemented draconian public health policies, such as closures of schools and businesses, curfews, and travel restrictions, to try to stop the spread of the disease and save lives (Resnik 2021). It soon became clear that COVID-19 posed a far greater threat to people who were elderly or infirm than to those who were young and healthy. While the case fatality rate for people 80 years old or older was nearly 15%, the case fatality rate for children was less than 0.5%, and most of the children who died from the disease were already seriously ill (Wu and McGoogan 2020). Most children and young adults who contracted COVID-19 experienced mild symptoms or none at all (Resnik 2021).

Although vaccine clinical trials began in the spring of 2020, they were not expected to be completed until later in the year, and vaccines were not likely to be widely available until early 2021. Vaccine trials normally take from 12 to 18 months to complete because investigators must wait until enough people have been exposed to the pathogen as a result of normal life activities so that statistically valid comparisons can be made between experimental and control groups (Resnik 2021).

In light of these circumstances, some scientists and ethicists argued that researchers should accelerate the vaccine development process by conducting studies that intentionally expose healthy, young volunteers to COVID-19. The volunteers, who would not have previously had COVID-19, would receive a vaccine or placebo before being inoculated with the virus. The purpose of these studies would be to identify promising vaccine candidates and cut months off the time needed to develop an effective vaccine (Eyal et al. 2020). Proponents of these studies argued that the risks imposed on volunteers would be justified in terms of the potential benefits. The risks of death or serious hospitalization for the study population, though not trivial, would still be orders of magnitude lower than the risks for general population or vulnerable groups. Proponents argued that the trials could potentially save thousands of lives, even if they shortened the length of vaccine development by only a few weeks (Eyal et al. 2020). When this proposal became widely known, thousands of people said they would volunteer for these studies to help in the fight against COVID-19. A non-profit organization, 1Day Sooner (2024), was formed to promote participation in vaccine challenge trials to accelerate the development of vaccines and treatments for infectious diseases. The World Health Organization (2020) developed criteria for ethical acceptability of COVID-19 challenge studies.

However, there was—and still is—significant opposition to these studies.[12] Critics argued that these challenge studies were too risky, especially since there were no effective treatments for COVID-19 at the time; that the benefits of the studies were speculative; and that exposing healthy volunteers to a dangerous virus as part of a medical experiment could undermine public trust in research if people died in the study (Kahn et al. 2020).

Although vaccine developers, such as Pfizer, Moderna, Johnson and Johnson, shied away from sponsoring COVID-19 challenge studies, the UK government did not. In the fall of 2020, the UK Vaccine Taskforce approved the funding of COVID-19 challenge studies. When the first COVID-19 vaccines were approved in early 2021, many argued that the COVID-19 challenges studies were no longer needed. Others argued, however, that there were still compelling reasons to do challenge studies, since these experiments could provide useful information about the pathophysiology of the disease and its clinical manifestations under controlled conditions (Eyal 2024).

In May of 2022, researchers from the Imperial College of London published the results of the first COVID-19 challenge study (Killingsley et al. 2022). In the study, 34 volunteers (age 18–30) were inoculated with COVID-19. 53% of the participants became infected, with a peak viral load occurring within five days of exposure. Of these, 89% experienced mild to moderate symptoms and 11% had no symptoms. There were no serious adverse events (Killingsley et al. 2022). According to the

[12] In Chap. 7, I will examine the ethics of these and other challenge studies in greater depth.

investigators, the study helped to better establish "viral kinetics over the course of primary infection with SARS-CoV-2" which has "implications for public health recommendations and strategies to affect SARS-CoV-2 transmission (Killingsley et al. 2022: 1031)." In September 2024, the investigators published the results of cognition and memory tests performed on the volunteers at 30, 90, 180, 270, and 360 days. The investigators found that infection with SARS-CoV-2 was associated with small deficits in memory and cognition that persisted for at least a year. The participants earned up to $6100 for completing the study (Trender et al. 2024). Investigators at the Imperial College of London (2024) are currently recruiting healthy volunteers for COVID-19 challenge studies. Some have argued that the debate about the COVID-19 challenge studies represents an important turning point in vaccine development policy and that, in the future, challenge trials may become an accepted method part of the vaccine development process (Eyal 2024).

1.7 Xenotransplantation Experiments

On January 7, 2022, surgeons at the University of Maryland School of Medicine performed the world's first pig-to-human heart transplant on David Bennett, a 57-year-old male with heart disease. Although the operation was initially deemed a success, Bennett's condition deteriorated, and he died on March 9, 2022 (Wang et al. 2022; Associated Press 2022). The heart Bennett received came from a genetically modified, "humanized" pig. The pig's genome had been modified to code for human cell-surface proteins, which would make its organs less likely to be attacked by the human immune system (Wang et al. 2022). The FDA authorized the procedure under its expanded access, compassionate use provisions, which allow the agency to approve an experimental device that has not undergone clinical trials for a patient with a life-threatening medical condition for which there are no available alternative treatments (Singh et al. 2022; Food and Drug Administration 2024). This was not the first time that doctors had performed cardiac xenotransplantation on a human being. In 1984, an infant in California known as Baby Fae lived for 21 days after receiving a baboon heart (Associated Press 2022).

Since about 5600 people in the US die each year while waiting for a new organ (Donate Life America 2024), research on the use of humanized pig organs for transplantation offers substantial benefits for society. However, the research raises several ethical concerns (Johnson 2022). First, pigs carry endogenous viruses that could infect organ recipients and enter the larger human population. Although these viruses might not cause the pigs any health problems, because the pigs have developed immunity to them and/or have evolved to live with them, the viruses could be very dangerous to human beings, who probably lack immunity (Rollin 2020). As the world learned during the COVID-19 pandemic, zoonoses (i.e., animal diseases that can infect humans) pose a major threat to humanity. Although researchers are taking various measures to minimize the risk of zoonoses, such as testing pigs for viruses and rearing them in sterile environments, these risks cannot be eliminated entirely. The U.S. research regulations (discussed in Chap. 2) focus on risks to subjects and do not provide the IRB with guidance on how to address societal risks related to research. In Chap. 7, we will consider the issue of whether (and how) IRBs should address these risks.

Second, to protect the human population from risks of zoonoses, organ recipients must agree to lifelong monitoring, which raises issues concerning the validity of consent, since recipients may not fully understand the implications of this monitoring, and the enforceability of the consent agreement, since it may be difficult to locate or control recipients who decide to shirk their responsibilities (Johnson 2022).

Third, the research raises animal welfare concerns because the pigs will be killed to harvest their organs. People who oppose killing pigs for food may also oppose killing pigs for medical purposes (Rollin 2020). Animal welfare concerns are not addressed by the US human research regulations and there are not unique to xenotransplantation. Most new drugs and medical devices are tested in animals before they are tested on people (Shamoo and Resnik 2022).

Fourth—and this issue is also not unique to xenotransplantation—the research raised issues about experimenting on brain-dead people, since pig organs were tested on brain-dead patients first (Associated Press 2023).[13] Research on brain dead people raises challenging ethical and regulatory questions because U.S. research regulations (discussed in Chap. 2) apply only to living individuals. Since research on brain-dead people falls outside the jurisdiction of the IRB, there are questions about how it should be overseen. In the absence of a clear legal framework for research on brain dead people, some institutions have formed special committees to review this research or asked IRBs to review it (Madhusoodanan 2024).

1.8 Overview of This Book

As one can see from the cases discussed above, research with human subjects raises complex and controversial ethical, social, and legal issues. In most of these cases, conscientious and well-informed scientists and bioethicists reached different conclusions concerning ethical and regulatory questions being raised. The researchers, as far as we know, were acting thoughtfully and responsibly and were not egregiously violating laws or ethical standards. They were trying to do what they considered to be ethical research to obtain important results for science and society. Moreover, with exception of the Lacks case, an extensive system of regulating and managing human research with human subjects was already in place when these controversies erupted.

Disagreements occurred at several levels. At one level, the disagreements were about whether a particular study was or was not ethical or would be ethical if it were

---

[13] A brain-dead person is an individual whose brain that no longer functions. In the U.S. and many other countries, brain-dead people are legally dead. However, their bodies can be kept alive by means of medical technology.

modified in a particular way. At another level, the disagreements revolved around the interpretations of regulations and guidelines and their applicability to specific studies. At a third level, the disagreements were also about the kinds of policies and oversight systems that should be in place to protect human research subjects. The overarching issue—the good of individual versus the common good—impacted these disputes explicitly or implicitly. Scientists and bioethicists came down on different sides of the issue, with some emphasizing the importance of protecting research subjects and others the importance of advancing scientific knowledge and benefitting society. Looming in the background were questions about how decisions made by investigators, IRBs, institutions, or oversight officials would impact the public's trust in the research enterprise.

Questions concerning the resolution of moral conflicts in research with human subjects will take center-stage in this book. In most of the ethical controversies related to research with human subjects, both sides agree that it is important to protect individual rights and wellbeing and promote scientific research. The disagreements usually have to do with emphasis or priority-setting. The aim of this book is to develop a philosophical framework for thinking about the ethics of research with human subjects that can be applied to ethical and policy dilemmas like those discussed above. The central insight of this framework is that the goal of promoting trust should play an essential role in our thinking about the ethics of research with human subjects.

Some argue that there is no need to develop a philosophical framework for thinking about the ethics of research with human subjects because one already exists, namely, the three ethical principles—respect for persons, beneficence, and justice—articulated in the *Belmont Report* (National Commission 1979). While I think that the *Belmont* principles provide useful guidance for the conduct of research with human subjects, I will argue in Chap. 2 that they are not able to adequately address difficult ethical dilemmas because they do not include a method for prioritizing the principles when they conflict, and most controversial issues involve such conflicts. Likewise, the seven ethical principles articulated in a widely cited article by Emanuel et al. (2000) provide useful guidance for research with human subjects but also fall short of the mark because they do include a way of adjudicating conflicts among principles.

Some philosophers are skeptical of the entire project of developing a philosophical framework for thinking about the ethics of research with human subjects. According to Alan Wertheimer:

> Research ethics is a practical discipline that has developed in response to specific historical events. It is not built on any general or overarching theory. The reining principles…respond to the desire to square the genuine need for biomedical research with the protection of human subjects in the context of a history that contains several episodes of serious abuse and exploitation of human subjects (Wertheimer 2011, p. 3).

While I agree with Wertheimer that the current oversight system largely reflects ad hoc responses to historical events, I think that one can develop a philosophical framework that helps us to understand and critique the current system. While the framework need not be an "overarching theory" in any deep sense, it should provide us with some guidance concerning ethical and policy dilemmas.

When I wrote the first edition of this book in 2017, there were very books on the ethics of research with human subjects that developed a comprehensive, philosophical framework for thinking about the issues and dilemmas.[14] There were articles, anthologies, and guidance documents that discussed the foundations of research ethics, but there were no books on the subject.[15]

Four years after my book first appeared in print, Alex London (2022) published a well-received book on the philosophical foundations of ethical research with human subjects.[16] In this book, London argues that (1) scientific research is a cooperative enterprise that should promote common societal goals and (2) the current system of regulatory oversight should be reformed to better reflect a commitment to the common good. According to London:

> Research is a scheme of social cooperation that serves a public purpose grounded in considerations of justice. One such consideration of justice concerns the claims that community members have on the goals and ends that are advanced by the research enterprise. Following the egalitarian research imperative, the public purpose of research is to generate the knowledge necessary to bridge gaps in the capacity of the basic social institutions of a community—such as its system of public health and clinical medicine—to safeguard and advance the basic interests of that community's members (London 2022: 251–252).

While I think London's book provides readers with some valuable insights, I believe my book still has considerable value as an alternative to London's for two reasons. First, London's book is highly idealistic and abstract and does not adequately deal with the historical, economic, political, and social aspects of modern science, or practical matters related to IRB review, research oversight, informed consent, study design, methodology, and research integrity. My book is more pragmatic than London's and encompasses both theory and applications. Second, the theory London defends in his book is based on an egalitarian approach to questions of morality and political philosophy. While egalitarianism is a highly respectable and influential philosophical theory (see discussion in Chap. 3), it is also controversial, because it entails restrictions on individual rights and redistribution of wealth to achieve egalitarian social goals, such as equality of opportunity and social justice (Arneson 2013). As a result of his strong commitment to egalitarianism, London does not

---

[14] Some books worth mentioning include books by Veatch (1987), Levine (1988), Brody (1998), and Wendler (2010). Each of these books has strengths and limitations. Veatch's depicts research as a partnership between the subject and investigator but does not say much about what that partnership involves. Levine provides useful guidance related to interpreting and applying ethical rules and the U.S. federal regulations but does not provide a philosophical foundation for these rules and regulations. Wendler describes and defends a philosophical framework for thinking about the ethics research on children, but it is not clear how the framework does not extend far beyond pediatric research.

[15] Three excellent collections of articles include Emanuel et al. (2008), Coleman et al. (2015), and Iltis and Mackay (2024).

[16] Elsewhere (Resnik 2022), I have published a review and critique of London's book.

adequately consider opposing views of the common good, which weakens his analysis of contested topics in research ethics, such as the commercialization of research, using placebos in research, and ancillary care obligations. The approach I defend in this book is likely to have greater appeal than London's because it is grounded on widely-accepted ethical notions, such as the *Belmont* principles (discussed in Chap. 2), and the idea of trust (discussed on Chap. 4).

My plan for the books is as follows. In Chap. 2, I will describe the historical events that have shaped ethical rules and policies concerning research with human subjects. I will argue that the social and political response to these events has been to develop an oversight system to restore and maintain public trust in research. An important question that emerges from this chapter is whether the current system provides too much or too little protection for human research subjects. To answer this question, we require a philosophical framework that can help us deal with ethical and regulatory dilemmas in research with human subjects.

In Chap. 3, I will consider whether moral theories can provide us with the guidance we are seeking. I will argue that since no single theory satisfactorily accounts for our moral intuitions and addresses crucial objections, some form of moral pluralism is the most reasonable approach. The type of pluralism I adopt, Beauchamp and Childress' (2019) four principles approach, will be familiar to many readers. Since pluralism still leaves us with questions concerning priority-setting when basic principles or values conflict in research with human subjects, we require some additional guidance for dealing with ethical and policy dilemmas.[17]

In Chap. 4, I will describe and defend my trust-based approach. The key premise of the framework is that research is founded on trust: trust between participants and investigators and institutions; between communities and investigators and institutions; among investigators; and between the public and the research enterprise. I will argue that the goal of promoting trust augments other moral principles, and that reflecting on the nature and importance of trust can often help us to decide how resolve ethical and policy dilemmas in research with human subjects. Although promoting trust does not solve all ethical dilemmas in research, it can be an essential tool in helping us to develop workable solutions.

In Chaps. 5, 6, 7, 8, and 9, I will apply the trust framework to ethical issues in research with human subjects, including:

- Obtaining and documenting informed consent
- Determining whether a person has the capacity to consent
- Conducting research without consent

[17] A moral principle is a general rule for conduct, e.g., "do not lie" or "keep your promises." A moral value is something that is morally good or worthwhile. Things have value for their own sake and not as a means to something else are intrinsically valuable. For example, most people would view happiness as intrinsically valuable. Things that have value as a means to something else are extrinsically valuable. For example, most people would regard money as valuable not in itself but for what we buy with it. Some things, such as knowledge and health, may be intrinsically and extrinsically valuable (Timmons 2002).

- Using general (or blanket) consent forms
- Using opt-out consent procedures
- Deceiving research subjects
- Paying people for research participation
- Safeguarding privacy and confidentiality
- Sharing data and biospecimens
- Assessing and balancing risks and benefits
- Minimizing risks
- Defining minimal risk
- Choosing the appropriate clinical trial design (e.g., whether to use a placebo control group)
- Conducting challenge studies on healthy volunteers
- Upper limits for risks to healthy volunteers
- Dealing with risks to third parties, communities, and society
- Sharing individualized results with research participants
- Sharing research benefits with subjects, communities, and populations
- Providing ancillary care to subjects and post-trial access to medications
- Enrolling and protecting vulnerable subjects in research
- Issues with special populations, such children, pregnant people, and adults with impaired decision-making capacity
- Promoting research integrity
- Investigating and reporting misconduct and non-compliance
- Disclosing and managing conflicts of interest in research

In Chap. 10, I will summarize the book's main arguments and conclusions and discuss some proposals for reforming the current oversight system and areas of further study.

Before proceeding further, two caveats are in order. First, policy discussions in the book will focus, for the most part, on U.S. regulations, and, to a lesser extent, on regulations in other countries and international guidelines. Although some may regard the U.S.-focus as narrow-minded, the ethical and legal standards for research with human subjects in the U.S. are similar to those found in other countries (Office of Human Research Protections 2024). Moreover, the U.S. and other countries are dealing with similar issues, and the standards adopted in the U.S. have considerable international influence (Brody 1998). Second, although I will discuss legal and regulatory issues in this book, nothing I say should be taken as legal advice. I am presenting my views to provide investigators and ethicists with some additional tools and perspectives for thinking about research with human subjects. Readers who are interested in legal advice should consult an attorney. Third, the opinions expressed in this book are my own and do not represent the views of the NIEHS, NIH, DHHS, or U.S. government.

References

1Day Sooner. 2024. About 1Day Sooner. Available at: https://www.1daysooner.org/about/. Accessed 3 May 2024.

Angell, M. 1997. The ethics of clinical research in the third world. *New England Journal of Medicine* 337 (12): 847–849.

Arneson, R. 2013. Egalitarianism. Stanford Encyclopedia of Philosophy. Available at: https://plato.stanford.edu/entries/egalitarianism/. Accessed 6 May 2024.

Associated Press. 2022. A man who got the 1st pig heart transplant has died after 2 months. NPR, 9 March. Available at: https://www.npr.org/2022/03/09/1085420836/pig-heart-transplant. Accessed 5 May 2024.

———. 2023. Pig kidney works in a brain-dead man for over a month, a step toward animal-human transplants. NBC News, 16 August. Available at: https://www.nbcnews.com/health/health-news/pig-kidney-transplant-human-rcna100098. Accessed 12 May 2024.

Beauchamp, T.L., and J. Childress. 2019. *Principles of biomedical ethics*. 8th ed. New York: Oxford University Press.

Blair, R., and R. Taylor. 2007. *Biostatistics for the health sciences*. Englewood Cliffs: Prentice-Hall.

Brody, B.A. 1998. *The ethics of biomedical research: An international perspective*. New York: Oxford University Press.

Carlo, W.A., E.F. Bell, M.C. Walsh, and SUPPORT Study Group of the Eunice Kennedy Shriver NICHD Neonatal Research Network. 2013. Oxygen-saturation targets in extremely preterm infants. *New England Journal of Medicine* 368 (20): 1949–1950.

Chalmers, I. 1999. People are "participants" in research. *British Medical Journal* 318 (7191): 1141.

Coleman, C.H., J.A. Menikoff, J.A. Goldner, and N.N. Dubler, eds. 2015. *The ethics and regulation of research with human subjects*. 2nd ed. Durham: Carolina Academic Press.

Dabs, F., P. Msellati, N. Meda, C. Welffens-Ekra, B. You, O. Manigart, V. Leroy, A. Simonon, M. Cartoux, P. Combe, A. Ouangré, R. Ramon, O. Ky-Zerbo, C. Montcho, R. Salamon, C. Rouzioux, P. Van de Perre, and L. Mandelbrot. 1999. 6-month efficacy, tolerance, and acceptability of a short regimen of oral zidovudine to reduce vertical transmission of HIV in breastfed children in Côte d'Ivoire and Burkina Faso: A double-blind placebo-controlled multicentre trial. DITRAME Study Group. DIminution de la Transmission Mère-Enfant. *Lancet* 353 (9155): 786–792.

Department of Health and Humans Services. 2017. Protection of Human Subjects. Code of Federal Regulations, Title 45, Part 46.

Department of Homeland Security, Department of Agriculture, Department of Energy, National Aeronautics and Space Administration, Department of Commerce, Social Security Administration, Agency for International Development, Department of Justice, Department of Labor, Department of Defense, Department of Education, Department of Veterans Affairs, Environmental Protection Agency, Department of Health and Human Services, National Science Foundation, and Department of Transportation. 2015. Notice of proposed rulemaking. Federal policy for the protection of human subjects. *Federal Register* 80 (173): 53933–54061.

Donate Life America. 2024. Organ, eye and tissue donation statistics. Available at: https://donatelife.net/donation/statistics/#:~:text=5%2C600%20people,on%20the%20transplant%20waiting%20list. Accessed 6 May 2024.

Drazen, J.M., C.G. Solomon, and M.F. Greene. 2013. Informed consent and SUPPORT. *New England Journal of Medicine* 368: 1929–1931.

Dye, T.D., and J.A. Menikoff. 2024. Biospecimen research and the law. *Science* 385 (6710): 695.

Emanuel, E.J., D. Wendler, and C. Grady. 2000. What makes clinical research ethical? *Journal of the American Medical Association* 283 (20): 2701–2711.

Emanuel, E.J., C. Grady, R.A. Crouch, R.K. Lie, F.G. Miller, and D. Wendler, eds. 2008. *The Oxford handbook of clinical research ethics*. New York: Oxford University Press.

Eyal, N. 2024. Research during pandemics and epidemics. In *The Oxford handbook of research ethics*, ed. A.S. Iltis and D. MacKay, 425–444. New York: Oxford University Press.

Eyal, N., M. Lipsitch, and P.G. Smith. 2020. Human challenge studies to accelerate coronavirus vaccine licensure. *The Journal of Infectious Diseases* 221 (11): 1752–1756.

Facebook. 2017. Terms of service. Available at: https://www.facebook.com/legal/terms. Accessed 3 May 2024.

Food and Drug Administration. 2024. Expanded access for medical devices. Available at: https://www.fda.gov/medical-devices/investigational-device-exemption-ide/expanded-access-medical-devices#compassionate. Accessed 5 May 2024.

Goel, V. 2014. As data overflows online, researchers grapple with ethics. New York Times, August 14, 2014:B1.

Henrietta Lacks Foundation. 2017. Available at: http://henriettalacksfoundation.org/. Accessed 3 May 2024.

Hudson, K.L., A.E. Guttmacher, and F.S. Collins. 2013. In support of SUPPORT – A view from the NIH. *New England Journal of Medicine* 368 (25): 2349–2351.

Iltis, A., and D. Mackay, eds. 2024. *The Oxford handbook of research ethics*. New York: Oxford University Press.

Imperial College of London. 2024. COVID-19 human challenge study: COVHIC002. Available at: https://www.imperial.ac.uk/infectious-disease/research/human-challenge/covhic002/. Accessed 5 May 2024.

Johnson, L.S. 2022. Xenotransplantation: Three areas of concern. Hastings Bioethics Forum. Available at: https://www.thehastingscenter.org/xenotransplantation-three-areas-of-concern/#:~:text=Using%20pigs%20and%20nonhuman%20primates,shortage%20is%20a%20formidable%20problem. Accessed 5 May 2024.

Kahn, J.P., L.M. Henry, A.C. Mastroianni, W.H. Chen, and R. Macklin. 2020. Opinion: For now, it's unethical to use human challenge studies for SARS-CoV-2 vaccine development. *Proceedings of the National Academy of Sciences USA* 117 (46): 28538–28542.

Kass, N., R. Faden, S. Goodman, S. Tunis, P. Pronovost, and T. Beauchamp. 2013. The research-treatment distinction: A problematic approach for determining which activities should have ethical oversight. *Hastings Center Report, Special Report* 43 (1): S4–S15.

Khan, T., S. Abimbola, C. Kyobutungi, and M. Pai. 2022. How we classify countries and people-and why it matters. *BMJ Global Health* 7 (6): e009704.

Killingley, B., A.J. Mann, M. Kalinova, A. Boyers, N. Goonawardane, J. Zhou, K. Lindsell, S.S. Hare, J. Brown, R. Frise, E. Smith, C. Hopkins, N. Noulin, B. Löndt, T. Wilkinson, S. Harden, H. McShane, M. Baillet, A. Gilbert, M. Jacobs, C. Charman, P. Mande, J.S. Nguyen-Van-Tam, M.G. Semple, R.C. Read, N.M. Ferguson, P.J. Openshaw, G. Rapeport, W.S. Barclay, A.P. Catchpole, and C. Chiu. 2022. Safety, tolerability and viral kinetics during SARS-CoV-2 human challenge in young adults. *Nature Medicine* 28 (5): 1031–1041.

Klitzman, R.L. 2015. *The ethics police? The struggle to make human research safe*. New York: Oxford University Press.

Klitzman, R.L., and P.S. Appelbaum. 2014. Facebook's emotion experiment: Implications for research ethics. The Hastings Center, *Bioethics Forum*, 21 July. Available at: https://www.thehastingscenter.org/facebooks-emotion-experiment-implications-for-research-ethics/. Accessed 3 May 2024.

Kramer, A.D., J.E. Guillory, and J.T. Hancock. 2014. Experimental evidence of massive-scale emotional contagion through social networks. *Proceedings of the National Academy of Science U.S.A.* 111 (24): 8788–8790.

Levine, R.J. 1988. *Ethics and the regulation of clinical research*. 2nd ed. New Haven: Yale University Press.

Lie, R.K., E. Emanuel, C. Grady, and D. Wendler. 2004. The standard of care debate: The declaration of Helsinki versus the international consensus opinion. *Journal of Medical Ethics* 30 (2): 190–193.

London, A.J. 2022. *Philosophical foundations of research ethics*. New York: Oxford University Press.

Lurie, P., and S.M. Wolfe. 1997. Unethical trials of interventions to reduce perinatal transmission of the human immunodeficiency virus in developing countries. *New England Journal of Medicine* 337 (12): 853–856.

Macklin, R., and L. Shepherd. 2013. Informed consent and standard of care: What must be disclosed. *American Journal of Bioethics* 13 (12): 9–13.
Macklin, R., L. Shepherd, A. Dreger, A. Asch, F. Baylis, H. Brody, L.R. Churchill, C.H. Coleman, E. Cowan, J. Dolgin, J. Downie, R. Dresser, C. Elliott, M.C. Epright, E.K. Feder, L.H. Glantz, M.A. Grodin, W. Hoffman, B. Hoffmaster, D. Hunter, A.S. Iltis, J.D. Kahn, N.M. King, R. Kraft, R. Kukla, L. Leavitt, S.E. Lederer, T. Lemmens, H. Lindemann, M.F. Marshall, J.F. Merz, F.H. Miller, M.E. Mohrmann, H. Morreim, M. Nass, J.L. Nelson, J.H. Noble Jr., E. Reis, S.M. Reverby, A. Silvers, A.C. Sousa, R.G. Spece Jr., C. Strong, J.P. Swazey, and L. Turner. 2013. The OHRP and SUPPORT–another view. *New England Journal of Medicine* 369 (2): e3.
Madhusoodanan J. 2024. The allure and dangers of experimenting on brain-dead bodies. Undark, 25 March. Available at: https://undark.org/2024/03/25/experimenting-with-brain-dead-bodies/. Accessed 12 May 2024.
Magnus, D.M., and A.L. Caplan. 2013. Risk, consent, and SUPPORT. *New England Journal of Medicine* 368 (20): 1864–1865.
Merelli, A. 2023. Henrietta lacks settlement hailed by experts as step toward correcting medicine's racist history. Stat News, 1 August. Available at: https://www.statnews.com/2023/08/01/henrietta-lacks-family-settlement-hela-cells/. Accessed 3 May 2024.
Miller, F.G., and H. Brody. 2002. What makes placebo-controlled trials unethical? *American Journal of Bioethics* 2 (2): 3–9.
Miller, F.G., and E.J. Emanuel. 2008. Quality improvement research and informed consent. *New England Journal of Medicine* 358 (8): 765–767.
National Commission for the Protection of Human Subjects of Biomedical or Behavioral Research. 1979. *The Belmont report: Ethical principles and guidelines for the protection of human subjects of research*. Washington, DC: Department of Health, Education, and Welfare.
National Institutes of Health. 2013. NIH, Lacks family reach understanding to share genomic data of HeLa cells. Available at: http://www.nih.gov/news-events/news-releases/nih-lacks-family-reach-understanding-share-genomic-data-hela-cells. Accessed: January 22, 2025.
Office of Human Research Protections. 2013. Non-compliance determination letter to the University of Alabama at Birmingham, March 7, 2013. Available at: http://www.hhs.gov/ohrp/detrm_letrs/YR13/mar13a.pdf. Accessed 3 May 2024.
———. 2014. Draft guidance on disclosing reasonably foreseeable risks in research evaluating standards of care. Available at: http://www.hhs.gov/ohrp/newsroom/rfc/comstdofcare.html#. Accessed 3 May 2024.
———. 2024. International compilation of human research standards. Available at: https://www.hhs.gov/ohrp/international/compilation-human-research-standards/index.html. Accessed 5 May 2024.
Pronovost, P., D. Needham, S. Berenholtz, D. Sinopoli, H. Chu, S. Cosgrove, B. Sexton, R. Hyzy, R. Welsh, G. Roth, J. Bander, J. Kepros, and C. Goeschel. 2006. An intervention to decrease catheter-related bloodstream infections in the ICU. *New England Journal of Medicine* 355 (26): 2725–2732.
Resnik, D.B. 1998. The ethics of HIV research in developing nations. *Bioethics* 12 (4): 286–306.
———. 2008. Closing loopholes in the federal research regulations: Some practical problems. *American Journal of Bioethics* 8 (11): 6–8.
———. 2013. What is a reasonably foreseeable risk? The SUPPORT study controversy. *Journal of Clinical Research Best Practices* 9 (8) Available at: Ttps://www.researchgate.net/publication/258822641_What_Are_Reasonably_Foreseeable_Risks. Accessed 3 May 2024.
———. 2021. *Precautionary reasoning in environmental and public health ethics*. Cham: Springer.
———. 2022. Philosophical foundations of human research ethics: Review of London's philosophical foundations of research ethics. *Perspectives in Biology and Medicine* 65 (3): 499–513.
Rollin, B. 2020. Ethical and societal issues occasioned by xenotransplantation. *Animals* 10: 1695.
Shamoo, A.E., and D.B. Resnik. 2022. *Responsible conduct of research*. 4th ed. New York: Oxford University Press.

Shamoo, A.E., and J. Schwartz. 2008. Universal and uniform protections of human subjects in research. *American Journal of Bioethics* 8 (11): 3–5.
Singh, A.K., B.P. Griffith, C.E. Goerlich, D. Ayares, and M.M. Mohiuddin. 2022. The road to the first FDA-approved genetically engineered pig heart transplantation into human. *Xenotransplantation* 29 (5): e12776.
Skloot, R. 2010. *The immortal life of Henrietta lacks*. New York: Crown Publishing.
SUPPORT Study Group of the Eunice Kennedy Shriver NICHD Neonatal Research Network. 2010. Target ranges of oxygen saturation in extremely preterm infants. *New England Journal of Medicine* 362 (21): 1959–1969.
Timmons, M. 2002. *Moral theory: An introduction*. Lanham: Rowman and Littlefield.
Trender, W., P.J. Hellyer, B. Killingley, M. Kalinova, A.J. Mann, A.P. Catchpole, D. Menon, E. Needham, R. Thwaites, C. Chiu, G. Scott, and A. Hampshire. 2024. Changes in memory and cognition during the SARS-CoV-2 human challenge study. *eClinicalMedicine* 76: 102842.
United Nations. 2023. Least developed nations. Available at: https://www.un.org/development/desa/dpad/wpcontent/uploads/sites/45/publication/ldc_list.pdf. Accessed 11 May 2024.
Varmus, H., and D. Satcher. 1997. Ethical complexities of conducting research in developing countries. *New England Journal of Medicine* 337 (14): 1003–1005.
Veatch, R.M. 1987. *The patient as partner: A theory of human experimentation ethics*. Bloomington: Indiana University Press.
Verma, I.M. 2014. Editorial expression of concern: Experimental evidence of massive-scale emotional contagion through social networks. *Proceedings of National Academy of Science USA* 111 (29): 10779.
Wadman, M. 2023. What does the historic settlement won by Henrietta Lacks's family mean for others? Science 7 August. Available at: https://www.science.org/content/article/what-does-historic-settlement-won-henrietta-lacks-s-family-mean-others. Accessed 3 May 2024.
Wang, W., W. He, Y. Ruan, and Q. Geng. 2022. First pig-to-human heart transplantation. *Innovations* 3 (2): 100223.
Wendler, D. 2010. *The ethics of pediatric research*. New York: Oxford University Press.
Wendler, D., E.J. Emanuel, and R.K. Lie. 2004. The standard of care debate: Can research in developing countries be both ethical and responsive to those countries' health needs? *American Journal of Public Health* 94 (6): 923–928.
Wertheimer A. 2011. *Rethinking the ethics of clinical research: Widening the lens*. New York: Oxford University Press.
Wilfond, B.S., D. Magnus, A.H. Antommaria, P. Appelbaum, J. Aschner, K.J. Barrington, T. Beauchamp, R.D. Boss, W. Burke, A.L. Caplan, A.M. Capron, M. Cho, E.W. Clayton, F.S. Cole, B.A. Darlow, D. Diekema, R.R. Faden, C. Feudtner, J.J. Fins, N.C. Fost, J. Frader, D.M. Hester, A. Janvier, S. Joffe, J. Kahn, N.E. Kass, E. Kodish, J.D. Lantos, L. McCullough, R. McKinney Jr., W. Meadow, P.P. O'Rourke, K.E. Powderly, D.M. Pursley, L.F. Ross, S. Sayeed, R.R. Sharp, J. Sugarman, W.O. Tarnow-Mordi, H. Taylor, T. Tomlinson, R.D. Truog, Y.T. Unguru, K.L. Weise, D. Woodrum, and S. Youngner. 2013. The OHRP and SUPPORT. *New England Journal of Medicine* 368 (25): e36.
World Health Organization. 2020. Key criteria for the ethical acceptability of COVID-19 human challenge studies. https://www.who.int/publications/i/item/WHO-2019-nCoV-Ethics_criteria-2020.1. Accessed 3 May 2024.
———. 2024. Coronavirus disease of 2019. Available at: https://www.who.int/news-room/fact-sheets/detail/coronavirus-disease-(covid-19). Accessed 3 May 2024.
World Medical Association. 2013. Declaration of Helsinki: Ethical principles for medical research involving human subjects (2013 revision). Available at: https://www.wma.net/policies-post/wma-declaration-of-helsinki-ethical-principles-for-medical-research-involving-human-subjects/. Accessed 3 May 2024.
Wu, Z., and J.M. McGoogan. 2020. Characteristics of and important lessons from the coronavirus disease 2019 (COVID-19) outbreak in China: Summary of a report of 72,314 cases from the Chinese Center for Disease Control and Prevention. *Journal of the American Medical Association* 323 (13): 1239–1242.

Chapter 2
Historical and Regulatory Background

The basic approach to the ethical conduct of research and approval of investigational drugs was born in scandal and reared in protectionism.

Carol Levine (1988: 167).

Abstract This chapter describes key historical events that have shaped ethical rules and policies concerning research with human subjects, including the Tuskegee Syphilis Study, the Nazi experiments on concentration camp prisoners, the Willowbrook hepatitis experiments, the US government's secret human radiation experiments, Stanley Milgram's obedience to authority experiments, the Stanford prison experiment, and Jesse Gelsinger's tragic death in a gene therapy study. The chapter also discusses the Belmont Report and its impact; provides an overview of the US federal regulations, including the Common Rule, the FDA regulations, and the IRB system; and reviews international guidelines, such as the Nuremberg Code and the Helsinki Declaration. The chapter argues that the social and political response to historical events has been to develop an oversight system to restore and maintain public trust in research. An important question that emerges at the end of this chapter is whether the current system provides too much or too little protection for human research subjects. To answer this question satisfactorily, one needs a philosophical framework that can address ethical and regulatory dilemmas in research with human subjects.

Keywords Ethics · Human subjects · History · Federal regulations · Oversight · Trust

D. B. Resnik, *The Ethics of Research with Human Subjects*, The International Library of Bioethics 111, https://doi.org/10.1007/978-3-031-82757-0_2

To understand the ethics of research with human subjects, it is important to have some familiarity with important events and trends that have entered the public's consciousness and influenced legislation, regulation, and policy. (See Table 2.3 for a summary of important events.) Examining this historical background will help to frame the issues and provide us with the social, political, economic, and legal context for the current oversight system. The main thesis of this chapter is that the current system has emerged in response to unethical—and in some cases horrific—treatment of human research subjects. According to Alexander Morgan Capron (1989: 127), "the darkest moments in the medical annals have involved abuses of human research subjects." Existing guidelines, regulations, principles, and policies are understandably protectionistic because they seek to prevent the kinds of abuses that have generated moral outrage and undermined the public's trust in the research enterprise (Levine 1988).

Before reviewing this history, I would like to counteract an impression the reader may draw from the narrative. It may appear to the reader that I think that research with human subjects is always ethically problematic and often morally corrupt. Nothing could be further from the truth. It is my opinion, based on my review of the literature and my experiences as an IRB chair and committee member, researcher, and research subject, that most research with human subjects is ethically sound and socially valuable. Research with human subjects has led to the development of medications, vaccines, surgical procedures, medical devices, and public health practices that have saved millions of human lives, reduced disability, and alleviated pain and suffering. Research has also provided us with important knowledge concerning human biology, psychology, sociology, and anthropology which has helped us gain a better understanding of ourselves and has led applications in medicine, nursing, public health, psychology, and other practical disciplines. Research has also provided us with knowledge that has informed public policies (Porter 1999).

It seems that we do not hear enough about the good things that happen in research with human subjects and that we hear too much about the bad. Journalists and ethicists—and I include myself in this group—tend to be more interested in bringing scandals and controversies to the public's attention than in reporting on the daily interactions between investigators and research subjects where nothing goes wrong and much goes right. Although the historical narrative concerning the ethics of research with human subjects includes many troubling episodes, it is important to come to terms with this storyline and understand its implications for practice and policy.

2.1 Research with Human Subjects Before World War II

Hippocrates, engraving by Peter Paul Reubens, public domain. https://commons.wikimedia.org/wiki/File:Hippocrates_rubens.jpg

I shall begin my review of history in ancient Greece. Hippocrates (460–377 BCE) is widely recognized as the father of scientific medicine, due to his careful attention to observation and his view that diseases should be explained in terms of natural, rather than supernatural, causes (Porter 1999). Hippocrates believed that diseases result from imbalances of different substances (such as blood, phlegm, and mucous) in the body and that medicine should attempt to restore the body's natural balance and help the body heal itself. Hippocrates took detailed case histories, observed signs and symptoms in his patients, and kept records on the effectiveness of different cures. Hippocrates also developed a medical ethos known as the Hippocratic Oath. One of the key tenets of the Oath is: "I will use those dietary regimens which will benefit my patients according to my greatest ability and judgement, and I will do no harm or injustice to them (National Library of Medicine 2024)." Others have simplified this part of the Oath to the principle of non-maleficence ("do no harm") and the principle of beneficence ("benefit your patient") (Beauchamp and Childress 2019). Hippocratic medicine therefore tended to be very nonaggressive and observational rather than experimental. Common remedies prescribed by Hippocratic physicians included special diets, herbal medicines, exercise, baths, and massage. The Hippocratic Oath still plays an important role in medical practice, and many medical students take the Oath when they begin their studies (Porter 1999).

The Hippocratic tradition dominated scientific medicine for 2000 years but yielded to experimentally-oriented approaches during the Scientific Revolution (circa 1500–1700). For example, Paracelsus (1493–1542), Andreas Vesalius (1514–1564), and William Harvey (1578–1657) sought to apply the experimental method to medicine. They dissected animal and human bodies, which allowed them to develop a better understanding of human anatomy and physiology. Although Harvey performed vivisections on animals (to demonstrate, for example, that the heart pumps blood), his experiments on humans tended to be low-risk. Harvey measured pulse and blood pressure and applied tourniquets to human research subjects to demonstrate the direction of blood flow (Porter 1999).

Edward Jenner, public domain. https://upload.wikimedia.org/wikipedia/commons/3/32/Eduard_Jenner.jpg

By the eighteenth century, human experimentation had become more commonplace and more risky. For example, on May 14, 1796, Edward Jenner (1749–1823) inoculated James Phipps, an eight-year-old boy, with fluid from a cowpox pustule to determine whether this would prevent him from developing smallpox. Jenner had observed that dairymaids exposed to cowpox seemed to have resistance to smallpox and hypothesized that exposure to cowpox would provide protection against smallpox. Phipps suffered no adverse effects from the inoculation other than a slight fever. Six weeks after the inoculation, Jenner exposed Phipps to smallpox, and he did not develop the disease. Jenner also tested this procedure on his children, who also did not contract smallpox (Porter 1999).

Louis Pasteur, photo by Félix Nadar Crisco. Licensed under Public Domain via Wikimedia Commons. https://upload.wikimedia.org/wikipedia/commons/a/a6/Louis_Pasteur%2C_foto_av_Paul_Nadar%2C_Crisco_edit.jpg

Louis Pasteur (1822–1896), who developed the germ theory of disease, also tested experimental vaccines on human subjects (Porter 1999). In 1885, Pasteur gave an experimental rabies vaccine, which had been tested on animals, to a nine-year old boy, Joseph Meister. Meister had been mauled by a dog, and Pasteur gave the boy the vaccine in the hope of preventing the disease. The vaccine worked, and other parents asked Pasteur to administer it to their children (Geison 1978). After observing that carbolic acid could reduce wound infections in cattle, Joseph Lister (1827–1912) hypothesized that this compound had antiseptic properties, and he tested it on human subjects. One of the first patients to receive this treatment was James Greenless, an eleven-year-old boy who had a compound fracture of the leg. The boy recovered from his treatment without infection or amputation (Porter 1999).

Other experiments placed subjects at greater risk and caused more suffering. For example, in 1874 Robert Bartholow (1831–1904) noticed that his 30-year-old mentally disabled patient, Mary Rafferty, had a two-inch hole in her skull caused by a cancerous ulcer. He took advantage of this opportunity to study her brain and inserted electrodes into the hole. Bartholow stimulated her brain with an electric current and recorded her responses. He noticed that small amounts of current caused different bodily movements, depending on the site of the stimulation, but that larger amounts caused pain. His patient fell into a coma and died in a few days, after experiencing distress, convulsions, and seizures (Lederer 1995).

In 1897, Giuseppe Sanarelli (1864–1940) claimed to have isolated the bacterium which causes yellow fever from the blood of patients with the disease. To test his hypothesis, he injected the bacteria into five patients without their consent. The patients all developed yellow fever symptoms and three died. William Osler (1849–1919) and many other physicians sharply criticized Sanarelli's experiments (Lederer 1995).

Although documentation of the use of slaves in research is not as extensive as the evidence for other abuses, there is evidence that African slaves were used as research subjects in the Caribbean and Southern U.S. in the 1800s. For example, new surgical techniques were often tried out on slaves, usually without anesthesia, and slaves were used in medical schools to train doctors (Savitt 1982). James Marion Sims (1813–1883), the father of American gynecology, developed surgical techniques used to repair childbirth complications on enslaved women. Although Sims wrote in his journals that women "clamored" for these operations it is unclear whether they genuinely consented (Holland 2018).

Walter Reed, public domain. https://en.m.wikipedia.org/wiki/File:WalterReed.jpeg

An important medical experiment that helped to pioneer the use of informed consent documents took place in Cuba in the early 1900s, when U.S. Army physician Walter Reed (1851–1902) sought to determine the cause of yellow fever. Yellow fever had been a major public health problem in the Caribbean and Central America that killed thousands of people each year and threatened commerce and U.S. military operations. Reed's experiments helped to establish that *Aedes aegypti* mosquitos carry yellow fever, which was a major breakthrough in controlling the spread of the disease. To determine whether the mosquito was the cause, Reed asked participants who had never had yellow fever to allow themselves to be bitten by mosquitos that had fed on patients with active yellow fever, or to be injected with blood from a yellow fever patient (Lederer 2008). Reed asked the volunteers to sign informed consent documents stating that they understood the risks of the experiment (including the possibility of death) and agreed to participate. The documents were translated into Spanish. Participants received $100 in gold and additional $100 and free medical care if they contracted yellow fever (Lederer 2008). Family members of participants who died also received $100. Two of Reed's collaborators, James Carroll and Jesse Lazear, volunteered for the experiment. They both developed yellow fever and Lazear died. Reed had been planning to also volunteer for the experiment but Carroll talked him out of it due to his age (the disease was more deadly for patients over 40 years old, such as Reed). A total of 33 volunteers participated in the experiment, including 18 Americans (2 civilians and 16 soldiers) and 15 Spanish immigrants. Six people died from yellow fever (Lederer 1995). The participants in the experiments were regarded as martyrs and heroes. The surviving military personnel received medals and government pensions, and the Army's Medical Center in Washington, DC was named after Reed (Lederer 1995).

It is not entirely clear why Reed asked the volunteers to sign consent documents, but the medical community's harsh reaction to Sanarelli's experiments probably influenced his decision-making (Lederer 2008). In any case, it is noteworthy that Reed used consent documents long before they became commonplace in medical practice or research in the 1970s. In 1822, William Beaumont (1785–1853) asked his patient Alexis St. Martin, to sign a document which some have regarded as a consent form, but it was more like an employment contract (Lederer 2008). St. Martin had been shot in the abdomen at close range. Beaumont tried to close the wound but did not succeed. Beaumont, who saw a unique opportunity to observe digestive processes, asked St. Martin's permission to take samples of his digestive fluids, place food in the opening, and record internal body temperatures. In exchange for his cooperation, St. Martin received $150 for one year's employment as a Beaumont's personal servant. Beaumont also provided St. Martin with food, clothing, and lodging (Lederer 2008).

By the early twentieth century, physicians had mixed views about medical research. Physicians still accepted the Hippocratic ethos, but they were also interested in advancing medical science and testing new therapies. Although some physicians, such as Reed, sought consent from their patients, medical practice was highly paternalistic. Physicians often treated their patients without explaining what they were doing or why. They were more concerned with helping their patients than with obtaining consent, and they often viewed medical experimentation as a necessary part of treatment. Indeed, physicians practicing medicine during this era did not have a clear sense of the difference between research and treatment and many mixed the two together (Lederer 1995).

In 1900, Prussia became the first nation to adopt laws concerning medical research. The Prussian Directive required medical researchers to obtain informed consent from their subjects and barred physicians from conducting research on children. The Prussian Directive was adopted in reaction to abuses of research subjects, such as an experiment in which a physician pricked the eye of a patient with a needle contaminated by nodules from a patient with leprosy to determine whether the bacteria he was studying cause the disease (Shamoo and Resnik 2022). The American Medical Association also recognized that physicians encountered some ethical dilemmas concerning medical research. For several decades the organization deliberated about developing a code of ethics for medical research on human subjects but did not formally adopt a code until 1946 (Lederer 1995).

During this same period, the eugenics movement spread throughout Europe and the U.S. Eugenics was a social and political ideology based on Darwin's theory of evolution by natural selection (or "survival of the fittest"). The founder of eugenics, Francis Dalton (1822–1911) argued that "defective" individuals should be prevented from breeding and "superior" ones encouraged to breed to improve the fitness of the human population. By the 1930s, 28 U.S. states had passed laws mandating involuntary sterilization of "unfit" individuals, including those who were mentally ill or disabled, criminally insane, or psychopathic. Approximately 100,000 people were sterilized under U.S. eugenics laws (Proctor 1988). Prominent proponents of eugenics included British Prime Minister Winston Churchill (1874–1965), U.S. President Theodore Roosevelt (1858–1919), science fiction writer H.G. Wells (1866–1946), and even African American scholar and civil rights activist W.E.B. Dubois (1868–1963), who opposed racism but believed in encouraging the most talented members of all races to interbreed.

German Chancellor and leader of the Nazi Party Adolf Hitler (1889–1945), who spoke favorably of America's eugenics laws in *Mein Kampf*, led Germany's eugenics movement (Proctor 1988). Inspired by the ideology that Aryans constitute a "master race," Germany began in the 1930s a program of racial/ethnic purification to rid the country of "undesirable" members of the population, including Jews, mentally or physically disabled people, Gypsies, and mixed-race children. The program began after Hitler came to power in 1933, with the enactment of mandatory sterilization laws and it soon included "euthanasia" of mentally or physically disabled people. Approximately 250,000 people died in Germany's "euthanasia" program. Racial purification expanded to include segregation, forced labor, confiscation of property, imprisonment of Jews and others in concentration camps, and eventually to systematic genocide. Approximately six million people were killed in Germany's racial purification program, most of whom were Jewish concentration camp prisoners (Weindling 2008).

2.2 Research with Human Subjects During World War II

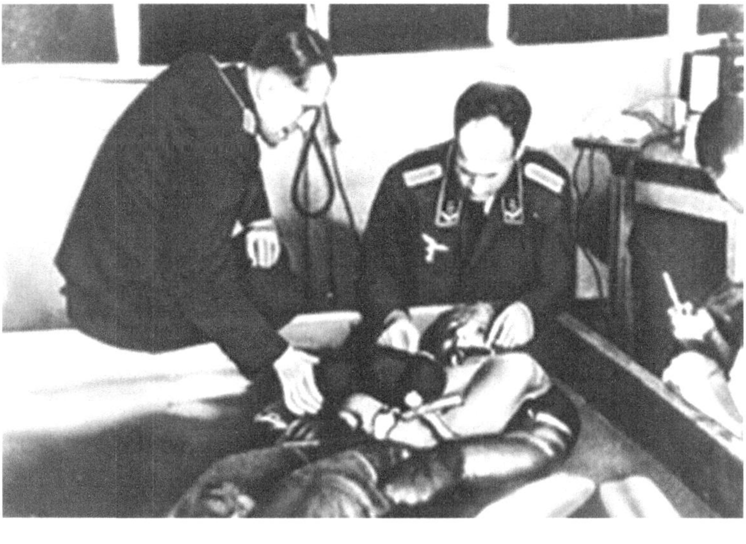

Cold water immersion experiment on a Dachau concentration camp prisoner. (Source: Wikipedia. Licensed under Fair use via Wikipedia—https://en.wikipedia.org/wiki/File:Dachau_cold_water_immersion.jpg#/media/File:Dachau_cold_water_immersion.jpg)

Coerced experimentation on concentration camp prisoners, mostly Jewish people and other "undesirables", was a natural outgrowth of the ideology of racial purification. In Nazi Germany, Jews did not have the same rights as other people, such as the right to refuse to participate in dangerous experiments. The Nazi experiments are among the worst ever performed on human subjects. Some of these included (Proctor 1988; Weindling 2008):

- Exposing research subjects to freezing cold temperatures, extremely low pressures, and high levels of electricity and radiation to study effects on the human body;
- Inflicting wounds on human subjects to study wound healing (subjects were shot, stabbed, or injected with metal or glass);
- Intentionally exposing human subjects to malaria, staphylococcus, tetanus, and other infectious diseases to study the effectiveness of vaccines;
- Injecting dye into human eyes to try to change eye color (many subjects went blind);
- Josef Mengele's (1911–1979) twin studies, which included forcing fraternal twins to have sex, mixing the blood of identical twins, sewing twins together to produce conjoined twins, and placing one member of a pair of identical twins in isolation from birth to study the role the environment in human development.

Over 15,000 people were killed, maimed, or permanently disabled in these experiments. The experiments began in the 1930s, rapidly increased around 1942, and continued until the end of the war (Weindling et al. 2016). Although most of these experiments were conducted by German military physicians or scientists, some were conducted by civilians. Most of these experiments did not produce useful results, due to their poor design; however, some did, and researchers have debated about whether to use or cite Nazi data (Post 1991). Even today, the full extent of these experiments is not known, and many have come to light in recent years due to historical scholarship (Weindling 2008).

Scientists working at Unit 731 of the Imperial Japanese Army conducted chemical and biological weapons experiments on thousands of Chinese prisoners of war from 1932 to 1945. The Japanese researchers also exposed human subjects to extremes of temperature and pressure, and conducted wound-healing, vaccination, and surgical studies involving people, including vivisections. The experiments, which took place on mainland China, did not gain much notoriety until the 1990s because the U.S. government was interested in using the data from these experiments. The government agreed to refrain from prosecuting the researchers for war crimes in exchange for the data. (Tsuchiya 2008). Victims also included Korean civilians, as well as Mongolians, Manchurians (distinct from Chinese), and Russians.

At the end of World War II, the Allied Powers put German military and political leaders on trial at Nuremberg, Germany for war crimes. Among these were the Nazi scientists and physicians who conducted experiments on human beings. Since the actions of the Nazi researchers were legal in Germany, the Nuremberg judges needed a source of international law to serve as a basis for prosecuting the defendants. The War Crimes Tribunals adopted the Nuremberg Code (1949) to serve as a basis for prosecuting Nazi researchers for war crimes. The Code became the first internationally recognized ethical standard for research with human subjects (Capron 1989). The Code consists of ten principles. The first principle states that "The voluntary consent of the human subject is absolutely essential (Nuremberg Code 1949)." Other principles address the importance of minimizing risk and ensuring that the research is well-designed and serves socially worthwhile purpose.

Doctors' trial, Nuremberg, 1946–1947, public domain. https://en.wikipedia.org/wiki/List_of_Nazi_doctors#/media/File:Doctors'_trial,_Nuremberg,_1946%E2%80%931947.jpg

2.3 Research with Human Subjects After World War II

After the Second World War, a seminal event involving research with human subjects was the development of the polio vaccine. Polio (poliomyelitis) had been one the most feared childhood diseases, killing or permanently disabling thousands of people in the U.S. each year. The incidence of the disease had risen to 28.3 per 100,000 people by 1949 and 37.2 per 1000,000 by 1952. In 1952, polio killed 2500 people. Polio outbreaks led to school closings, quarantines, and other measures designed to prevent the spread of the disease (Meldrum 2008). Research on a polio vaccine took place in the 1930s, with little success. In the late 1940s, the National Foundation for Infantile Paralysis (NFIP), nicknamed the March of Dimes for millions of contributors giving a dime to the cause, began sponsoring research on a polio vaccine. In 1953, the Foundation announced it would conduct field trials of a vaccine, led by Jonas Salk (1914–1995). The NFIP decided to conduct a blinded randomized controlled trial (RCT), with an experimental group of second-grade children receiving the vaccine and first- and third-grade children receiving a placebo. The researchers collected data on the incidence of polio in these two populations. This was the first RCT conducted on healthy children (Meldrum 2008). In 1953, the field trials began. Informed consent was obtained for the study via a letter sent home to the children's parents. U.S. counties with the highest incidence of the disease with more than 50,000 people were targeted for recruitment. 455,474 children (60.8% of the eligible population) participated in the study and 401,974 completed the full series of injections. On April 12, 1955, the researchers reported the results of the study in the *American Journal of Public Health*. The study was an overwhelming success: the vaccine was 80–90% effective at preventing the disease. Soon a massive vaccination campaign began. By 1956, 75% of children, ages 5–12, had received the vaccine. In 1957, the incidence of polio dropped to 3 per 100,000 (Meldrum 2008).

Salk headlines. March of Dimes, public domain via Wikimedia Commons—https://commons.wikimedia.org/wiki/File:Salk_headlines.jpg#/media/File:Salk_headlines.jpg

Although the polio vaccine clinical trial was a major victory in the fight against childhood diseases, it did raise some ethical concerns. The NFIP solicited a great deal of advice on scientific and ethical issues from investigators, physicians, and public health officials. Some were concerned about the safety of vaccine. While the vaccine used a dead virus, it was possible that in some batches could contain live viruses. Some argued that the vaccine should be tested on a smaller group before launching a large field trial. However, the investigators were responding to an urgent need to develop a vaccine to deal with the epidemic, and they knew that testing it a smaller group first would have delayed the launch of larger study. Others worried that consent process was inadequate because the letter that came home from school did not fully inform parents about the risks of the vaccine, and fear of polio compromised parents' ability to carefully weigh risks and benefits (Meldrum 2008).

In 1964, the World Medical Association (WMA) met in Helsinki, Finland to draft an ethics code which is known as the Helsinki Declaration. The Declaration has been amended nine times since then, most recently in 2013 (World Medical Association 2013). The original version of the Declaration expanded upon the principles of Nuremberg Code and included 22 principles for ethical medical research. The Declaration addressed topics not covered by the Code, such as obtaining informed consent for legally incompetent subjects (such as children) and the use of control groups in clinical research. The Declaration prohibited the use placebo control groups when there is a known effective therapy (World Medical Association 1964). As noted in Chap. 1, the most recent version of the Declaration allows placebo control groups when there is a known effective therapy if subjects receiving a placebo do not face a risk of serious or irreversible harm (World Medical Association 2013).

The preamble of the Declaration emphasized the importance of acting in the patient's best interests and stated that "the interests of science and society should **never** take precedence over considerations related to the well-being of the subject (World Medical Association 1964, emphasis added)." The most recent version of the Declaration includes a similar statement: "While the primary purpose of medical research is to generate new knowledge, this goal can **never** take precedence over the rights and interests of individual research subjects (World Medical Association 2013, emphasis added)." Given that there is an inherent conflict between the good of the research subject and the good of society (see cases in Chap. 1), "never" may be too strong of a word, especially if the subjects are willing to place the interests of society ahead of their own interests. We will return to this issue when discuss the risks of research on health volunteers in Chap. 7.

After the Clinical Center of the National Institutes of Health (NIH) opened in 1953, researchers began informally reviewing human studies to minimize risks and unusual hazards. In 1965, the NIH formalized this process and required prior review of intramural human studies. In 1966, the NIH extended this prior review requirement to include extramural research. In 1971, the Food and Drug Administration (FDA) issued similar rules for human studies that generate data submitted to the agency. These peer review committees were precursors to institutional review boards (IRBs) (Capron 1989).

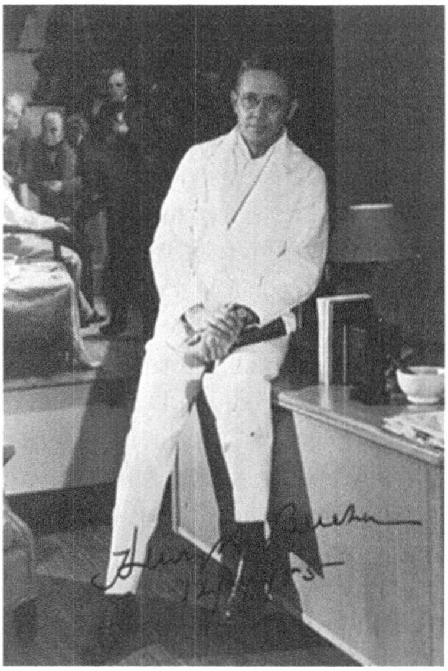

Henry K. Beecher. (Source: National Library of Medicine (public domain))

In 1966, Henry Beecher (1966) published an article in the *New England Journal of Medicine* describing 22 studies involving human subjects which he categorized as unethical because they violated informed consent, withheld treatment, or exposed subjects to unacceptable risks. Studies mentioned took place in hospitals, medical schools, private clinics, and government, military, and industry research centers, and included two infamous cases. This groundbreaking article helped to raise awareness among clinical researchers about ethical problems that plagued medical practice but did not lead to any immediate changes in government policy. According to Beecher (1966: 1354): "American medicine is sound, and most progress is soundly attained. There is, however, a reason for concern in certain areas, and I believe the type of activities to be mentioned will do great harm to medicine unless soon corrected."

One of the studies mentioned in Beecher's article, the Willowbrook hepatitis experiment, took place from 1956 to 1980 at the Willowbrook State School for Mentally Retarded Children in New York. Saul Krugman (1911–1995) supervised this research, which involved infecting healthy children with a mild form of viral hepatitis which was endemic at Willowbrook to determine whether this would provide them with immunity. The study included an experimental group, which received the virus prior to being admitted to the institute, and a control group, which did not receive the virus. The researchers obtained written informed consent from the parents. Several committees, including the New York State Department of Mental Health, approved the study. Over 700 children participated in the research.

The investigators argued that the experiments were ethical because (1) they were important for developing a vaccine that could benefit many children and (2) the research subjects would benefit from their participation since they were likely to develop the disease within 6–12 months of staying at Willowbrook, and they would receive better medical care if they were intentionally infected and treated quickly, instead of becoming infected naturally, with treatment possibly delayed. Also, infected subjects would likely have immunity to the disease. Critics of the studies, including Beecher, argued that the experiments were unethical because they imposed unacceptable risks on children and because consent process was deficient (Robinson and Unruh 2008).

Another study mentioned by Beecher took place at the Jewish Chronic Disease Hospital in Brooklyn, New York, in 1963. In this study, Chester Southam and Deogracias Custodio injected live cancer cells into 22 debilitated patients. The purpose of this research, which was funded by the American Cancer Society and the Public Health Service, was to better understand whether the immune deficiencies of cancer patients are due cancer or its effects on their immune system. The investigators did not tell the patients they were injecting them with cancer cells and they did not obtain written informed consent. They argued that there was no need to inform the patients that they were receiving cancer cells because their immune systems were likely to reject the foreign tissue, and so this posed very little risk to them. As it turned out, all patients, except those who died from their disease, rejected the cancer cells (Arras 2008).

During the 1960s, Harvard University psychology Stanley Milgram conducted behavioral experiments that generated ethical controversy. The experiments included three types of participants: the learner, the punisher, and the researcher. Electrodes were attached to the learner. If the learner answered a question incorrectly, the researcher instructed the punisher to press a button on a machine that would deliver an electric shock to the learner. The learners would wince or cry in pain upon receiving the shock. Punishers continued to administer electric shocks even when the meter on the machine indicated that the shocks were at a dangerous level. In some cases, punishers continued to administer shocks despite learners' demands to stop the experiment. In reality, no one received a shock, but the punishers did not know this until the experimental session concluded. The purpose of the study was to determine the extent to which human beings would obey an authority figure (the researcher), even when they were ordered to do something they considered to be morally wrong. Milgram argued that this was an important topic to study, since many German soldiers and prison guards during World War II defended their immoral actions by claiming that they were only following orders. The experiment would not work if the learners had been told what was really happening, because this would not test their willingness to obey authority. During the debriefing after the experimental sessions, many of the punishers expressed a deep sense of regret that they were willing to do something they considered to be wrong. Some also objected to the deception and the experimental manipulation of their behavior (Milgram 1974).

Two other social science studies that have had a major impact on the public's perception of the trustworthiness of research include the Tearoom Trade and the Stanford Prison Experiment. The Tearoom Trade is part of the title of a Laud Humphreys' Ph.D. dissertation, which he completed in 1968. In the dissertation, Humphreys reported findings based on his observations of homosexual encounters in public bathrooms conducted during. To do his research, Humphreys (1976) misrepresented his identity and intent to the subjects to avoid response bias. He also used their license plates numbers to ascertain their identities. Although the Tearoom Trade helped to debunk stereotypes about male homosexuality, for example, by showing that gay men come from diverse backgrounds, it also angered many people in the gay community and the public more broadly, because it involved deception and possibly endangered the subjects.

The Stanford Prison Experiment, which was depicted in a film in 2015, took place from August 14 to 20, 1971, when a team of researchers, led by Stanford University psychology professor Philip Zimbardo, asked 24 students to participate in a role-playing activity designed to simulate the prison environment. The students, who received $15 a day for their participation, were randomly divided into two groups, the "prisoners" and the "guards," with Zimbardo serving as warden. The "prison" was a boarded-off area in the basement of the psychology department building. Zimbardo instructed the students on how to perform their roles, but he did not anticipate the degree to which they would become embrace them. Several of the guards implemented authoritarian rules and subjected prisoners to psychological torture, and many of the prisoners accepted or encouraged this mistreatment. Zimbardo stopped the experiment after six days to protect the participants from further abuse from their peers. The experiment raised issues about anticipating and assessing psychosocial risks in research (Zimbardo 2024).

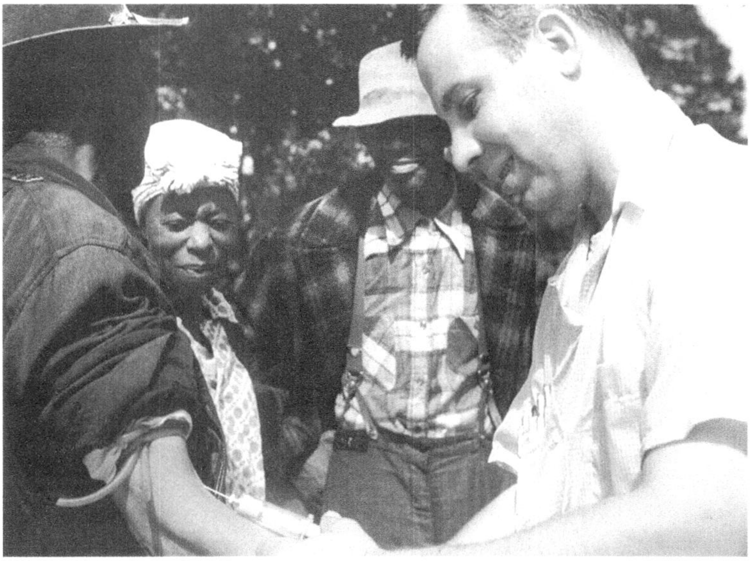

Doctor injecting subject in the Tuskegee Syphilis Study, Wikimedia Commons. https://en.wikipedia.org/wiki/Tuskegee_Syphilis_Study#/media/File:Tuskegee-syphilis-study_doctor-injecting-subject.jpg

A study that began in 1932, the Tuskegee Syphilis Study, did not gain national attention until 1972, when it was stopped. The purpose of this research, which was funded by the Department of Health, Education, and Welfare (DHEW), was to better understand the etiology of late-stage syphilis in African American men. The scientific rationale for this research was that this disease had been well-studied in Whites but not in Blacks, and it was important to be able to identify the signs and symptoms of this disease in this population. Because treatment for syphilis would interfere with observation of its natural course, 399 subjects in the "experimental" group did not receive treatment for syphilis. Subjects without syphilis constituted the "control" group for the study. The study was initially planned to last only a year, but it went on for decades. Subjects were not told that they were in a research study nor offered treatment for syphilis when it became available in the 1940s (Jones 1981). Subjects thought they were receiving treatment for "bad blood." The "treatment" involved nothing more than medical examinations and routine care. Investigators steered subjects away from clinics where they might receive treatment for syphilis. The participants received free lunches and burials. The study was stopped when Peter Buxton, a Public Health Service employee, informed the Associated Press about the research, and newspapers around the country ran the story. Surviving research subjects and families of the deceased filed a class-action lawsuit against the U.S. government, and the parties agreed to a settlement in 1973. President William Clinton issued an official apology on behalf of the U.S. government to survivors and the families in 1997 (Shamoo and Resnik 2022).

The Belmont Report, Wikimedia Commons, https://en.m.wikipedia.org/wiki/File:The_Belmont_report_-_ethical_principles_and_guidelines_for_the_protection_of_human_subjects_of_research_%28IA_belmontreporteth01unit%29.pdf

2.4 The Belmont Report

Public outrage over the Tuskegee study led to Congressional hearings on the ethics of research with human subjects and passage of the National Research Act in 1973 (signed into law in 1974). The Act authorized federal agencies to draft regulations for research involving human subjects and created the National Commission for the Protection of Human Subjects of Biomedical and Behavioral Research (Shamoo and Resnik 2022). In 1978, DHEW (a precursor to DHHS) announced revisions to its regulations for research with human subjects. In 1979, the National Commission issued the *Belmont Report*, which provided an ethical framework for research involving human subjects that served as a conceptual foundation for a major revision of federal regulations in 1981 (National Commission 1979).

The *Belmont Report* described three principles for ethical research involving human subjects: respect for persons (which requires informed consent from research subjects as well as additional protections for those who cannot consent); beneficence (which requires maximization of the benefits of research and minimization of risks); and justice (which requires fair distribution of benefits and burdens of research). The authors of the report did not say what makes a distribution "fair" other than to list some different ways of distributing benefits and burdens, such as on the basis of equality, merit, or desert. The authors acknowledged that the principles may sometimes conflict and that to make an ethical decision concerning research with human subjects one must balance these different principles in light of the relevant facts and circumstances (National Commission 1979).

The *Belmont Report* also distinguished between medical therapy and medical research. The goal of therapy, according to the authors of the *Report*, is to benefit the patient, whereas the goal of research is to produce generalizable knowledge (National Commission 1979). The authors of the *Report* argued that it is important to distinguish between research and therapy to determine what types of activities should be regulated as research. For example, medical practice is regulated by medical licensing boards and hospital committees, and is subject to malpractice liability, so it does not require additional forms of oversight. The authors also noted that research and therapy often occur concurrently but that if any part of a medical intervention involves research it should be regulated as such. The authors also noted that innovative medical practices do not need to be regulated as research, but they encouraged physicians to develop research protocols to evaluate them (National Commission 1979).

The authors of *Belmont Report* stressed the importance of protecting research subjects who are vulnerable, due to their compromised ability to provide consent or their dependency on others. The authors said care should be taken to avoid unjustly using vulnerable individuals as research subjects solely for administrative convenience or because they are easy to manipulate due to illness or their socioeconomic circumstances (National Commission 1979). The authors also stated that justice requires that there should be a preference in the order of selection of research subjects, i.e., adults should be preferred to children for research that is not on exclusively pediatric conditions and some subjects, such as prisoners or mentally disabled people, should be used only under special circumstances (National Commission 1979).

The twelve-page *Belmont Report* has become one of the most influential documents pertaining to research with human subjects. It is frequently cited in books, articles, government reports, and official guidance. Also, many institutions agree to follow the ethical principles of the report when they sign an agreement (known an as an assurance) to receive federal funding for research with human subjects (Shamoo and Resnik 2022).

Despite its considerable influence, the *Report* has some flaws. First, the principles summarize some key insights from ethical theories and traditions, but the *Report* itself does not examine these principles or their underlying philosophical justification in any depth (Beauchamp 2005). Second, as noted earlier, the *Report* does not include a decision-procedure for prioritizing the principles when they conflict. Third, although the *Report* includes some applications of the principles, none of these provide details concerning real or hypothetical studies. Consequently, one is left wondering exactly how to apply the principles to situations and resolve conflicts (Veatch 2005).

2.5 The U.S. Federal Regulations

As mentioned earlier, the NIH and FDA have had rules pertaining to research within human subjects since the 1960s, but in 1981, U.S. federal agencies implemented major revisions to their regulations for protecting human subjects in research. The U.S. regulations include the Common Rule, i.e., rules which apply to research supported by different 19 federal agencies; and the FDA regulations, which apply to research submitted to the agency pertaining to products under its jurisdiction, including new drugs, biologics, medical devices, food additives, and cosmetics. The regulations have been revised several times since 1981, most recently in 2017 (effective 2018) (Department of Homeland Security et al. 2017).

The Common Rule[1] includes four subparts: Subpart A, which include general requirements for research with human subjects; Subpart B, which includes requirements for pregnant women, fetuses, or neonates involved in research; Subpart C, which include requirements for prisoners; and Subpart D, which includes requirements for children. Although the FDA has regulations which differ in some from the Common Rule, it has adopted Subpart D for its research. Subparts B, C, and D will be discussed in Chap. 9 (Table 2.1).

Institutions the receive funding from Common Rule agencies are required to sign a written assurance (essentially, a contract with the agency) stating that they will follow the regulations; designate one or more IRBs to review research; provide space and staff support for the IRB; provide the agency with a list of IRB members and their qualifications; establish standard operating procedures (SOPs) for IRBs to follow; and establish procedures for reporting to IRBs, institutional officials, and

Table 2.1 U.S. federal regulations (sample)

Agency	Regulations
DHHS (includes NIH and CDC)	45 CFR 46
DOD	32 CFR 219
ED	34 CFR 97
EPA	40 CFR 26
FDA	21 CFR 50; 21 CFR 56
NASA	14 CFR 1230
NSF	45 CFR 690

[1] Appendix B include a decision tree for applying the Common Rule to studies.

the agency unanticipated problems involving risks to subjects or others and serious or continuing noncompliance with federal regulations or IRB determinations (Department of Health and Human Services 2017). Although not required by law, most U.S. universities apply the Common Rule to all human research on campus, regardless of the source of funding (Shamoo and Resnik 2022).

The Office of Human Research Protection (OHRP) oversees research supported by DHHS, including NIH and CDC research, and the FDA oversees research it regulates. OHRP (2024) and FDA publish guidance documents to help investigators, IRBs, institutions, and sponsors interpret the regulations. OHRP and FDA both have the authority to audit studies and research sites, issue determination letters of non-compliance, and suspend research. For example, in 1999, the Office of Protection from Research Risks (a precursor to OHRP) suspended all DHHS-funded research conducted at Duke University after an audit found the institution was not complying with record-keeping and continuing review requirements (Weiss 1999). In 2000, the FDA suspended all gene therapy trials at the University of Pennsylvania after the agencies found glaring deficiencies in risk monitoring (Marshall 2000). (See discussion of gene therapy below.) (Table 2.2).

The Common Rule applies to research involving human subjects. A human subject is defined as:

> [A] living individual about whom an investigator (whether professional or student) conducting research:
>
> (i) Obtains information or biospecimens through intervention or interaction with the individual, and uses, studies, or analyzes the information or biospecimens; or
> (ii) Obtains, uses, studies, analyzes, or generates identifiable private information or identifiable biospecimens (45 CFR 46.102e).

Research is defined as:

> [A] systematic investigation, including research development, testing, and evaluation, designed to develop or contribute to generalizable knowledge. Activities that meet this definition constitute research for purposes of this policy, whether or not they are conducted or supported under a program that is considered research for other purposes. For example, some demonstration and service programs may include research activities (45 CFR 46.102l).

Some types of activities are classified as <u>not</u> research, including:

> Scholarly and journalistic activities (e.g., oral history, journalism, biography, literary criticism, legal research, and historical scholarship), including the collection and testing of information or biospecimens, conducted, supported, requested, ordered, required, or authorized by a public health authority....Public health surveillance activities... Collection and analysis of information, biospecimens, or records by or for a criminal justice agency for activities authorized by law or court order solely for criminal justice or criminal investigative purposes...Authorized operational activities (as determined by each agency) in support of intelligence, homeland security, defense, or other national security missions (45 CFR 46.102l)

The FDA regulations apply to clinical investigations, which are defined as:

> [A]ny experiment that involves a test article and one or more human subjects and that either is subject to requirements for prior submission to the Food and Drug Administration under section 505(i) or 520(g) of the act, or is not subject to requirements for prior submission to

> the Food and Drug Administration under these sections of the act, but the results of which are intended to be submitted later to, or held for inspection by, the Food and Drug Administration as part of an application for a research or marketing permit (21 CFR 50.3c).

The FDA defines a human subject as:

> [A]n individual who is or becomes a participant in research, either as a recipient of the test article or as a control. A subject may be either a healthy human or a patient (21 CFR 50.3g)

The Common Rule, but not the FDA, exempts some types of research with human subjects from IRB review. The exemptions do not apply to research covered under Subpart C (prisoners). Some common types of exemptions under the Common Rule include (45 CFR 46.104 emphasis added):

(1) **Research, conducted in established or commonly accepted educational settings,** that specifically involves normal educational practices that are not likely to adversely impact students' opportunity to learn required educational content or the assessment of educators who provide instruction. This includes most research on regular and special education instructional strategies, and research on the effectiveness of or the comparison among instructional techniques, curricula, or classroom management methods.

(2) **Research that only includes interactions involving educational tests (cognitive, diagnostic, aptitude, achievement), survey procedures, interview procedures, or observation of public behavior (including visual or auditory recording)** if at least one of the following criteria is met:

 (i) The information obtained is recorded by the investigator in such a manner that the identity of the human subjects cannot readily be ascertained, directly or through identifiers linked to the subjects;

 (ii) Any disclosure of the human subjects' responses outside the research would not reasonably place the subjects at risk of criminal or civil liability or be damaging to the subjects' financial standing, employability, educational advancement, or reputation; or

 (iii) The information obtained is recorded by the investigator in such a manner that the identity of the human subjects can readily be ascertained, directly or through identifiers linked to the subjects, and an IRB conducts a limited IRB review to make the determination required by

(3) (i) **Research involving benign behavioral interventions** in conjunction with the collection of information from an adult subject through verbal or written responses (including data entry) or audiovisual recording if the subject prospectively agrees to the intervention and information collection…

(4) **Secondary research for which consent is not required: Secondary research uses of identifiable private information or identifiable biospecimens**, if at least one of the following criteria is met:

 (i) The identifiable private information or identifiable biospecimens are publicly available;

 (ii) Information, which may include information about biospecimens, is recorded by the investigator in such a manner that the identity of the human subjects cannot readily be ascertained directly or through identifiers linked to the subjects, the investigator does not contact the subjects, and the investigator will not re-identify subjects;

 (iii) The research involves only information collection and analysis involving the investigator's use of identifiable health information when that use is regulated under 45 CFR parts 160 and 164, subparts A and E..

Deciding whether a proposed activity qualifies as "not human subjects research" or "exempt" is a complex determination that can be difficult for even seasoned researchers or IRB staff to make. To ensure adequate oversight of human subjects research, most institutions require investigators to submit a form containing information that the IRB office can use to determine whether their proposed activity is "not human subjects research" or "exempt." Investigators can share this form with journals, most of which require documentation of ethics oversight of human subjects research (International Committee of Medical Journal Editors 2024).

Under the U.S. federal regulations, the IRB is the locus of institutional oversight of human subjects research. IRBs must have at least five members of varying backgrounds, including at least one member not affiliated with the institution and at least one non-scientist member. IRBs should include members who are qualified to provide the board with advice concerning scientific, professional, ethical, and legal issues and community perspectives. IRBs should include members of different genders and different races/ethnicities. If an IRB regularly reviews research on vulnerable subjects, it should include members who have knowledge and expertise concerning these categories of subjects. An IRB may invite outside, non-voting experts to assist with its decision-making. If an IRB member has a conflict of interest related to a study (for example, they are an investigator on the study), they may provide information about it to the board but cannot vote (45 CFR 46.107).

IRBs have the authority to approve, disapprove, require changes to, audit, suspend, or terminate research (45 CFR 46.108). IRBs are required to notify investigators in writing of their decisions; conduct continuing reviews of previously approved studies no less than once a year; and keep records of meeting minutes, research proposals, consent forms, SOPs, and communications with investigators; (45 CFR 46.109, 45 CFR 46.115).

To avoid duplication of effort, institutions that engage in cooperative research within the U.S. (for example, by conducting a multisite clinical trial) are required to

Table 2.2 Common rule subparts and topics

Subpart A 45 CFR 46.100	Applicability 46.100 Definitions 46.102 Assurances 46.103 Exemptions 46.104 IRB Membership 46.107 IRB Functions 46.108 IRB review 46.109 Expedited review 46.110 Criteria for approval 46.111 Cooperation research 46.114 IRB records 46.115 Consent 46.116 Consent documentation 46.117
Subpart B 45 CFR 46.200	Additional protections for pregnant women, fetuses, and neonates
Subpart C 45 CFR 46.300	Additional protections for prisoners
Subpart D 45 CFR 46.400	Additional protections for children

enter into agreements (known as reliance agreements) which designate a single IRB as the reviewing IRB. There are some exceptions to the single IRB requirement, however: more than one IRB is permitted when required by local or tribal law or the department or agency head has determined that more than one IRB is appropriate (45 CFR 46.114). Some institutions contract out IRB review and oversight to private IRBs (Emanuel et al. 2006).

To approve a proposed study, an IRB must determine that it meets all the following criteria (quoting 45 CFR 46.111, emphasis added; 21 CFR 56.109 has similar language):

1. **Risks to subjects are minimized**: (i) By using procedures which are consistent with sound research design and which do not unnecessarily expose subjects to risk, and (ii) whenever appropriate, by using procedures already being performed on the subjects for diagnostic or treatment purposes.
2. **Risks to subjects are reasonable** in relation to anticipated benefits, if any, to subjects, and the importance of the knowledge that may reasonably be expected to result. In evaluating risks and benefits, the IRB should consider only those risks and benefits that may result from the research (as distinguished from risks and benefits of therapies subjects would receive even if not participating in the research). The IRB should not consider possible long-range effects of applying knowledge gained in the research (for example, the possible effects of the research on public policy) as among those research risks that fall within the purview of its responsibility.
3. **Selection of subjects is equitable**. In making this assessment the IRB should take into account the purposes of the research and the setting in which the research will be conducted and should be particularly cognizant of the special problems of research involving vulnerable populations, such as children, prisoners, pregnant women, mentally disabled persons, or economically or educationally disadvantaged persons.
4. **Informed consent will be sought** from each prospective subject or the subject's legally authorized representative, in accordance with, and to the extent required by §46.116.
5. **Informed consent will be appropriately documented**, in accordance with, and to the extent required by §46.117.
6. When appropriate, the research plan makes adequate provision for **monitoring the data** collected to ensure the **safety of subjects**.
7. When appropriate, there are adequate provisions to protect the **privacy of subjects and to maintain the confidentiality of data**.
8. When some or all of the subjects are likely to be **vulnerable** to coercion or undue influence, such as children, prisoners, pregnant women, mentally disabled persons, or eco nomically or educationally disadvantaged persons, **additional safeguards** have been included in the study to protect the rights and welfare of these subjects (45 CFR 46.111).

An IRB must also determine that a study continues to meet these requirements when it conducts continuing review or approves an amendment to an approved study.[2]

The Common Rule includes provisions for expedited review of research. Expedited review can be conducted by the IRB Chair or a designated IRB member for new studies that are deemed to be minimal risk or minor changes (amendments) to previously approved studies. The person doing the expedited review can

[2]The Common Rule also includes provisions for approving a study that only involves collections of biospecimens for research. Biospecimen collections protocols do not need to meet all of the requirements of 45 CFR 46.111 but must meet requirements set forth on 45 CFR 46.111a8.

approve research but cannot disapprove research (45 CFR 46.110). Minimal risk is defined as:

> [T]he probability and magnitude of harm or discomfort anticipated in the research are not greater in and of themselves than those ordinarily encountered in daily life or during the performance of routine physical or psychological examinations or tests (45 CFR 46.102j).

OHRP (1998) publishes a list of types of procedure it considers to be approvable under expedited review, i.e., qualifying as minimal risk. Although the agency has said it will periodically revise the list to take into account changes in science and technology, it has not done so since 1998. Some types of research on the list include collection of urine, saliva, hair, nail clippings, and blood (from a healthy adult weighing at least 110 pounds and no more than 550 ml in an eight-week period); non-invasive medical imaging, such x-rays, magnetic resonance imaging, or sonography; audio or voice recordings; surveys, interviews, or focus groups; hearing or visions tests; moderate exercise testing; and physical measurements, such blood pressure monitoring, electrocardiography, and electroencephalography; and continuing review of research "where (i) the research is permanently closed to the enrollment of new subjects; (ii) all subjects have completed all research-related interventions; and (iii) the research remains active only for long-term follow-up of subjects; or where no subjects have been enrolled and no additional risks have been identified; or where the remaining research activities are limited to data analysis (OHRP 1998)."

The Common Rule also includes numerous informed consent requirements. Some of the key ones include (quoting 45 CFR 46.116, emphasis added):

(1) Before involving a human subject in research covered by this policy, an investigator shall **obtain the legally effective informed consent of the subject or the subject's legally authorized representative.**[3]

(2) An investigator shall seek informed consent only under circumstances that provide the prospective subject or the legally authorized representative sufficient opportunity to discuss and consider whether or not to participate and that **minimize the possibility of coercion or undue influence.**

(3) The information that is given to the subject or the legally authorized representative shall be in **language understandable to the subject** or the legally authorized representative.

(4) The prospective subject or the legally authorized representative must be provided with the **information that a reasonable person would want to have in order to make an informed decision** about whether to participate, and an opportunity to discuss that information.

(5) Except for broad consent obtained in accordance with paragraph (d) of this section:

  (i) **Informed consent must begin with a concise and focused presentation of the key information** that is most likely to assist a prospective subject or legally authorized representative in understanding the reasons why one might or might not want to participate in the research.

  (ii) Informed consent as a whole must present information in sufficient detail relating to the research, and must be organized and presented in a way that does **not merely provide lists of isolated facts, but rather facilitates the prospective subject's or legally authorized representative's understanding of the reasons why one** might or might not want to participate.

[3] A legal representative could include a parent, guardian, or next of kin, depending on local laws.

(6) No informed consent may include any exculpatory language through which the subject or the legally authorized representative is made to waive or appear to waive any of the subject's legal rights, or releases or appears to release the investigator, the sponsor, the institution, or its agents from liability for negligence.
(b) Basic elements of informed consent. Except as provided in paragraph (d), (e), or (f) of this section, in seeking informed consent the following information shall be provided to each subject or the legally authorized representative:

(1) **A statement that the study involves research, an explanation of the purposes of the research and the expected duration of the subject's participation, a description of the procedures to be followed, and identification of any procedures that are experimental;**
(2) **A description of any reasonably foreseeable risks or discomforts to the subject;**
(3) **A description of any benefits to the subject or to others that may reasonably be expected from the research;**
(4) **A disclosure of appropriate alternative procedures or courses of treatment,** if any, that might be advantageous to the subject;
(5) **A statement describing the extent, if any, to which confidentiality of records identifying the subject will be maintained**;
(6) For research involving more than minimal risk, an explanation as to whether any compensation and an explanation as to whether any medical treatments are available if injury occurs and, if so, what they consist of, or where further information may be obtained;
(7) An explanation of **whom to contact** for answers to pertinent questions about the research and research subjects' rights, and whom to contact in the event of a research-related injury to the subject;
(8) A statement that **participation is voluntary, refusal to participate will involve no penalty or loss of benefits to which** the subject is otherwise entitled, and the subject may discontinue participation at any time without penalty or loss of benefits to which the subject is otherwise entitled; and
(9) **One of the following statements about any research that involves the collection of identifiable private information or identifiable biospecimens:**

(i) **A statement that identifiers might be removed** from the identifiable private information or identifiable biospecimens and that, after such removal, the information or biospecimens **could be used for future research studies or distributed to another investigator for future** research studies without additional informed consent from the subject or the legally authorized representative, if this might be a possibility; or
(ii) A statement that the subject's information or biospecimens collected as part of the research, even if identifiers are removed, will not be used or distributed for future research studies.

(c) Additional elements of informed consent. Except as provided in paragraph (d), (e), or (f) of this section, one or more of the following elements of information, when appropriate, shall also be provided to each subject or the legally authorized representative…

(5) A statement that significant new findings developed during the course of the research that may relate to the subject's willingness to continue participation will be provided to the subject;

(7) **A statement that the subject's biospecimens (even if identifiers are removed) may be used for commercial profit and whether the subject will or will not share in this commercial profit;**

(8) **A statement regarding whether clinically relevant research results, including individual research results, will be disclosed to subjects, and if so, under what conditions; and**

The Common Rule also includes provisions for waiving or altering informed consent requirements. For example, an IRB can waive or alter consent requirements if it finds that (45 CFR 116.f):

(i) The research involves no more than minimal risk to the subjects;
(ii) The research could not practicably be carried out without the requested waiver or alteration;
(iii) If the research involves using identifiable private information or identifiable biospecimens, the research could not practicably be carried out without using such information or biospecimens in an identifiable format;
(iv) The waiver or alteration will not adversely affect the rights and welfare of the subjects; and
(v) Whenever appropriate, the subjects or legally authorized representatives will be provided with additional pertinent information after participation.

The Common Rule also allows investigator to "obtain information or biospecimens for the purpose of screening, recruiting, or determining the eligibility of prospective subjects without the informed consent of the prospective subject" if certain conditions are met (45 CFR 46.116 g).

The FDA regulations include provisions, which will be discussed in Chap. 5, for conducting research without informed consent for life-threatening conditions for which there are no alternative treatments (emergency research) (21 CFR50.24).

The Common Rule also includes consent documentation requirements. Documentation can take two forms: the long form; that is, a written document that includes all the required consent elements and is given to the subject (or their representative); and the short form, that is, a document stating that the consent elements have been presented to the subject orally. The short form must be accompanied by a written summary of the research that is given to the subject (45 CFR 46.117). Although not required by the regulations, must IRBs restrict the use of the short form to unanticipated enrollment of a subjects who have limited English proficiency or are blind or illiterate (Resnik and Jones 2006).

The IRB can waive documentation if one of the following apply (45 CFR 46.117c):

(i) That the only record linking the subject and the research would be the informed consent form and the principal risk would be potential harm resulting from a breach of confidentiality. Each subject (or legally authorized representative) will be asked whether the subject wants documentation linking the subject with the research, and the subject's wishes will govern;
(ii) That the research presents no more than minimal risk of harm to subjects and involves no procedures for which written consent is normally required outside of the research context; or
(iii) If the subjects or legally authorized representatives are members of a distinct cultural group or community in which signing forms is not the norm, that the research presents no more than minimal risk of harm to subjects and provided there is an appropriate alternative mechanism for documenting that informed consent was obtained.

The FDA regulations do not include provisions for waiving documentation.

2.6 Continuing Scandals and Controversies

Although the federal regulations helped establish ethical and legal standards for conducting research with human subjects, controversies and scandals continued to surface. Some of these have involved exposés of past research. Congress held hearings in 1975 and 1977 on Central Intelligence Agency (CIA) secret mind control research, which took place from 1951 to 1973. The agency conducted human behavioral experiments using lysergic acid diethylamide (LSD), sodium pentothal, hypnosis, radiation, and electric shocks. The purpose of the research was to develop methods for obtaining information from someone, planting suggestions in someone's mind, and other forms of controlling thought or behavior (Goliszek 2003). These experiments, which often occurred without consent, involved dozens of human subjects. On November 19, 1953, CIA researchers gave LSD to unwitting human subjects. One of these victims, civilian army employee Frank Wilson, became paranoid, agitated, and severely depressed. Several days after taking LSD, Wilson died after jumping out a tenth-floor window. Some of the "voluntary" CIA LSD experiments involved drug addicts who were serving prison sentences for drug violations.[4] The researchers gave the subjects a variety of hallucinogenic drugs and observed them for several days. To pay them for their participation, the researchers provided the subjects with morphine, or any other drug they wanted, after the experiment was over (Goliszek 2003). Ironically, declassified documents have revealed that research ethics pioneer, Henry Beecher, whose article was discussed earlier, conducted secret experiments for the CIA on the use of LSD as an interrogation tool on unwitting human subjects from 1951 to 1952 (McCoy 2007).

On October 26, 1984, in Loma Linda Hospital near Riverside, CA, an infant born with a defective heart, known as Baby Fae, received a heart transplant from a female baboon named Goobers, who was about 7–10 months old (Pence 1995). Fae's doctors determined that she would die very soon if she did not receive a new heart. No human infant hearts were available. On November 12, Baby Fae began rejecting the baboon heart and died on November 15. Baby Fae's plight attracted a great deal of national attention. Her family received thousands offers of emotional and financial support. Critics argued, however, that the procedure was an ill-conceived, costly, and exploitative experiment that had little chance of benefitting the patient. Although the hospital's IRB approved the procedure, there was considerable confusion among IRB members and Baby Fae's physicians as to whether the procedure was therapy or research. Critics also questioned whether Baby Fae's parents gave effective informed consent for the procedure, especially given the stress they were under and the lack of viable alternatives. Critics also argued that the use of animals was unjustified, given the operation's minute chances of success (Pence 1995).

In 1994, the Clinton administration began declassifying thousands of documents pertaining to secret experiments funded or conducted by the U.S. government during the Cold War period (1944–1974). Many of these were human radiation

[4] Voluntariness will be discussed in Chap. 5.

experiments sponsored by the Department of Energy (DOE) or Department of Defense (DOD) involving approximately 4000 civilians or military personnel. The Clinton Administration appointed a national commission to review these experiments and determine whether they violated legal rules or ethical standards. Most of these experiments occurred after the international community had adopted the Nuremberg Code but before the U.S. had developed any human research regulations (Advisory Committee on Human Radiation Experiments 1995). Some of the survivors of these experiments and families of the deceased filed a class-action lawsuit against researchers, institutions, and the U.S. government (*In Re Cincinnati Litigation* 1995).

Investigators, government officials, and military leaders invoked a utilitarian rationale for this research: the U.S. needed to conduct radiation experiments to have a better understanding of how radiation affects human health to obtain information that could be used to formulate national security strategies related to a potential conflict with the Soviet Union (Goliszek 2003). Most of these experiments involved exposure to low doses of radiation, and did not entail any serious, long-term harm to subjects. However, some of the studies increased the lifetime cancer risk of the subjects, and some used children or pregnant women. Also, consent was not obtained in most of the studies (Advisory Committee on Human Radiation Experiments 1995). Some noteworthy studies include (Advisory Committee on Human Radiation Experiments 1995; Goliszek 2003):

- In the 1940s, researchers at Vanderbilt University gave radioactive iron pills to pregnant women to understand how radiation impacts fetal development;
- From the 1940s to the 1960s, the military exploded nuclear weapons or released radioactive fallout clouds in the South Pacific Islands, Utah, New Mexico, Nevada, and Washington State to study the impact of radiation on ecosystems, organisms, and human beings;
- From the 1940s to the 1960s, investigators injected radium into the noses of about 1500 military personnel, many of whom developed headaches and nosebleeds;
- From 1951 to 1956, researchers at the Texas M.D. Anderson Cancer Center administered total body radiation to 263 male cancer patients, many of whom experienced nausea, vomiting, or bone marrow depletion, to understand how radiation affects motor skills needed for flying aircraft;
- In the 1950s, researchers at Columbia University administered radioactive isotopes of calcium and strontium to patients with terminal cancer to understand how human bones and tissues absorb radioactive material;
- From 1963 to 1971, researchers at Oregon State Prison x-rayed the testicles of male prisoners (most of whom were African American) to understand how radiation affects sperm function.

In 2010, an experiment hidden from view for over 50 years came to light while Wellesley College professor Susan Reverby was conducting research on the Tuskegee Syphilis Study. When she was examining Public Health Service (PHS) records on the study she discovered another PHS study that took place between

1946 and 1948 in Guatemala. The study exposed 696 subjects to syphilis to determine whether taking penicillin prophylactically can prevent the disease. 427 of the subjects developed syphilis. To expose human subjects to syphilis, the researchers first asked prisoners to have sex with prostitutes who had the disease. When this method proved to be ineffective, they inoculated subjects' cheeks, forearms, and penises with syphilis. The researchers selected Guatemala for the study site, in part, because they knew they would have trouble getting permission for some of their research activities (such as subjects having sex with prostitutes) in the U.S. After Reverby published her findings, President Barack Obama issued an official apology to the Guatemalan government and asked his Presidential Commission for the Study of Bioethical Issues to investigate the incident and review U.S. human research protections to ensure that nothing like this could happen again (Semeniuk and Reverby 2010).

During the 1990s, pharmaceutical companies began conducting clinical trials in developing nations. Although some companies were genuinely interested in helping address public health problems that were of concern to local populations, others moved studies to these venues to save money on labor and facilities and reduce regulatory burdens. At that time, many developing nations had no laws, regulations, or guidelines concerning the conduct of research with human subjects. Ethicists and human rights activists became concerned that (1) pharmaceutical companies were conducting studies in these nations because they knew these studies would not be approved in developed nations but would pass muster in developing nations (a practice sometimes referred to as "ethics dumping") and (2) they were exploiting local participants and populations by placing them at significant risk without providing substantial benefits (Hawkins and Emanuel 2008; Nordling 2018).[5]

A vaccine trial conducted by Pfizer in 1996 illustrates these concerns. Pfizer claimed in was testing the vaccine in response to an epidemic of meningitis in Africa, which would eventually kill over 15,000 people. The company enrolled nearly 100 children in a clinical trial of Trovan, an experimental meningitis drug that had never been tested in children. The children were randomly assigned to two groups, one which received Trovan and another which received a medication commonly used to treat the disease. While there is no evidence that the children suffered additional, significant harm from participating in the study—five died in the Trovan arm of the study and six died in the comparator arm—there is also no evidence that the parents provided consent for the study or knew their children were participating in an experiment. Also, Pfizer did not obtain proper ethics approvals for the study. Parents of the children sued Pfizer and eventually won an out-of-court settlement. In 1997, the FDA had approved the Trovan for use in adults but not children and restricted use of the drug, based on evidence that it increased the risk of liver damage and death (Stephens 2006).

[5]The Guatemala syphilis research uncovered by Reverby (see discussion above) is an example of ethics dumping that predates the concerns about this practice that emerged in the 1990s.

In 1998, the Environmental Working Group published a report accusing chemical companies of conducting unethical pesticide experiments on human subjects. The experiments involved administering low doses of pesticides to healthy human subjects to determine whether the chemicals produce adverse at effects at minimal exposure levels. In most of these experiments, human subjects ingested small quantities of pesticides. The researchers collected biological samples from the subjects to measure pesticide metabolite levels in the blood and urine and biochemical indicators of toxicity. Many of the subjects were college students who were paid as much as $470 (about $900 in 2024 dollars) for their participation, but some were company employees (Resnik 2012). Critics said that the studies were scientifically and ethically flawed because: (1) sample sizes were too small to generate valid data; (2) some of research subjects were company employees, which was potentially coercive;[6] (3) appropriate measures to minimize risks were not implemented; (4) data gathered on adults were not relevant to pediatric risks; and (5) the results were not likely to yield useful information for society because the health risks of the pesticide exposures were already well-understood (Lockwood 2004).

The companies conducted these experiments to generate data to submit to the Environmental Protection Agency (EPA) to convince the organization to increase allowable levels for these chemicals on foods. Congress passed the Food Quality Protection Act (FQPA) in 1996 to require the EPA to enhance safety standards for pesticide residues on food. Prior to the adoption of the FQPA, the EPA regulated allowable pesticide residue levels on foods using two safety factors of ten, one to account for variations between animals and humans and another to account for variations among humans. The FQPA mandated an additional safety factor of ten to protect children from exposure risks, i.e., an allowable exposure of 1/1000 of the level which produces no adverse effects in rodents. Pesticide companies and representatives from the agricultural industry objected to these enhanced safety standards because they believed the standards would interfere with effective pesticide use and could produce pesticide resistance. The companies conducted the studies to convince the EPA to reduce its animal-human safety factor (Resnik 2012).

The EPA had adopted the Common Rule for studies sponsored by the agency but did not have any regulations for studies conducted by private companies (third-parties). The agency had accepted third-party data on a case-by-case basis and applied relevant ethical and scientific criteria to the research. In 1998, the EPA asked the National Research Council (NRC) to study the ethical and scientific issues related to testing pesticides and other toxic chemicals on human subjects and announced that it would not accept any third-party human data until the NRC released its report. Pesticide manufacturers sued the EPA on the grounds that it had engaged in unlawful rulemaking because it had not given proper notice or solicited public comments prior to making the decision to no longer accept third party human data. In 2003, a federal court ruled in favor of the plaintiffs and ordered the EPA to

[6] Coercion will be discussed in Chap. 5.

engage in appropriate rulemaking for research third-party with human subjects (Resnik 2012).

By 2004, environmental groups, environmental researchers, ethicists, and members of the public had raised serious concerns about testing pesticides on human subjects. Some argued that privately funded human pesticide testing was inherently unethical and should never be performed (Stokstad 2005). In February of 2004, the NRC issued its report on experiments that involve testing pesticides or other toxic chemical on human subjects. The report said that these types of experiments were acceptable if they meet stringent scientific and ethical standards, such as good study design, informed consent, risk minimization, social value, and protection of vulnerable subjects. The NRC recommended that the EPA adopt the Common Rule for third-party research (National Research Council 2004). The EPA began to take steps to revise its human subjects regulations in accordance with the NRC's recommendations (Resnik 2007a, b).

In the fall of 2004, an observational study of children's pesticide exposures generated a controversy that would have a dramatic impact on the EPA's human subjects regulations. The study, named Children's Environmental Exposure Research Study (CHEERS), was sponsored by the EPA, Centers for Disease Control and Prevention (CDC), and Duvall County, Florida Health Department. The goal of the study was to measure and observe children's exposures to pesticides and household chemicals in the home. The investigators planned to recruit sixty families with children living in the home, including a group with high pesticide use and a control group of families with low pesticide use (or no pesticide use). The study would include five home visits over a two-year period. During the home visits, the investigators would collect biological samples from the children (blood and urine) and soil and dust samples from the home. Families would keep a journal of their pesticide and chemical use and videotape their children's activities. For their participation, parents would receive $970 (about $1600 in 2024 dollars) and t-shirts, and they would be allowed to keep the camera used for videotaping. Investigators and the sponsors regarded the study as important for promoting children's health because little was known about children's exposures to pesticides and other chemicals in the home. The American Chemistry Council (ACC) contributed $2 million to the study, but it had no role in designing the research or interpreting the results. IRBs and peer review committees from the EPA, CDC, and Duvall County Department of Health approved the study and classified it as minimal risk (Resnik and Wing 2007).

Before the study could get off the ground, a controversy erupted. Environmental groups, human research ethics groups, and journalists portrayed CHEERS as an intentional exposure study that would use children as "guinea pigs." This impression was mistaken because the study had an observational, rather than experimental design (see Appendix A). However, some were concerned that parents might start using pesticides to qualify for enrollment in the high-use group in the study. To prevent this from happening, the investigators planned to make an initial screening visit to the home immediately after enrolling a family in the study to determine the family's level of pesticide use. They also told families they were free to withdraw from the study at any time and could remain in the study even if they were in the high-use group and stopped using pesticides (Resnik and Wing 2007).

Another criticism of the study is that it was targeting low-income minority groups living in Duvall County. However, investigators selected the site because they knew that many families in the area were using large amounts of pesticides to control roaches and other pests. Race, ethnicity, and income were not part of the study's inclusion or exclusion criteria. Although the study did not target racial/ethnic minorities or low-income participants, it probably would have enrolled a disproportionate number of families from these groups (Resnik and Wing 2007).

Critics also argued that the amount of money offered to the parents was excessive and could induce them to enroll in the study against their better judgment, especially since it was likely that most of the parents in the study would be economically disadvantaged. However, the amount of money offered to the parents was probably a fair rate for their participation since the study involved considerable inconvenience and about 150 hours of the parents' time. The parents would receive $6.47 per hour, slightly above the federal minimum wage at the time of $5.15 per hour (Resnik and Wing 2007).[7]

Although the EPA and CHEERS investigators attempted to defend the study against these criticisms, the controversy persisted. In the spring of 2005, California Representative Barbara Boxer and Florida Senator William Nelson threatened to block Stephen Johnson's nomination as EPA administrator if the agency did not study the study. The EPA surrendered to this political pressure and cancelled the concerns study. Later that year, Congress passed a law prohibiting the EPA from sponsoring pesticide experiments on children or pregnant or nursing women. In 2005, the EPA adopted regulations that went beyond the requirements of the law. The regulations prohibit the agency from sponsoring any research that intentionally exposes children or pregnant or nursing women to chemicals, allergens, cosmetics, air pollution, or other environmental agents. These regulations are more restrictive than the Common Rule's protections for children and pregnant women, which allow such research if the IRB categorizes it as minimal risk (Resnik 2007a, b).

2.7 Concerns About Integrity in Biomedical Research

In the 1980s and 1990s there were growing concerns about integrity in biomedical research (and to a lesser extent in other types of research), due to some highly publicized cases of data fabrication and falsification in studies funded by the U.S. government as well as allegations of bias and conflict of interest (COI) in privately-funded research.[8] During that period, government agencies, including the National Institutes of Health (NIH) and National Science Foundation (NSF) developed policies

[7] Payments to human subjects will be discussed in Chap. 5.

[8] Research integrity will be discussed in more depth in Chap. 9.

pertaining to misconduct and conflicts of interest in research. Academic institutions were obliged to comply with these policies as condition of receiving funding. The FDA also developed COI policies for research it regulates and for its advisory committees. In 2001, the federal government finalized a policy that defines research misconduct as fabrication, falsification, or plagiarism. In 2005, the NIH strengthened its COI policies for intramural researchers, and in 2011, the NIH and NSF strengthened their policies for extramural researchers.[9] Scientific journals developed policies pertaining to research misconduct and disclosure of financial or other conflicting interests (Shamoo and Resnik 2022).

One of the misconduct cases with human subjects which made national headlines involved the NIH-funded National Surgical Adjuvant Breast and Bowel Project (NSABP). In 1993, University of Montreal professor of Surgery Roger Poisson admitted to altering medical records from 117 of his patients between 1977 and 1990 to enable them to qualify for enrollment. Poisson admitted wrongdoing after being investigated by his institution and the Office of Research Integrity (ORI). In defending his actions, Poisson claimed that he was trying to help his patients participate in a study which could benefit them. The ORI ordered Bernard Fisher, a University of Pittsburgh cancer researcher and NSABP chairman, to reanalyze the data to determine whether the falsification impacted the study's outcome. Fortunately, the reanalysis showed that Poisson's misconduct had no effect on the study's overall results. The study found that lumpectomies are just as effective as mastectomies at treating breast cancer tumors less than 4 cm wide. Although Fisher was never accused of misconduct, he resigned his position to appease Congress (which was investigating the incident) and the NIH. Fisher sued the government for damaging his reputation because the NIH had labelled 93 of his papers with a warning that misconduct was suspected (Resnik 2008).

In 1998, British surgeon Andrew Wakefield and 12 coauthors published a paper in *Lancet* in 1998 speculating that exposure to the measles, mumps, and rubella (MMR) vaccine can cause autism in children. The paper claimed that 12 healthy children developed gastrointestinal disease and developmental regression after receiving the vaccine. The paper had a negative impact on vaccination rates in the U.K. and other countries because the anti-vaccination community cited Wakefield's results as proof that childhood vaccinations cause autism. British journalist Brian Deer began investigating Wakefield's research. In 2004, Deer published an article in the *Sunday Times* claiming that Wakefield had not disclosed a significant financial interest and had not obtained ethics board approval for the study. Wakefield's research had been supported by a law firm that was suing vaccine manufacturers, and a lawyer for the firm had helped Wakefield recruit patients. Wakefield did not disclose his relationship to the law firm in the 1998 paper. In 2010, the U.K.'s General Medical Council (GMC) revoked Wakefield's license to practice medicine

[9] Intramural researchers are employed by the NIH; extramural researchers are employed by academic institutions that receive funding from the NIH for their research.

following an investigation which concluded that he had not disclosed a significant financial interest and had performed risky procedures, such as colonoscopies and lumbar punctures, without appropriate pediatric qualifications or ethics committee approval. *Lancet* retracted the paper shortly thereafter. Deer continued pursuing Wakefield. In 2011 he published an article in the *British Medical Journal* accusing the surgeon of fabricating and falsifying data in the 1998 paper. Deer reviewed the medical records of the children involved in the study and found that five of the children already had developmental problems before entering the study and that four of the children which Wakefield said had autism or developmental regression were healthy. Deer also found that the researchers had changed nine normal pathology results to colitis. Wakefield has denied these allegations and continues to advise anti-vaccine groups (Shamoo and Resnik 2022).

On March 17, 2005, University of Montreal professor Eric Poehlman admitted to fabricating or falsifying data on 15 federal grant applications (worth $2.9 million in funding) and 17 papers over a ten-year period as part of a comprehensive settlement dealing with criminal, civil, and administrative law charges against him. For several years, the University of Vermont (where Poehlman worked from 1987–1993 and 1996–2001) and the ORI had been investigating allegations of research misconduct against Poehlman made by Walter Denino, one of his research assistants. Denino found that Poehlman had changed some data points on Excel spreadsheets to support his hypotheses. In one study, most of the research subjects did not exist. During the investigation, Poehlman destroyed evidence and gave false testimony to investigators. Under the terms of the settlement, Poehlman agreed to pay $180,000 to the government and $16,000 to Denino's attorney. He also agreed to a lifetime ban on receiving federal research funding and to retract papers affected by his misconduct. Poehlman was also sentenced to serve one year and one day in federal prison for defrauding the government. He had published highly-cited papers on menopause, metabolism, and aging. In his defense, Poehlman said he broke the rules of science because he felt significant pressure to obtain funding to keep his lab operating and support the salaries of his students and staff (Shamoo and Resnik 2022).

Another episode of research misconduct involving human subjects made international headlines in December 2005, when Seoul University professor Woo Suk Hwang admitted to fabricating and falsifying data in two papers published in *Science* in 2004 and 2005, which described the derivation of human embryonic stem cell lines by therapeutic cloning.[10] Hwang had become a national hero in South Korea for bringing recognition to his country for its scientific accomplishments. Earlier that year, Hwang had been the first person to clone a dog (named Snuppy). Problems with the research began to surface when one of Hwang's collaborators, University of Pittsburgh stem cell researcher Gerald Schatten, said that Hwang had

[10] In therapeutic cloning, researchers transfer a nucleus from a patient's somatic cells into a zygote which has had its nucleus removed to produce embryonic stem cells which are virtually genetically identical to cells taken from the patient.

misled him about the source of human eggs used in the experiments. Contrary to what Hwang had told Schatten, some of the eggs had come from Hwang's laboratory technicians. Though this practice was legal in South Korea, many regarded it as unethical because it was potentially coercive. Hwang admitted that he helped the donors fill out some paperwork and that he paid them up to $1400 (about $2200 in 2024 dollars) for their eggs. Allegations of research fraud emerged when Sung Rho, one of Hwang's collaborators, informed the media that Hwang had fabricated 9 out of 11 images of cell lines in the 2005 paper. An investigation by Seoul University found that all of the images of cell lines in the papers had been fabricated or falsified, and that Hwang had used 273 eggs, not the 185 he had reported. In May 2005, a South Korean court convicted Hwang and five collaborators of fraud and embezzlement (totaling $3 million), but the court suspended their sentence (Shamoo and Resnik 2022).

During the 1990s, financial ties between academic scientific researchers and private companies expanded. Private companies (especially pharmaceutical and biotechnology companies) boosted their funding for academic scientific research, continuing medical education, and other forms of financial support for universities and medical centers. Researchers increasingly reported financial relationships to private sponsors, such as stock ownership, and receipt of fees for consulting or speaking. Researchers also began more aggressively pursuing intellectual property related to their work (such as patents) and formed start-up companies (often with institutional support) to commercialize their discoveries and inventions. Academic institutions also became more entangled with industry via ownership of stock or intellectual property, licensing of intellectual property, and receipt of donations from private companies. Academic institutions formed technology transfer offices to manage their intellectual property and help researchers commercialize research and created private foundations to hold stock and engage in corporate fundraising (Krimsky 2003; Resnik 2007c).As these financial ties between academia and industry strengthened, scientists, ethicists, policymakers, and members of the public became increasingly concerned that the financial interests of private sponsors could threaten the objectivity and integrity of biomedical research. Researchers began to publish studies demonstrating relationships between sources of funding and finan cial interests and research results (Sismondo 2008a, b). Researchers also published articles reporting that a significant percentage of industry-sponsored studies have "ghost authors," i.e., individuals who have made contributions to the study which meet authorship criteria but who are not named as authors. For example, Gøtzsche et al. (2007) found that 75% of studies reporting industry-sponsored clinical trials had ghost authors. Corporate sponsors sometimes use ghost authors to conceal financial interests and industry ties related to published research (Resnik 2007c).

Jesse Gelsinger (age 18) with Rocky Balboa statue. (Used with permission from Paul Gelsinger)

In 1999, a high-profile tragedy occurred which gave a human face to concerns about COIs in biomedical research.[11] Jesse Gelsinger was an eighteen-year-male with ornithine transcarbamylase (OT) deficiency, a genetic disease in which the individual lacks a functional copy of a gene that codes for OT, which plays an important role in protein metabolism. Most children born with this condition die in infancy, but Gelsinger had a mild form of the illness which had been well-managed through with drugs to help him metabolize proteins and dietary restrictions. Gelsinger enrolled in a Phase I (first in human) gene therapy trial at the University of Pennsylvania. The goal of the study was to determine whether it was possible to use an adenovirus vector to transfer copies of the OT gene into the subject's liver. The investigators had planned to perform the experiments on children with the disease but decided to focus on adults, based on the advice of bioethicist Art Caplan, who argued that parents might face self-imposed emotional pressures to consent for the experiment. Although the study had a small chance of benefiting Gelsinger, its

[11] See Chap. 9 for further discussion of COIs.

main purpose was to demonstrate whether the procedure would work and be safe. Unfortunately, Gelsinger developed a severe immune response to the infusion of the adenovirus and died shortly thereafter (Resnik 2007c). Gelsinger's family sued the study's principal investigator, James Wilson, Caplan (who was later dropped from the lawsuit), the university, and Genovo, the company which sponsored the study. The lawsuit, which was settled of our court, alleged that the investigators mislead Gelsinger about the benefits and risks of the study and had failed to inform him about adverse effects (including death) in prior animal experiments and the investigator's and university's financial interests. Wilson had 20 patents on gene therapy techniques, some of which had been transferred to the university. Wilson also helped to start Genovo and held 30% of its stock even though university policies required that equity interests be 5% or less. The university also owned stock in Genovo.

The FDA and OHRP investigated the incident and found that there were problems with adverse event reporting in gene therapy research and that the consent document that Gelsinger signed was inadequate. The NIH and Association of American Medical Colleges revised their COI guidelines for research with human subjects in response to this episode. Gelsinger's death undermined the public's trust in gene therapy research and set the field back several years (Resnik 2007c; Gelsinger and Shamoo 2008). In an editorial commenting on the incident, DHHS Secretary Donna Shalala observed that "[T]he public's confidence in our work, our competence, and our ethics has been seriously shaken (Shalala 2000: 8)." Jesse Gelsinger's father, Paul Gelsinger, had this to say about the COIs in this case:

> The financial conflicts of interest for both the sponsor (James Wilson) and the University were basically ignored by the University's Conflict of Interest Committee. This led to a blindness on the part of all parties to the dangers they were seeing. These were not bad men doing evil. They were men blinded by ambition and greed. They have told me that they could not have foreseen Jesse's death, yet their data was screaming at them to stop. We trusted a system that was untrustworthy, one that didn't even pay attention to its own stops...So, my son, doing the right thing, was killed by a system and people rife with conflicts of interest, and real justice has been found to be very lax. It's essentially business as usual (Gelsinger 2008).

In June 2001, not long after the Gelsinger incident, another significant death of a human research subject occurred when Ellen Roche, a healthy 24-year-old technician at Johns Hopkins University (JHU), died from respiratory failure a month after inhaling hexamethonium as part of study to better understand the causes of asthma. In July of that year OHRP temporarily halted most of JHU's federally-funded human subjects research. An investigation into the incident found that study investigators had not adequately reviewed the literature on the risks of inhaling the drug. Most of the participants in the experiment were JHU employees (Keiger and De Pasquale 2002).

2.8 Concerns About Research Involving Human Biospecimens

During the 1990s, there were growing concerns about the ethics of research involving human biospecimens (e.g., blood, tissues, DNA). Biospecimen are playing an increasingly important role in biomedical research. Scientists can use biospecimens to understand basic physiologic and metabolic processes and mechanisms, as well as relationships between genetics, environment, and disease, and clinical investigators can use them to measure the effects of medical treatments. Many biomedical researchers collect and store human biospecimens for their own work or to share with other scientists. Some studies involve the creation of biobanks which store biospecimens and associated clinical or genetic data. For example, NIH's All of Us (2021) study, aims to collect biospecimens and health data from over one million U.S. adults to better understand the relationship between genes, the environment, and disease. There are many different types of biobanks ranging in size from a few hundred samples to millions of samples. Some are non-profit organizations operated by academic institutions or government agencies, while others are for-profit organizations run by private companies. Some have governing boards, while others do not (Weir and Olick 2004; Budimir et al. 2011).

Most researchers today obtain informed consent from human subjects before collecting research with biospecimens from them. In thinking about the ethics of research with biospecimens, it is useful to distinguish between three types of biospecimens:

- Personally identified biospecimens: the specimens include the subject's name, medical record number, address, or other information that can personally identify the person;
- Coded biospecimens: the investigator has removed personally identifying information from the specimens but marked them with a code linked to personally identifying information;
- Anonymous biospecimens: the specimens include no identifying information; if the specimens were coded at one point, the key to the code has been destroyed.

The Common Rule (45 CFR 46) does not require consent from human subjects when researchers are conducting research on biospecimens that do not include information that can identify participants because the specimens are anonymous, or they are coded and the researcher does not have access to the code.

Long before the Henrietta Lacks case entered the public spotlight (see discussion in Chap. 1), leftover tissues and blood from medical procedures have long been an important resource for researchers. Until fairly recently, doctors rarely obtained their patients' consent to use leftover blood or tissues for research. In 1990, the California Supreme Court decided a case with implications for this practice of using tissues without consent. In 1976, John Moore had a splenectomy at the University of California at Los Angeles (UCLA) medical center to treat his leukemia. David Golde, the physician who had recommended the surgery, discovered that Moore's tumor was overproducing proteins that regulate immune cell functions known as lymphokines. Golde decided to develop a cell line from Moore's tissue and asked him to make several additional visits to the medical center to provide blood and tissue samples, without telling him the true purpose of the visits. Golde reached an agreement with UCLA and a pharmaceutical company to commercialize the cell line. Golde patented the cell line and transferred it to UCLA, which sold the patent to company. The cell line generated several billion dollars in revenue (Resnik 2007a, b, c). Moore eventually found out about this clandestine affair and sued the university, the company, and the investigator for conversion (wrongfully taking or using someone's property) and lack of informed consent. The court did not recognize Moore's conversion claim because it said that he did not have a property interest in tissue which he had abandoned, but it did recognize his informed consent claim because it said that Golde should have told Moore about his true reasons for collecting the additional samples and his plans to commercialize the cell line (*Moore v. Regents of the University of California* 1990).

Several courts have revisited the property issues addressed in *Moore*, but so far none have recognized that patients or human subjects have property rights pertaining to tissues that they have abandoned (e.g., tissues leftover from surgery) or voluntarily given away (e.g., tissues donated to research institutes) (Shuster 2017).

Consent for use biospecimens and data was an important issue in 1998 when Iceland's government created a database of the health care records of its citizens which could be linked to genealogical and genetic data. The government granted an exclusive license to deCODE genetics, a for-profit company, to create and manage the database. Only de-identified data would be shared with biomedical researchers. The database was controversial because it used an opt-out consent process: citizens' records would be included in the database unless they requested to not include their data. Critics argued that an opt-out process did not grant sufficient respect for citizens' rights to control access to their medical records, because some people might not know about the opt-out procedure or understand how to opt-out (Weir and Olick 2004).[12]

A dispute between Arizona State University (ASU) and the Havasupai Native American tribe, which unfolded in the early 2000s, also raised important issues concerning consent for the use of biospecimens and data. In 1990, ASU investigators collected 200 blood samples from members of the tribe for a study they told tribal leaders would focus on the genetics of diabetes. However, the consent form indicated that samples and data would also be used from research on mental illness. Members of the tribe became upset when they learned that researchers had shared samples and data with other scientists without their permission and used them to study schizophrenia, tribal in-breeding, and the tribe's evolutionary history. They filed a $50 million lawsuit against the university and the investigators, which was settled out-of-court in April 2010. Under the terms of the settlement, ASU agreed to return the blood samples to the tribe and pay the 41 study participants $17,000 each. This incident soured the university's relationship with the tribe and many members

[12] Opt-in vs. opt-out consent will be discussed in Chap. 5.

vowed to never participate in ASU studies again (Mello and Wolf 2010). In her testimony before the President's Commission for the Study of Bioethical Issues, Havasupai Tribal Council member Carletta Tilousi said: "The reason why I'm here today was a lot of our blood samples were misused. The people's trust in the institution was shattered (Tilousi 2011)."

Consent for access to blood samples was also an issue in a U.S. law passed on December 14, 2014, which required investigators to obtain parental consent to conduct research on de-identified blood samples from newborn screening programs (Bayefsky et al. 2015). All states have mandatory newborn screening programs supported by funds from the state and the Health Resources and Services Administration, a federal agency. States test for a varietal of congenital diseases and disorders including deafness, cystic fibrosis, phenylketonuria, sickle cell anemia, and severe combined immune deficiency. Prior to the enactment of the law, parental consent was not necessary to conduct research on de-identified leftover newborn blood samples, because this would not be classified as human subjects research under the Common Rule. The new law defined research on de-identified newborn blood samples as human subjects research, so consent would be required. Some investigators objected to this change in federal policy because would impede research, since many parents might refuse to consent. They also argued that consent should not be necessary because research on de-identified blood samples poses virtually no risk to newborns (Bayefsky et al. 2015).

Research involving the analysis of biospecimens or data poses no physical risks to the subjects, other than the small risks associated with sample collection, such as bleeding or bruising from a venipuncture. The main risk from this research is the potential loss of privacy if confidentiality is breeched and personal information is inadvertently disclosed, but this risk can be minimized by implementing security measures, such as removing personal identifiers from the specimens or data, limiting access to specimens and data, implementing cybersecurity measures (such as encryption and firewalls) and storing specimens securely (Shamoo and Resnik 2022). If research on biospecimens or data is low risk, why have people raised ethical concerns about it?

One reason is that some people would like to control how their biospecimens or data are used. In the ASU case, members of the Havasupai tribe did not want their blood samples or data to be used from research on mental illness, in-breeding in the tribe, or tribal origins. Some people may object to having their biological specimens or data used for commercial purposes. Women who donate eggs for research may not want researchers to use them for cloning human embryos or creating human animal chimeras (Shamoo and Resnik 2022).

Another reason is that de-identification may no longer guarantee confidentiality because statisticians have developed methods for re-identifying individuals in de-identified genomic databases from a sample of their DNA or their phenotypic information, such as medical diagnoses, height, weight, gender, and so on (Lowrance and Collins 2007; Homer et al. 2008). Statisticians have also shown how to identify individuals in de-identified databases based on their some of their demographic and medical information (see discussion in Chap. 6).

The NIH revised its genomic data-sharing policy in 2010 due to concerns about re-identification. Previously, the NIH had required funded investigators to make their human genomic data available to the public by depositing it one of two publicly accessible NIH websites, the Genetic Markers of Susceptibility database or the database of Genotypes and Phenotypes. Under the new policy, most of the human genomic data on these websites will be accessible only by restricted access, such as signing a data use agreement in which one promises not to share data with others without permission or to try to re-identify individuals. Some limited human genomic data is available by unrestricted access, e.g., on a publicly available website (Resnik 2010; National Institutes of Health 2024).

Another concern with biospecimens that emerged during the 1990s was sharing individualized results of laboratory tests or procedures with research subjects, such as the result of blood tests, genetic or genomic tests, urinalysis, radiologic imaging, and clinical evaluations. Subjects may want to have this information because it can be useful in making decisions related to the prevention or treatment of disease or they are curious and want to learn more about themselves. Researchers may be hesitant to share this information because they are concerned it may have little relevance to medical decision-making and could cause subject unnecessary confusion and distress (see discussion in Chap. 8).[13]

In 2001, the Maryland Supreme Court ruled that investigators have legal duties to inform research subjects about significant health risks discovered during the course of a study (*Grimes v. Kennedy Krieger Institute, Inc.* 2001). The litigation in this case pertained to an EPA-funded study on lead abatement methods conducted by the Kennedy Krieger Institute, a research center associated with JHU, in the 1990s. The aim of the study was to determine whether less expensive forms of lead abatement are as effective at preventing exposure to lead as full lead abatement, which could cost as much as $10,000 per home. The study enrolled 25 Baltimore, MD families with small children living in the home. Twenty families living in homes with lead paint were randomly assigned to groups receiving the full level of lead abatement, or $1650, $3500, or $6000–$7000 worth a lead abatement. The study also included a control group of five families living in homes without lead paint. Investigators measured lead levels in children's blood and in dust, soil, and waters samples from the homes. They told the families that they would inform them of dangerous lead levels in their children's blood or in the home. Two families sued the investigators and KKI on the grounds that the investigators had acted negligently by not informing them of dangerous lead levels in a timely fashion. The investigators did not inform Viola Hughes that her daughter, Erica Grimes, had dangerous lead levels in her blood until nine months after they had detected this hazard. The defendants made a motion to dismiss the case on the grounds that they did not have a legal duty to inform the research subjects of dangerous lead levels. However, the court ruled in favor of the plaintiffs. The court found that the investigators have legal duties to the participants based on a special relationship implied by the federal

[13] Return of results will be discussed in Chap. 5.

research regulations and the contractual obligations in the informed consent document (*Grimes v. Kennedy Krieger Institute, Inc. 2001*). Although other cases had addressed clinical investigators' legal obligations to research subjects, this was the first case in which a court found that investigators had legal obligations to research subjects who are not their patients.

2.9 Conclusion: Public Trust, Government Oversight, and Protectionism

For much of the history of science and medicine, society has entrusted investigators with protecting the rights and welfare of human research subjects. The protection of human subjects depended, in large part, on the integrity of the investigators, not on institutional or governmental oversight (Moreno 2001). While investigator integrity is still a vital part of research, in the 1970s protection of human subjects shifted toward external oversight by means of regulations and guidelines. These rules and guidelines emerged as a reaction to egregious abuses of human subjects and have continued to evolve in response to the need to assure the public the research enterprise is trustworthy (Levine 1988; Moreno 2001; Marshall 2002). As incidents have come to the public's attention, government officials, scientists, bioethicists, and policymakers have taken steps to restore and maintain the public's trust. This chapter has catalogued some of the events which have alarmed the public, such as: The Nazi experiments, the Tuskegee study, the Willowbrook hepatitis experiments, the secret human radiation experiments, the CIA's mind control research, the deaths of Jesse Gelsinger and Ellen Roche, the Havasupai case, the CHEERS study, and reports of fraud, bias, and COI in biomedical research. Each step of the way, the trend has been to adopt new regulations and guidelines or revise existing ones to protect the rights and welfare of human subjects in research (Marshall 2002; Rhodes 2005).

The result of this shift from internal control toward external oversight is that research with human subjects is one of the most heavily regulated—and some would say overregulated—activities in the U.S. (Schrag 2010; Schneider 2015). Despite the presence of numerous legal rules and ethical guidelines pertaining to research with human subjects, controversies and scandals still arise. Controversies continue to emerge despite ample regulation and guidance because research with human subjects involves ethical conflicts between the rights and welfare of the individual and the goal of advancing scientific knowledge that can benefit society. Regulations and guidelines, by themselves, cannot resolve these tensions because issues can reemerge when interpreting and applying rules. For example, the SUPPORT study and the Johns Hopkins infection control research, discussed in Chap. 1, both involved disputes about how to interpret the federal regulations.

Given the historical events and societal responses that shaped these regulations and guidelines, the question one needs to ask is whether the current oversight system is appropriate reaction to a crisis of confidence or an overreaction. Should human subject research, generally, have more or less regulation? Should specific domains of research have more or less regulation? These questions, which are moral abd philosophical in nature, are just as important today as they were ten years or fifty years ago. Thus, there is an ongoing need for meaningful reflection on the rules and standard for research, which is a principal aim of this book.

Some authors, such as Rhodes (2005), Schrag (2010), Wertheimer (2011), Sachs (2011), and Schneider (2015) argue that some of these regulations and guidelines have gone too far and impede valuable research. One of the main reasons why federal agencies revised their regulations in 2017 was to reduce regulatory burden on researchers (Emanuel and Menikoff 2011). For example, the 2017 version of the Common Rule reduced regulatory burdens by excluding certain activities, such as public health surveillance and oral history, from the definition of research, expanding the scope of activities that can be treated as exempt research, and eliminating requirements for continuing review of certain types of studies. However, the 2017 version of the Common Rule also increased regulatory burdens by including additional provisions in the required elements of informed consent (Office of Human Research Protections 2021).

Various writers have observed that human research subjects have more protections than most people have outside of the research context (Edwards et al. 2004; Rhodes 2005; Miller and Wertheimer 2007; Wilson and Hunter 2010; Resnik 2015; Jansen and Wall 2018). For example:

- The regulations allow healthy adults to take risks in research only if an IRB determines that the risks are reasonable in relation to the expected benefits. Outside of the research context, healthy adults are free to make their own decisions concerning risks and are free to perform many risky activities, such skydiving, mountain climbing, smoking tobacco (and marijuana in some states), riding motorcycles, and so on (see discussion or risks in Chap. 7).
- The regulations place limits on the risks that children, pregnant women, and prisoners may be exposed to in research that does not provide direct benefits. Outside of the research context, parents are free to expose their children to various risks beyond what would be allowed by the regulations. For example, parents can allow their children to participate in contact sports, ride horses, work on farms, scuba dive, and so on. Pregnant women are free to ride horses, smoke, drink alcohol, and so on. Prisoners are routinely exposed to far greater risks than they would be allowed to face in non-beneficial research (see discussion in Chap. 8).
- The regulations allow human subjects to withdraw from a study at any time without penalty. Outside of the research context, a person may incur a penalty for breaching a contract.
- The regulations include numerous types of information that must be conveyed to the subject (or his or her representative) and documented for consent to be valid. Outside of the research context, people are free to decide what to include in a contract if they reach some agreement on what each party will do under the contract.

- The regulations prohibit human subjects from waiving or appearing to waive legal rights during consent. Outside of the research context, people are free to waive various kinds of legal rights. For example, one could agree not to sue a skydiving company as a condition of using its services.
- Ethical guidelines place limits on the amount of money someone may be paid to participate in a study. These guidelines have been implemented to ensure that the consent process is free from coercion or undue inducement. Outside of the research context, we usually do not place any limits on the amount of money that someone may be paid to perform a job (See discussion of payments in Chap. 5).

Most writers refer to these additional protections as paternalistic because they involve controlling people's decision-making for their own good. An important distinction pertaining to paternalism is the distinction between hard and soft paternalism (Miller and Wertheimer 2007; Dworkin 2020). Hard paternalism involves restricting the actions or decisions of a person who is fully autonomous (e.g., a competent adult) for their own good, whereas soft paternalism involves restricting the actions or decisions of someone who is not fully autonomous (e.g., a child or mentally disabled adult) for their own good. A law requiring all automobile passengers, including adults, to wear seatbelts is an example of hard paternalism, whereas a law requiring children to wear helmets when bicycling is an example of soft paternalism. Most people would accept some form of soft paternalism because they agree that people who have compromised decision-making, due to mental incapacity or lack of information, deserve some protection from harm. In a famous example, nineteenth century British philosopher John Stuart Mill (1806–1873), an ardent defender of liberty, conceded that it would be acceptable to stop a competent adult from crossing a dangerous bridge if they don't know the bridge is dangerous and there is no time to warn them. Mill also argued that it is acceptable to restrict the liberty of mentally disabled people for their own good (Mill 1978).

Hard paternalism is more controversial than soft because most people value personal freedom and do not want outside interference in their decisions or actions. People would like to make their own choices, even choices which many would consider to be too risky or ill-advised. While the additional protections for children, neonates, fetuses, and prisoners and mentally ill people could be viewed as soft paternalism, most of the additional protections for competent adults would be hard paternalism (Miller and Wertheimer 2007). For example, limiting the amount of money that competent adults can receive for research participation to prevent them from making decisions against their better judgment would be hard paternalism.

Soft paternalism can also be controversial if it provides too much protection for people with compromised autonomy. For example, each year in the U.S. about 40,000 children have injuries requiring a visit to the emergency department while participating in gymnastics (Tisano et al. 2022). A law banning children from participating in gymnastics would protect them from harm, but parents would probably object to it because it would deny children valuable benefits (such as exercise and socialization) and would interfere with parental decision-making. Although we should do our best to protect vulnerable people from harm, their lives cannot—and should not—be risk-free.

Although I consider paternalism to be an important issue in research with human subjects, I prefer to frame the ethical and regulatory issues in terms of questions concerning protectionism, i.e., what justifies a decision or policy that increases or decreases protections for human research subjects? It may seem to some readers that the difference between paternalism and protectionism is a terminological issue of little consequence. However, I think it is important to make this distinction because protectionism goes beyond mere paternalism, which focuses on the external control of an individual's actions or choices for their own good (Dworkin 2020). Protectionism could include policies that promote the welfare of individuals without directly controlling their decision-making. For example, requiring a clinical trial to have an independent data and safety monitoring committee would be protectionistic but not paternalistic. Likewise, providing medications to human subjects in a clinical trial in a developing country after the trial ends would be protectionistic but not paternalistic. Paternalism is a type of protectionism.

Another reason why I prefer to focus on protectionism rather than paternalism is that paternalism can be difficult to distinguish from other doctrines, such as the public harm principle or the public welfare principle, when it comes to social policy (Feinberg 1986). For example, consider a law that prohibits gambling. On the one hand, it could be viewed as a form of strong paternalism because it controls the choices of competent adults for their own good. On the other hand, it could be justified on the grounds that it protects society from the harmful effects of gambling, such as gambling addiction and the organized crime that is often associated with gambling. So, it is not paternalism but a form of public harm prevention. Paternalism applies must clearly to cases where the harm occurs only to the individual whose rights are being restricted, but those cases tend to be rare, because our actions almost always affect others.

Determining what constitutes the appropriate level of ethical and legal protection for human subjects in research involves a conflict between the good of the individual and the good of society, because protecting participants from harm or exploitation often hinders the advancement of science. The historical record clearly indicates that human subjects were under-protected in the past. Today, many researchers believe that society has erred in the opposite direction and that the current regulatory regime is overprotective (Schrag 2010; Schneider 2015). To make some headway in addressing the under-protection vs. overprotection dilemma, we need to make use of a decision-making framework that balances the competing values at stake.[14] Moral theories may be able to provide us with such a framework or help us to craft one. The next chapter will consider whether moral theories offer some useful guidance for answering fundamental questions concerning ethical and regulatory issues in research with human subjects (Table 2.3).

[14] I will return to this dilemma again in Chap. 10.

Table 2.3 Human research ethics timeline

400 BCE	Hippocratic oath
1796	Edward Jenner performs a test of a small pox inoculation on eight-year-old James Phipps
1874	Robert Bartholow inserts electrodes into a two-inch hole in the skull of Mary Rafferty, a mentally disabled cancer patient, and stimulates her brain with electricity
1885	Louis Pasteur administers an experimental rabies vaccine to nine-year-old Joseph Meister without first testing it on animals
1900–1901	Walter Reed's yellow fever experiments
1900	Prussia adopt human research regulations
1932–1972	Tuskegee syphilis study (sponsored by the DHEW)
1932–1945	Japanese experiments on prisoners
1940–1945	Nazi experiments on concentration camp prisoners
1945–1971	The U.S. government sponsors secret human radiation experiments
1946	The AMA adopts ethical guidelines for research with human subjects
1946–1948	The PHS sponsors syphilis prevention experiments in Guatemala
1947	Nuremberg Code adopted
1951	Researchers at Johns Hopkins Hospital develop a cell line from Henrietta Lacks' leftover cervical cancer tissue without her consent
1953–1973	The CIA conducts secret mind control experiments
1953–1955	Polio vaccine field trials led by Jonas Salk
1956–1980	Willowbrook hepatitis experiments
1961–1962	Stanley Milgram's obedience to authority experiments
1963	Jewish Chronic Disease Hospital experiments
1964	Helsinki Declaration adopted (first version)
1966	Henry Beecher publishes an exposé on 22 unethical human studies in the *New England Journal of Medicine*
1966	NIH adopts human research regulations
1971	The Stanford Prison Experiment
1971	FDA adopts human research regulations
1972	The Associated Press breaks reports on the Tuskegee study
1972	Congressional hearings on biomedical research
1973	Congress passes the National Research Act
1979	The U.S. National Commission releases the *Belmont Report*
1981	U.S. federal agencies implement a major revision of human research regulations
1990	The California Supreme Court rules that physician David Golde and the University of California at Los Angeles were liable for failing to obtain John Moore's informed consent for research they were conducting on his leftover tissue from a splenectomy
1991	Federal agencies revise and unify their research regulations (known as the Common Rule)
1993	Roger Poisson admits to altering patients records in the NSABP clinical trial

(continued)

Table 2.3 (continued)

1994	President Clinton declassifies thousands of documents pertaining to secret human radiation experiments and creates an advisory committee to examine scientific, ethical, and legal issues related to this research.
1997	The U.S. government issues an official apology to participants in the Tuskegee syphilis study and their families
1997	Peter Lurie and Sydney Wolfe publish a paper in the *New England Journal of Medicine* alleging that HIV prevention trials conducted in developing nations were unethical
1998	The U.S. Government Accounting Office issues a report calling for reform of IRBs
1998	Andrew Wakefield publishes a paper in *The Lancet* claiming that the MMR vaccine can cause autism; the journal retracted the paper in 2010 due to ethics problems
1999	Jesse Gelsinger dies in a Phase I human gene therapy experiment; OHRP and the FDA investigate the University of Pennsylvania, where the study occurred
1999	OHRP temporarily halts all human studies at the Duke University Medical Center
2001	Ellen Roche dies in asthma experiment at Johns Hopkins University; OHRP temporarily halts federally-funded research at the university.
2004	The EPA cancels the Children's Environment Exposure Research Study in response to Congressional criticism
2004	Merck withdraws its drug Vioxx from the market, due to safety and liability concerns, including allegations of fraud
2005	Eric Poehlman admits to fabricating and falsifying data on 15 federal grant applications worth $2.9 million; Poehlman serves a year and a day in federal prison as part of a comprehensive legal settlement
2005	Woo Suk Hwang admits to fabricating and falsifying data pertaining to two papers on the derivation of human embryonic stem cells by therapeutic cloning
2006	The EPA revises its human research regulations
2007	OHRP issues a determination letter to Johns Hopkins University claiming that its IRB failed to adequately review a quality assurance research project conducted at 67 hospitals in Michigan
2010	Arizona State University settles a lawsuit with the Havasupai Native American tribe pertaining to improper use of blood samples by researchers
2013	OHRP issues a determination to the University of Alabama at Birmingham claiming the consent for the SUPPORT study was out of compliance with federal regulations
2017	Federal agencies publish major revisions to the Common Rule, which become effective in 2018
2022	Investigators publish findings from COVID-19 human challenge studies

References

Advisory Committee on Human Radiation Experiments. 1995. *Final report*. Washington, DC: U.S. Government Printing Office.

All of Us. 2021. FAQs. Available at: https://allofus.nih.gov/about/faq#:~:text=We%20aim%20to%20enroll%20one,the%20course%20of%20the%20program. Accessed 12 May 2024.

Arras, J.D. 2008. The Jewish chronic disease hospital case. In *The Oxford handbook of clinical research ethics*, ed. E.J. Emanuel, C. Grady, R.A. Crouch, R.K. Lie, F.G. Miller, and D. Wendler, 73–79. New York: Oxford University Press.

Bayefsky, M.J., K.W. Saylor, and B.E. Berkman. 2015. Parental consent for the use of residual newborn screening bloodspots: Respecting individual liberty vs ensuring public health. *Journal of the American Medical Association* 314 (1): 21–22.
Beauchamp, T.L. 2005. The origins and evolution of the Belmont report. In *Belmont revisited: Ethical principles for research with human subjects*, ed. J.F. Childress, E.M. Meslin, and H.T. Shapiro, 12–25. Washington, DC: Georgetown University Press.
Beauchamp, T.L., and J.F. Childress. 2019. *Principles of biomedical ethics*, 7th ed. New York: Oxford University Press.
Beecher, H. 1966. Ethics and clinical research. *New England Journal of Medicine* 274 (24): 1354–1360.
Budimir, D., O. Polasek, A. Marusić, I. Kolcić, T. Zemunik, V. Boraska, A. Jeroncić, M. Boban, H. Campbell, and I. Rudan. 2011. Ethical aspects of human biobanks: A systematic review. *Croatian Medical Journal* 52 (3): 262–279.
Capron, A.M. 1989. Human experimentation. In *Medical ethics*, ed. R.M. Veatch, 125–172. Boston: Jones and Bartlett.
Department of Health and Humans Services. 2017. Protection of Human Subjects. Code of Federal Regulations, Title 45, Part 46.
Department of Homeland Security, Department of Agriculture, Department of Energy, National Aeronautics and Space Administration, Department of Commerce; Social Security Administration, Agency for International Development; Department of Housing and Urban Development, Department of Labor, Department of Defense, Department of Education, Department of Veterans Affairs, Environmental Protection Agency, Department of Health and Human Services, and National Science Foundation; and Department of Transportation. 2017. Federal policy for the protection of human subjects. *Federal Register* 82 (12): 7149–7274.
Dworkin, G. 2020. *Paternalism*. Stanford Encyclopedia of Philosophy. Available at: https://plato.stanford.edu/entries/paternalism/. Accessed 17 May 2024.
Edwards, S.J., S. Kirchin, and R. Huxtable. 2004. Research ethics committees and paternalism. *Journal of Medical Ethics* 30 (1): 88–91.
Emanuel, E.J., and J. Menikoff. 2011. Reforming the regulations governing research with human subjects. *New England Journal of Medicine* 365 (12): 1145–1150.
Emanuel, E.J., T. Lemmens, and C. Elliot. 2006. Should society allow research ethics boards to be run as for-profit enterprises? *PLoS Medicine* 3 (7): e309.
Feinberg, J. 1986. *Harm to self*. New York: Oxford University Press.
Geison, G.L. 1978. Pasteur's work on rabies: Reexamining the ethical issues. *Hastings Center Report* 8 (2): 26–33.
Gelsinger, P. 2008. A comment from Paul Gelsinger on gene therapy and informed consent. Bioethics.net, 31 January 31. Available at https://bioethicstoday.org/blog/a-comment-from-paul-gelsinger-on-gene-therapy-and/. Accessed 12 May 2024.
Gelsinger, P., and A.E. Shamoo. 2008. Eight years after Jesse's death: Are human research subjects any safer? *Hastings Center Report* 38 (2): 25–27.
Goliszek, A. 2003. *In the name of science: A history of secret programs, medical research, and human experimentation*. New York: St. Martin's Press.
Gøtzsche, P.C., A. Hróbjartsson, H.K. Johansen, M.T. Haahr, D.G. Altman, and A.W. Chan. 2007. Ghost authorship in industry-initiated randomised trials. *PLoS Medicine* 4 (1): e19.
Hawkins, J.L., and E.J. Emanuel. (eds.). 2008. *Exploitation and developing countries: The ethics of clinical research*. Princeton, NJ: Princeton University Press.
Homer, N., S. Szelinger, M. Redman, D. Duggan, W. Tembe, J. Muehling, J.V. Pearson, D.A. Stephan, S.F. Nelson, and D.W. Craig. 2008. Resolving individuals contributing trace amounts of DNA to highly complex mixtures using high-density SNP genotyping microarrays. *PLoS Genetics* 4 (8): e1000167.
Holland, T. 2018. The father of modern gynecology performed shocking experiments on enslaved women. *History*, 4 December. Available at: https://www.history.com/news/the-father-of-modern-gynecology-performed-shocking-experiments-on-slaves. Accessed 7 May 2024.

Humphreys, L. 1976. *Tearoom trade: Impersonal sex in public places (observations)*. 2nd ed. New York: Routledge.
In Re Cincinnati Radiation Litigation. 1995. 874 F. Supp. 796 (S.D. Ohio 1995).
International Committee of Medical Journal Editors. 2024. Protectiono of research participants. Available at: https://www.icmje.org/recommendations/browse/roles-and-responsibilities/protection-of-research-participants.html. Accessed 9 Sept 2024.
Jansen, L.A., and S. Wall. 2018. Reconsidering paternalism in clinical research. *Bioethics* 32 (1): 50–58.
Jones, J.H. 1981. *Bad blood*. New York: Free Press.
Keiger, D., and S. De Pasquale. 2002. Trials & tribulation: A special report. *Johns Hopkins Magazine* 54 (1) Available at: http://pages.jh.edu/~jhumag/0202web/trials.html. Accessed 12 May 2024.
Krimsky, S. 2003. *Science in the private interest—Has the lure of profits corrupted biomedical research?* Lanham: Rowman and Littlefield.
Lederer, S.E. 1995. *Subjected to science: Human experimentation in America before the Second World War*. Vol. 24, 13. Baltimore: Johns Hopkins University Press.
———. 2008. Walter Reed and the yellow fever experiments. In *The Oxford handbook of clinical research ethics*, ed. E.J. Emanuel, C. Grady, R.A. Crouch, R.K. Lie, F.G. Miller, and D. Wendler, 11–17. New York: Oxford University Press.
Levine, C. 1988. Has AIDS changed the ethics of human subjects research? *Law, Medicine, and Health Care* 16: 167–173.
Lockwood, A.H. 2004. Human testing of pesticides: Ethical and scientific considerations. *American Journal of Public Health* 94 (11): 1908–1916.
Lowrance, W.W., and F.S. Collins. 2007. Ethics. Identifiability in genomic research. *Science* 317 (5838): 600–602.
Marshall, E. 2000. FDA Halts all gene therapy trials at Penn. *Science* 287 (5453): 565–567.
Marshall, M.F. 2002. Born in scandal: The evolution of clinical research ethics. *Science Careers*. 26 April. Available at: https://www.science.org/content/article/born-scandal-evolution-clinical-research-ethics. Accessed 16 May 2024.
McCoy, A.W. 2007. Science in Dachau's shadow: Hebb, Beecher, and the development of CIA torture and modern medical ethics. *Journal of the History of Behavioral Sciences* 43 (4): 401–417.
Meldrum, M.J. 2008. The Salk polio vaccine field trial of 1954. In *The Oxford handbook of clinical research ethics*, ed. E.J. Emanuel, C. Grady, R.A. Crouch, R.K. Lie, F.G. Miller, and D. Wendler, 61–72. New York: Oxford University Press.
Mello, M., and L. Wolf. 2010. The Havasupai Indian tribe case—Lessons for research involving stored biologic samples. *New England Journal of Medicine* 363 (3): 204–207.
Milgram, S. 1974. *Obedience to authority*. New York: Harper and Rowe.
Mill, J.S. 1978 [1859]. *On liberty*, ed. E. Rapaport. Indianapolis: Hackett.
Miller, F.G., and A. Wertheimer. 2007. Facing up to paternalism in research ethics. *Hastings Center Report* 37 (3): 24–34.
Moore v. Regents of the University of California. 1990. 793 P.2d 479 (Cal. 1990).
Moreno, J.D. 2001. Goodbye to all that. The end of moderate protectionism in human subjects research. *Hastings Center Report* 31 (3): 9–17.
National Commission for the Protection of Human Subjects of Biomedical or Behavioral Research. 1979. *The Belmont report: Ethical principles and guidelines for the protection of human subjects of research*. Washington, DC: Department of Health, Education, and Welfare.
National Institutes of Health. 2024. Accessing genomic data from NIH repositories. Available at: https://sharing.nih.gov/accessing-data/accessing-genomic-data/accessing-genomic-data-from-nih-repositories#:~:text=Open%20or%20Unrestricted%20Access%3A%20Some,to%20use%20the%20datasets%20responsibly. Accessed 25 Sept 2024.
National Library of Medicine. 2024. Oath of Hippocrates. Available at: https://www.nlm.nih.gov/hmd/topics/greek-medicine/index.html. Accessed 7 May 2024.

National Research Council. 2004. *Intentional human dosing studies for EPA regulatory purpose: Scientific and ethical issues*. Washington, DC: National Academy Press.

Nordling, L. 2018. Europe's biggest research fund cracks down on 'ethics dumping. *Nature* 559: 17–18.

Nuremberg Code. 1949. *Trials of war criminals before the Nuremberg military tribunals under Control Council Law 2 (10): 181–182*. Washington, DC: U.S. Government Printing Office. Available at: https://ori.hhs.gov/content/chapter-3-The-Protection-of-Human-Subjects-nuremberg-code-directives-human-experimentation. Accessed 7 May 2024.

Office of Human Research Protections. 1998. Expedited review: Categories of research that may be reviewed through an expedited review procedure 1998. Available: https://www.hhs.gov/ohrp/regulations-and-policy/guidance/categories-of-research-expedited-review-procedure-1998/index.html. Accessed 9 May 2024.

———. 2021. Common rule Q & A. Available at: https://www.hhs.gov/ohrp/education-and-outreach/revised-common-rule/revised-common-rule-q-and-a/index.html#:~:text=revised%20Common%20Rule.-,The%20three%20burden-reducing%20provisions%20of%20the%20revised%20Common%20Rule,research%3B%20and%20(3)%20the. Accessed 17 May 2024.

———. 2024. Guidance. Available at: https://www.hhs.gov/ohrp/regulations-and-policy/guidance/index.html. Accessed 10 May 2024.

Pence, G. 1995. *Classic cases in medical ethics*. 2nd ed. New York: McGraw-Hill.

Porter, R. 1999. *The greatest benefit to mankind: A medical history of humanity*. New York: W.W. Norton.

Proctor, R. 1988. *Racial hygiene: Medicine under the Nazis*. Cambridge, MA: Harvard University Press.

Resnik, D.B. 2007a. The new EPA regulations for protecting human subjects: Haste makes waste. *Hastings Center Report* 37 (1): 17–21.

———. 2007b. Are the new EPA regulations concerning intentional exposure studies involving children overprotective? *IRB* 29 (5): 15–19.

———. 2007c. *The price of truth: How money impacts the norms of science*. New York: Oxford University Press.

———. 2008. Fraud, fabrication, and falsification. In *The Oxford handbook of clinical research ethics*, ed. E.J. Emanuel, C. Grady, R.A. Crouch, R.K. Lie, F.G. Miller, and D. Wendler, 787–794. New York: Oxford University Press.

———. 2010. Genomic research data: Open vs. restricted access. *IRB* 32 (1): 1–6.

———. 2012. *Environmental health ethics*. Cambridge: Cambridge University Press.

———. 2015. Paternalism and utilitarianism in research with human participants. *Health Care Analysis* 23 (1): 19–31.

Resnik, D.B., and C.W. Jones. 2006. Research subjects with limited English proficiency: Ethical and legal issues. *Accountability in Research* 13 (2): 157–177.

Resnik, D.B., and S. Wing. 2007. Lessons learned from the Children's Environmental Exposure Research Study. *American Journal of Public Health* 97 (3): 414–418.

Rhodes, R. 2005. Rethinking research ethics. *American Journal of Bioethics* 5 (1): 7–28.

Robinson, W.M., and B.T. Unruh. 2008. The hepatitis experiments at the Willowbrook State School. In *The Oxford handbook of clinical research ethics*, ed. E.J. Emanuel, C. Grady, R.A. Crouch, R.K. Lie, F.G. Miller, and D. Wendler, 80–85. New York: Oxford University Press.

Sachs, B. 2011. Going from principles to rules in research ethics. *Bioethics* 25 (1): 9–20.

Savitt, T.L. 1982. The use of Blacks for medical experimentation and demonstration in the old south. *The Journal of Southern History* 48 (3): 331–348.

Schneider, C.E. 2015. *The Censor's hand: The misregulation of human-subject research*. Cambridge, MA: MIT Press.

Schrag, Z.M. 2010. *Ethical imperialism: Institutional review boards and the social sciences, 1965–2009*. Baltimore: Johns Hopkins University Press.

Semeniuk, L., and S. Reverby. 2010. A shocking discovery. *Nature* 467 (7316): 645.

Shalala, D. 2000. Protecting research subjects—What must be done. *New England Journal of Medicine* 343 (11): 808–810.

Shamoo, A.E., and D.B. Resnik. 2022. *Responsible conduct of research*. 4th ed. New York, NY: Oxford University Press.

Shuster, B. 2017. In the Wake of Henrietta Lacks: Current U.S. Law and policy on control and ownership of one's body tissues used in medical research. *The Journal of Healthcare Ethics & Administration* 2 (2). https://doi.org/10.22461/jhea.1.71614.

Sismondo, S. 2008a. Pharmaceutical company funding and its consequences: A qualitative systematic review. *Contemporary Clinical Trials* 29 (2): 109–113.

———. 2008b. How pharmaceutical industry funding affects trial outcomes: Causal structures and responses. *Social Science and Medicine* 66 (9): 1909–1914.

Stephens, J. 2006. Panel faults Pfizer in '96 clinical trial in Nigeria. *The Washington Post*, 7 May 2006:A1.

Stokstad, E. 2005. Pesticide testing. EPA draft rules for human subjects draw fire. *Science* 309 (5732): 232.

Tilousi, C. 2011. Testimony before the President's commission for the study of bioethical issues, August 30, 2011.

Tisano, B., A.J. Zynda, H.B. Ellis, and P.L. Wilson. 2022. Epidemiology of pediatric gymnastics injuries reported in US emergency departments: Sex- and age-based injury patterns. *Orthopaedic Journal of Sports Medicine* 10 (6): 23259671221102478.

Tsuchiya, T. 2008. The imperial Japanese experiments in China. In *The Oxford handbook of clinical research ethics*, ed. E.J. Emanuel, C. Grady, R.A. Crouch, R.K. Lie, F.G. Miller, and D. Wendler, 31–45. New York: Oxford University Press.

Veatch, R.M. 2005. Ranking, balancing, or simultaneity: Resolving conflicts among the Belmont principles. In *Belmont revisited: Ethical principles for research with human subjects*, ed. J.F. Childress, E.M. Meslin, and H.T. Shapiro, 184–204. Washington, DC: Georgetown University Press.

Weindling, P.J. 2008. The Nazi medical experiments. In *The Oxford handbook of clinical research ethics*, ed. E.J. Emanuel, C. Grady, R.A. Crouch, R.K. Lie, F.G. Miller, and D. Wendler, 18–30. New York: Oxford University Press.

Weindling, P., A. von Villiez, A. Loewenau, and N. Farron. 2016. The victims of unethical human experiments and coerced research under National Socialism. *Endeavour* 40 (1): 1–6.

Weir, R.F., and R.S. Olick. 2004. *The stored tissue issue: Biomedical research, ethics, and law in the era of genomic medicine*. New York: Oxford University Press.

Weiss, R. 1999. U.S. halts human research at Duke. *Washington Post*, 12 May: A1.

Wertheimer, A. 2011. *Rethinking the ethics of clinical research: Widening the lens*. New York: Oxford University Press.

Wilson, J., and D. Hunter. 2010. Research exceptionalism. *American Journal of Bioethics* 10 (8): 45–54.

World Medical Association. 1964. Declaration of Helsinki: Ethical principles for medical research involving human subjects.

———. 2013. Declaration of Helsinki: Ethical principles for medical research involving human subjects (2013 revision). Available at: https://www.wma.net/policies-post/wma-declaration-of-helsinki-ethical-principles-for-medical-research-involving-human-subjects/. Accessed 8 May 2024.

Zimbardo, P. 2024. The Stanford Prison Experiment. Available at: https://www.prisonexp.org/the-story. Accessed 11 July 2024.

Chapter 3
Moral Theory

Abstract This chapter provides an overview of moral concepts, theories, and principles that can be serve as a philosophical framework for making decisions involving dilemmas and issues in research with human subjects. The theories and ideas discussed in this chapter include the difference between science and ethics; moral justification and knowledge; reflective equilibrium; moral relativism and subjectivism; divine command theory; virtue ethics; natural law theory; natural rights theory; social contract theory; utilitarianism; Kantianism; Rawls' principles of justice; feminist ethics; and principlism. The chapter observes that each of the moral theories examined have strengths and weaknesses and emphasize different aspects of morality. The chapter argues that principlism is the most reasonable approach to moral theory because it is better able to accommodate conflicting intuitions concerning moral values than the other approaches and has a workable decision procedure. The principles defended in this chapter include respect for human dignity, nonmaleficence, beneficence, and justice. Principlism, like other moral theories, must still deal with the problem of priority-setting when conflicts arise among principles. In the following chapter, I argue that an additional principle, the principle of trust, can help resolve conflict among the other moral principles.

Keywords Moral theories · Moral concepts · Moral principles · Reflective equilibrium · Virtue ethics · Utilitarianism · Kantianism · Justice · Moral relativism

In the previous chapter I argued that the history of the ethics of research with human subjects indicates that the regulations and ethical guidelines have evolved in response to egregious abuses of people and blatantly unethical (or at least ethically questionable) research. Society has adopted regulations and guidelines to prevent these problems from occurring again and to restore and maintain public trust in research. The regulations and guidelines form a system of rules designed to protect the rights and welfare of human research subjects. Ethical dilemmas involving research with human subjects continue to emerge, however, because the rules do not completely resolve fundamental questions related to the conflict between the rights and welfare of human subjects and the goal of advancing human knowledge to

D. B. Resnik, *The Ethics of Research with Human Subjects*, The International Library of Bioethics 111, https://doi.org/10.1007/978-3-031-82757-0_3

benefit society. To be sure, regulations and guidelines can help investigators, IRBs, institutional leaders, policymakers, and concerned citizens address these issues, but questions remain because dilemmas occur when interpreting, applying, prioritizing, and revising regulations and guidelines. There are also questions about the rules themselves—are they overprotective or under-protective? Do they need to be changed? What is needed, I argued, is a moral decision-making framework which can be used to evaluate rules and policies and set priorities when ethical conflicts arise. This chapter will consider whether moral theories can provide such a framework.

3.1 What Is a Moral Theory?

To understand what a moral theory is, it will be useful to compare it to the more familiar concept of scientific theory. Scientific theories, such as Charles Darwin's (1809–1882) theory of evolution by natural selection, are groups of statements that can be used to describe, explain, and predict phenomena in a particular field of inquiry (Kitcher 1995). Darwin proposed his theory to explain the origin of species and variation in the Earth's biota and describe the relatedness of all living things on Earth. The core idea of Darwin's theory is that organisms which are best adapted to their environments will increase in prevalence in a population relative to those who are not as well adapted, i.e., survival of the fittest. As a result, the characteristics of the population will change and eventually a new species may emerge. The evidence that Darwin presented for his theory came from various sources, including the fossil record, the effects of selective breeding of animals and plants, and his studies of variation among finches and other species. While many of the predictions from the theory pertain to changes that occur over the course of geologic time, others, such as antibiotic resistance and pesticide resistance, can be observed in the span of a several years or months. Darwin's theory has been tested and confirmed many times since he first proposed it and has been modified to deal with problems and limitations. For example, the theory has had difficulty explaining the emergence of altruism in animal populations, because altruistic behavior seems to be opposed to the individual's fitness. To explain altruism, evolutionary biologists have expanded the notion of fitness to include fitness in genetically-related organisms (or kin selection) (Dennett 1995). Scientists could decide to reject the theory if enough contrary evidence emerges, but so far, that has not happened.

With this example in mind, we can think of a moral theory as a group of statements that can be used to describe, explain, and predict moral phenomena, which include our judgments, beliefs, reasons, and emotions concerning right and wrong, good and bad, and related concepts (Timmons 2012; Haidt 2013; Driver 2022). The judgments and beliefs may include general claims, such as "murder is wrong," or "people should not participate in risky experiments without their consent," or specific claims, such as "Martin Luther King was a brave man," or "the Tuskegee Syphilis Study was unethical." Moral reasons could include the thought processes we use to decide whether a particular act is right or wrong, for example, deciding whether it is wrong to tell a lie to avoid hurting someone's feelings. Moral emotions include such feelings like compassion, sympathy, love, guilt, reprobation, disgust, and shame (Greene 2013; Haidt 2013).

Moral theories provide us with an account of what we should do (i.e., duties or obligations), what is good or worthy of having or pursuing (i.e., moral values and virtues), and what is fair or just; or what is generally known as normative ethics. Normative ethics usually includes principles (or norms or standards) for making decisions in particular cases as well as definitions of morally relevant concepts. For example, Immanuel Kant's (1724–1804) moral theory (discussed below) includes a general principle for making decisions known as the Categorical Imperative and definitions of duty, dignity, autonomy, and good will (Timmons 2012).

Moral theories usually also include statements about moral knowledge, truth, metaphysics, and justification, or what is known as meta-ethics. For example, Kantians hold that a principle for action (or maxim) is morally justified only if it conforms to the Categorical Imperative. Kantians also hold that rational beings can have moral knowledge because they can understand the Categorical Imperative (Timmons 2012).

Moral theories are different from psychological, biological, anthropological, or sociological theories of ethical behavior because they include or imply prescriptive or normative claims, such as "one ought to do X," "X is good or worthwhile," and so on (Timmons 2012).[1] According to most moral theories, we ought not to kill innocent people. A scientific theory of human behavior could describe the psychological and social conditions associated with murder or explain why societies have prohibitions against murder, but it cannot tell us that one ought not to commit murder. Additionally, most moral theories hold that some things, such as human life, freedom, or justice, have intrinsic value or worth. Scientific theories can explain why people value human life, liberty, or justice but they cannot tell us why these things should have value (Harman 1977; Timmons 2012).

Moral theories (and moral statements) are also different from scientific theories because they refer to abstract properties and entities, such as goodness, rightness, wrongness, virtue, value, duty, responsibility, agency, and rights, which do not fit neatly into the natural world (Sayre-McCord 2015). Wrongness, for example, cannot be reduced to a single natural property or set of properties, because there are indefinitely many actions that we would judge as wrong. We cannot equate the wrongness in the Tuskegee Study, the Henrietta Lacks case, the Jesse Gelsinger case, or the Andrew Wakefield case, with a single observable natural property or set of properties that makes these situations wrong. It seems that wrongness is an emergent property that is not reducible to natural phenomena.

Now, it may be objected that scientific theories (and statements) also refer to abstract properties and entities that are not reducible to physical phenomena, but we

[1] To say that one ought, morally, do X, is equivalent to saying that one should do X, one has a duty, obligation, or responsibility to do X, that doing is X is the moral or right thing to do, that doing X is morally obligatory or that not doing X is not morally permissible.

still view these properties and entities as part of the natural world. For example, in evolutionary theory, adaptedness is not reducible to any single biological property or set of properties because what makes an organism adapted is different for each type of organism and environment. Likewise, the concept of a population cannot be equated with a single population because there are indefinitely many populations (Rosenberg 1985).

The abstract properties and entities described in scientific theories and statements are different from those included in moral theories in three important ways, however. First, the abstract properties and entities in scientific theories are descriptive or factual, not prescriptive or normative. Saying that an organism is well-adapted to its environment merely describes some facts and does prescribe any action or imply any normative judgement. However, saying that a person is murderer implies a great deal, morally, and legally. When we call someone a murderer, we are not merely stating a fact about that person but also making moral judgments about them.

Second, scientists have developed tests, methods, and procedures for measuring and detecting abstract properties and entities. For example, evolutionary theorists have developed measures of fitness and adaptedness (Brandon 1995; Peck and Waxman 2018). Although some philosophers, such as utilitarians (discussed below) have attempted to develop measurements of rightness and wrongness, these efforts have been largely unsuccessful because rightness and wrongness are not reducible to natural properties (Moore 1903).

Third, the tests, methods, and procedures used in science for measuring and detecting abstract properties and entities generate results that are reproducible. That is, scientists from vastly different parts of the world can use these tests, methods, and procedures to obtain same results. The same cannot be said for moral theorizing, however, which is marked by controversy. Although moral theorists (and ordinary people) often agree about some basic moral judgments (e.g., murder is wrong), they often disagree about specific cases (e.g., the morality of capital punishment or abortion) (Timmons 2012).

Because moral theories are different from scientific ones in several key respects, we cannot confirm moral theories in the same way that we can confirm scientific ones (Harman 1977). That is, we cannot use empirical methods, such as experiments, tests, or observations to confirm (or disconfirm) moral theories. Additionally, as noted above, it seems that morality refers to properties and entities, such as rightness or justice, that do not fit neatly into science's ontology.[2] Morality seems to be fundamentally different from science (Table 3.1).

Many philosophers who have reflected on these points view morality as a lesser form of human knowledge if they count it as knowledge at all (Tramel 2024).[3]

---

[2] Ontology is a branch of metaphysics that deals with questions of being or existence. For example, if you claim that "The Earth is the third planet from the Sun," you are making ontological claims about the Earth, the Sun, and planets.

[3] Most philosophers characterize knowledge as justified, true belief. There is little dispute about whether beliefs must be true to qualify as knowledge, but there is considerable debate about what constitutes appropriate justification.

Table 3.1 Some key difference between scientific and moral theories

Scientific theories	Moral theories
Do not include of imply prescriptive or normative claims.	Include or imply prescriptive or normative claims, e.g., "murder is wrong."
Refer to things that fit into the natural world, e.g., atoms, electrons, cells, societies, etc.	Refer to things that do not seem to easily fit into the natural world, e.g., rightness, goodness, justice, etc.
Refer to things that can measured or detected using various methods, tests, and procedures.	Refer to things that cannot be easily measured or detected.
Usually generate reproducible results.	Often do not generate reproducible results.

Several centuries ago, David Hume (1711–1776) argued that because moral claims cannot be tested empirically and are not true because of the meanings of words, then they are not a type of human knowledge and merely express our passions or emotions (Hume 1896).[4] Hume's view is a very pessimistic and unsatisfactory account of morality, however, because it implies that reasoning has little to do with moral decision-making and action (Haidt 2013). However, one might argue that even though emotion plays an important role in our morality, reasoning is not inert and can influence what we do (Richardson 2018).

Other philosophers have argued that we can have moral knowledge, but it is very different from scientific knowledge. In the twentieth century, G.E. Moore (1873–1958) and W.D. Ross (1877–1971) argued that we can acquire moral knowledge not by empirical methods but by intuition. That is, any rational, adult human being should be able to immediately grasp the truth of moral claims like "murder is wrong," "loyalty is good," and so on. Morality is self-evident (Tramel 2024). A key weakness with the intuitionist approach is that history shows us that our intuitions are often biased. For example, in 1958 only 4% of U.S. adults viewed interracial marriage as morally acceptable but only 94% feel that way today (McCarthy 2021). Also, intuitionism does not seem to account for the prevalence of moral controversy. If morality is self-evidently true, then why do we disagree so much about it?

In 1971, John Rawls (1921–2002) made in important breakthrough in moral theorizing that addressed these concerns about moral knowledge and justification. Rawls proposed that we can confirm (or justify) moral theories by means of a method he called reflective equilibrium (Rawls 1971). To follow the method, one begins with reflective judgments (or intuitions) of right/wrong, good/bad and so on, in particular cases. One then forms general principles (or theories) to systematize

[4] Hume would say that a statement like "Triangles have three sides" counts as knowledge because, even though it cannot be tested empirically, it is true by virtue of the meanings of words or what he called relations of ideas. See also Ayer (1952). Although W.V. Quine (1951) developed a powerful critique of the view that there is a distinction between statements that are true by virtue of meanings and those that are true by virtue of observations, this critique does not invalidate the view that moral knowledge is different from scientific knowledge.

those judgments and tests those principles against the original judgments and new ones, such as counterexamples. One may revise one's principles to accommodate those judgments, or one may reject some judgments, based on one's principles. As the process continues, one eventually reaches a point (known as reflective equilibrium) in which moral principles and judgments cohere (Rawls 1971). The method sets aside epistemological questions of whether moral statements are true and metaphysical questions about whether moral ontology fits into the natural world and simply asks us to focus on whether moral statements are justified or acceptable.

Some philosophers have argued that Rawlsian reflective equilibrium should be expanded to include scientific facts and highly confirmed scientific theories, so that moral theory will be consistent with the findings of modern science (Daniels 1996; Sayre-McCord 1996). This expansion of reflective equilibrium is known as wide reflective equilibrium (See Fig. 3.1). Requiring that moral theories be consistent with the findings of modern science does not imply that morality is derived from science; it only means that morality should not make claims which science has proven to be incorrect. For example, modern science has discredited the racist idea that some races are intellectually and morally inferior. Racism should not be assumed by or implied by any moral theory.

One of the potential shortcomings of reflective equilibrium is that our reflective judgments are susceptible to cultural, social, religious, political, or economic biases (Rawls 1971; Daniels 1996), which implies that reflective equilibrium might not help us achieve any moral progress but may simply reinforce existing biases.

Proponents of the reflective equilibrium respond to this problem by arguing that biases can be eliminated, in the long run, by repeatedly revising moral principles or

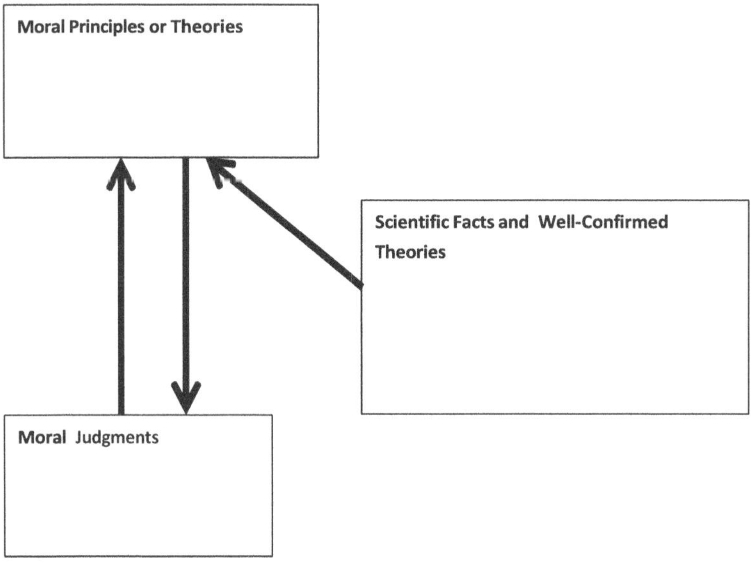

Fig. 3.1 Wide Reflective Equilibrium

theories in response to scientific findings and new/diverse moral judgments. To counteract the biases inherent in commonsense morality, one should caste a wide net and be open to diverse viewpoints and revisionary moral ideas (Greene 2013). Thus, moral principles and judgments in reflective equilibrium should represent a broad moral consensus, rather than the judgments or views of particular cultures, societies, or historical periods.

While this response to the possibility of bias seems plausible, it also creates a problem: if reflective equilibrium is reached in the long run, then how should we act in the short run? We must make moral decisions in the present moment and cannot wait for reflective equilibrium to provide us with the "correct" moral theory. How can we be assured that the decisions we make in the short run are not biased? While this is indeed an important problem, it is not unique to reflective equilibrium, since all moral theories must confront this issue. The most prudent course of action that one can take is to make the best decision possible, under the circumstances, realizing that one may need to revisit the decision in light of new scientific findings or moral judgments. When it comes to moral decision-making, we should accept the fact that we are fallible, and we should be prepared to correct mistaken ethical judgments.

Even if we accept reflective equilibrium as a method for justifying moral theories, an important question still remains: which criteria should we use to decide whether to accept, reject, or modify a particular moral theory during the reflective process (Driver 2022)? According to Timmons (2012), moral theories should be evaluated in terms of their ability to perform two fundamental tasks. The first is that a moral theory should be applicable. That is, it should provide us with a workable procedure for deciding right/wrong, good/bad, on so on, in particular cases. For our purposes, a theory should help us resolve complex ethical issues involving conflicts between protecting the rights and wellbeing or human subjects and the advancing the interests of science and society. The second is that a theory should provide a cogent explanation of our moral experience. That is, it should help to understand moral judgments, beliefs, decisions, reasons, and emotions. For our purposes, a theory should help understand why we ethical controversies concerning the protection of human subjects emerge and how we can resolve them.

With all of foregoing in mind, I will now survey some influential moral theories with an eye toward how they would approach the basic moral dilemma in research with human subjects, how to balance the good of the individual and the good of society.

3.2 Moral Subjectivism and Relativism

I will begin my survey by examining moral subjectivism and relativism, since if either view is correct, it may not be possible to develop a practical framework for addressing ethical dilemmas in research with human subjects, because accepting either theory would undermine any effort to develop widely accepted ethical guidance. Since I am optimistic about the prospect of developing an ethical framework that applies widely (i.e., globally), I need to address subjectivism and relativism before evaluating other moral theories.

Moral subjectivism is the view that what is right/wrong, good/bad is relative to a particular person: there is no difference between what a person believes is morally correct and what is morally correct. If I believe that it is morally acceptable for me to abuse animals, harm people, lie, cheat, steal, etc., then it is morally acceptable for me. Morality has to do with matters of taste, personal preferences, or subjective attitudes and opinions (Ayer 1952).

Subjectivists have developed several arguments for their position. According to the first argument, morality is subjective because we cannot prove or disprove moral statements by means of empirical observations, tests or experiments (Ayer 1952). However, as argued above, we can test moral statements by other methods, such as reflective equilibrium. These methods allow us to reach agreement on moral principles, even if they do not constitute empirical evidence for ethical theories.

The second argument is that morality is subjective because we cannot reach agreement on moral issues. Having a debate about morality is like debating about food, music, clothing, or other personal tastes. However, there is strong evidence that we can reach agreement on many moral topics even if we still disagree about others. Most people will agree that it is wrong to kill people, abuse animals, lie, cheat, steal, and so on, even if they disagree about the morality of capital punishment or abortion (Timmons 2012). Moreover, arguing about morality is not like debating about personal preferences, because we can often convince people to change their minds in response to facts and reasons.

According to the third argument, morality is subjective because moral statements are not capable of being true or false. Moral statements express emotions, feelings, or attitudes, but they do not express facts with truth-values (Ayer 1952). Saying that "murder is wrong" is like saying "murder, boo!" or "murder, yuck."

Fully responding to the argument involves examining complex issues of metaphysics and philosophy of language that are beyond the scope of this book (see Sayre-McCord 1996, 2015), but for our purposes, we can respond to the argument by asserting that what matters in morality is not whether moral statements are true or false but whether they can be appropriately justified. Most people will agree that the statement "murder is wrong" is justified even if they are not sure whether it is "true" in the sense that it states a fact or describes a mind-independent reality (Sayre-McCord 1996; Tramel 2024).[5] If we can agree that some moral statements are justified, then we can build on this common ground to develop principles that guide conduct and enable us to make moral decisions (Rawls 1971).

[5] The sense of truth alluded to here is known as the correspondence theory of truth, i.e., a statement is true if and only if corresponds a fact in the world (i.e., reality). The statement "snow is white is true" means that (a) there is a thing (or object) in the world that we call snow and (b) that thing is white, i.e. has the property of whiteness (David 2015). Moral statements pose a difficult for this notion of truth, however, because it is not clear that there are moral facts, objects, and properties (Sayre-McCord 1996).

Turning to moral relativism, many philosophers distinguish between meta-ethical relativism and normative relativism. Meta-ethical relativism is the view that the truth of moral statements is relative to a particular culture (or society). Normative relativism is the view that one ought to follow the moral standard of one's culture or society (Timmons 2012).

To see how these views are different, consider international variations in laws. Whether something is legal or illegal depends on what country you're in. The statement "It is illegal to drive on the left side of a two-way road" is true in the U.S. but not in the U.K. Accepting this meta-thesis about the law does not imply, however, that one ought to follow the laws of a country that one happens to live in, because the laws could be immoral. For example, homosexuality is illegal in Saudi Arabia but that does not mean that a homosexual person who moves to Saudi Arabia ought not engage in homosexual sex. They might decide to disobey the law because they view the law as immoral and they're willing to risk the consequences of being caught.

Although meta-ethical and normative relativism are conceptually distinct, in practice there may be little difference between the two positions since they both imply that there are no moral standards that apply across different cultures.

Philosophers have discussed moral relativism since ancient times (Gowans 2021), but the theory has increased in popularity during the twentieth century due to anthropological research on cross-cultural variations in social norms (Benedict 1946) and philosophical critiques of objectivism in science and epistemology (Harman 1977; Baghramian and Carter 2020).

One of the most influential arguments for relativism begins with the thesis of culturally diversity with respect to moral norms, i.e., different cultures have differing views on homosexuality, the status of women, capital punishment, children's rights, marriage, abortion, free speech, and many other moral topics (Benedict 1946). The next premise in the argument is that if two cultures disagree about some topic, then there is no correct view of the matter, because truth and justification are culturally-relative. From these two premises, the relativist concludes that there are no culturally-transcendent moral standards (Gowans 2021).

Let's consider the first premise. While there is considerable evidence to support the cultural diversity thesis, there is also evidence that weighs against it (Macklin 1999; Pojman 2005). For example, most cultures regard murder, battery, rape, theft, fraud, cheating, incest, and adultery as morally wrong. There is also considerable evidence from psychology that people from different cultures form similar moral judgments in response to decision-scenarios, and that the parts of the brain involved in moral judgment are the same throughout the human species (Greene 2013). People from different countries have come together many times throughout history to reach a consensus on moral topics, such as human rights, slavery, human trafficking, and torture. For example, in 1948 over 50 members of the United Nations drafted a Universal Declaration of Human Rights (United Nations 1948). As discussed in Chap. 2, nations around the world condemned the Nazi experiments on concentration camp prisoners, and the Nuremberg Code reflects an international consensus on the ethics of research with human subjects. Other international codes, such as the Helsinki Declaration, reflect a similar consensus. Sociobiologists argue that evolutionary biology provides an explanation for cross-cultural similarities in moral norms, because morality has evolved as an adaptation that promotes cooperation within a social group (de Waal 2009). Thus, this empirical premise in the argument is only partly true. At the very least, it should be restated more precisely.

What about the second premise? If culture A believes that the Earth is flat and culture B believes it is round, should we conclude that there is no correct view of the matter independent of these cultures? The problem with the second premise is that is leads to epistemological/scientific relativism, which seems nonsensical. People from different cultures have disagreed (and continue to disagree) about medicine, astronomy, biology, geology, and many other topics, yet we do not conclude that the correct view on those topics depends on what culture you are in. There are objective (culture-independent) truths in science and mathematics (Kitcher 1995).[6]

A defender of relativism might qualify the second premise by saying that it applies only to non-scientific areas of debate, such as morality, religion, politics, law, and not to scientific areas, such as physics, chemistry, and biology.

A key problem with this response is that it is not clear that we can always clearly distinguish between science and non-science (e.g., politics, religion, philosophy) with respect to moral questions, especially those that are related to human biology and psychology. For example, if a culture has "scientific" beliefs concerning the inferiority of certain races, it may be difficult to determine whether their racism comes from their science, ethics, or both. Likewise, it may be difficult to distinguish between scientific and moral or political beliefs in economics. For example, Marxists claim that their theory is scientific, but others view it essentially a type of politics or ideology (Rosenberg 2016).

An additional problem with moral relativism is that 'culture' is not easily defined. Is culture defined by a geographic region, such as nation, state, or city? Can culture be equated with race, ethnicity, religious affiliation, political allegiance, profession, or occupation? The degree of cultural variation in moral beliefs and norms may vary depending on the definition of 'culture' one uses. Without a clear definition of 'culture,' the argument for moral relativism commits the fallacy of ambiguity because

[6]The idea that knowledge is objective is regarded by some as incompatible with indigenous (or traditional) knowledge, but this need not be the case. Much of indigenous knowledge has been developed by observing local plants, animals, and ecologies for hundreds or thousands of years. For example, in 1993, elders from the Navajo Native American Tribe shared knowledge with Western epidemiologists that play a crucial role in discovering the cause of a hantavirus outbreak (McDonald 2018). According to the Navajo people, hantavirus outbreaks usually followed periods of heavy rain or snowfall. Western epidemiologists used this knowledge to determine that rats carry the disease because periods of increased rain or snowfall produce large crops of pinion nuts, which, in turn, increase rat populations that eat the nuts. Indigenous cultures have ways of acquiring knowledge that are different from Western, scientific methods but this do not mean that the patterns in nature that they observe are culturally-relative. The pattern observed by the Navajos (i.e., increased rain/snow➔increase disease) correspond to a fact in the world. Western scientists could have observed the same fact by different methods.

'culture' may mean different things in the premises and conclusion (Macklin 1999; Timmons 2012).

Some people may endorse moral relativism because they believe we should not judge or interfere with other cultures, and they believe that acceptance of relativism justifies tolerance. However, the connection between relativism and tolerance is not straightforward. Suppose that culture A is very peaceful and tolerant toward other cultures, but culture B is very aggressive and intolerant. The relativist cannot say which culture has the correct view, because the relativist holds that what is right/wrong depends on what culture you are in. If you are in culture A, you should be tolerant; if you are in culture B, you should be intolerant. But this is an incoherent form of tolerance. To provide a coherent defense of tolerance, one would need to appeal to moral principles that transcend any particular culture, which the relativist cannot do (Timmons 2012).

While tolerance is an important virtue, morality sometimes requires us to take a stand against actions or behaviors which are clearly wrong. For example, I think most people would agree that what the Nazis did to the Jews before and during World War II was morally abominable. We should also have similar reactions to the Hutu-led government's slaughter of nearly one million Tutsis during the Rwandan civil war, and other episodes of genocide. Tolerance does not require us to sit by idly when evil occurs; we should condemn it and do something about it if necessary (Macklin 1999). The relativist's condemnation of moral evil is very limited and weak, however. A relativist could say what the Nazis did to the Jews was wrong according to the cultural standards of the Jewish people, but that it was acceptable within Nazi culture, or perhaps the broader German culture. This type of view flies in the face of commonsense, since most people would agree that what the Nazis did was wrong without any qualifications: it was wrong not just from the perspective of Jewish culture but from any reasonable perspective.

In addition to the aforementioned problems, moral relativism appears to be incompatible with intra-cultural criticism and reform. Throughout U.S. history, moral visionaries have criticized the status quo practices and have inspired changes in attitudes, beliefs, practices, and laws. For example, leaders of the Civil Rights Movement criticized racial segregation and discrimination and helped to move the U.S. in the direction of racial equality. These visionaries appealed to moral ideals which transcend the ethical norms found in popular culture. If moral relativism were correct, then intra-cultural criticism and reform would be conceptually incoherent, since there are no moral ideals beyond what is contained within one's culture. But most of us would agree that intra-cultural criticism and reform is conceptually coherent and that moral visionaries have brought about important changes in the U.S. and many other countries. Moral relativism has no cogent explanation of this phenomenon.

While I think there are good reasons for rejecting moral relativism as a moral theory, I think that relativism raises important issues concerning cultural variation that moral theories should address in some fashion. It might be the case that different cultures interpret and apply common ethical standards differently (Macklin 1999).

For example, in some Islamic cultures a woman is not allowed to consent for medical treatment by herself and must have the consent of her husband (if married) or an older male relative (Afifi 2007). If we assume that informed consent is a universal moral standard, should we view this Islamic practice as a violation of the standard or an as acceptable variation of it? Does it make sense to say that informed consent can mean different things in different cultures or nations? At what point would we not consider a cultural practice to involve consent at all? We will take up these sorts of questions again in Chaps. 4 and 5. Other chapters will also address the topic of cultural variations in research ethics.

3.3 Classifying Moral Theories

If we assume that it is possible to develop a moral theory that applies to different individuals, cultures, communities, societies, and nations, we need to ask what such a theory should be founded upon. As we shall see below, different moral theories provide different foundations for morality: some base it on God's will; some on human nature; and some on human reason. Before we begin our examination of moral theories, it is importance to distinguish between two types of theories: teleological (or goal-directed) theories and deontological (or duty-based) theories. Teleological theories hold that morality consists in striving to achieve, acquire, or sustain things we regard as valuable, such as happiness, knowledge, virtue, or human life, whereas deontological theories hold that morality is concerned with obeying moral duties, irrespective of whether doing so helps us to promote things (or goals) we regard as valuable. Deontologists claim that whether an action is right or wrong depends on the nature of the act, and its relation to our duties or rights, rather than the goals, aims, or purposes served by the act (Timmons 2012). We shall now consider these different theories (See Table 3.2).

Table 3.2 Moral theories

Name of theory	Type of theory	Chief proponent
Divine command theory	Deontological	Various religions
Virtue ethics	Teleological	Aristotle
Natural law theory	Teleological	St. Thomas Aquinas
Utilitarianism	Teleological	John Stuart mill
Kantianism	Deontological	Immanuel Kant
Natural rights theory	Deontological	John Locke
Care ethics	Mixed	Carol Gilligan
Principlism	Mixed	W.D. Ross

3.4 Divine Command Theory

The first moral theory we will consider is known as the divine command theory. This is the oldest moral theory, since it coincides with the advent of religion in human societies thousands of years ago. According to this theory, morality is founded upon God's will. Something is right/wrong because God commands it or would command it (Timmons 2012). All of the world's major religions—Judaism, Christianity, Islam, Buddhism, Hinduism, and African and Native American belief systems—include ethical principles and teachings. The divine command theory accurately depicts how many people think about moral issues and understand their moral obligations, since many people look to their religion for moral guidance and inspiration. The divine command theory is usually construed as deontological because it holds that morality consists in following God's commands. For example, Jews, Christians, and Muslims all regard the Ten Commandments as a set of moral obligations or duties. Despite its widespread influence, the divine command theory has some problems which preclude it from serving as a viable framework for thinking about the ethics of research involving human subjects.

The first problem with the theory was raised by the Greek philosopher Plato (427–347 BCE) in the *Euthyphro* (Plato 2002). In this dialogue, Socrates asks Euthyphro whether something is pious (or moral) because it is loved by the Gods or is loved by the Gods because it is pious. On the one hand, if something is moral because it is loved by the Gods, then this would imply that whatever the Gods love would be moral, even something which we would consider to be wrong. For example, suppose that the Gods commanded a man to kill his only son. Would this make it the right thing to do? Morality would seem to be arbitrary and possibly cruel or unjust if it depends on what the Gods love. On the other hand, if the Gods love what is moral because it is moral, then this would mean that morality is independent of what the Gods love. The Gods do not create morality but they are bound by it (Timmons 2012). If this is the case, it appears that the Gods play no role in morality. We do not need to consult with the Gods to understand our moral obligations or make moral decisions.

One way for the theist to deal with the *Euthyphro* dilemma is to concede that God does not create morality, but that God follows it perfectly because that is part of God's nature. God's commands always conform to morality because God is perfectly good and omniscient. God understands and follows the moral law and apprehends what it the right thing to do in every situation because that is part of what God is. God would not command someone to do something wrong, because this would be against God's nature (Timmons 2012).

While proponents of the divine command theory may be able fend off the Euthyphro dilemma, a second problem presents greater difficulties. The second problem has to do with epistemological problems with the theory. Supposing that morality is based on God's will, how are we to know what is right or wrong, good or bad, and so on? There seem to be two choices: either moral knowledge comes from divine revelation (i.e. direct communication with God) or religious teachings.

Divine revelation is a problematic source of moral knowledge because it is subjective and therefore unverifiable and potentially erroneous. Suppose that a man says that God told him to kill his son. Should we believe that person or take steps to prevent him from committing murder? What if the man is mistaken and the divine revelation is really a voice he hears in his head as a result of mental illness? Divine revelation may be a useful source of inspiration for an individual, but it does not seem to be a reliable source of information for making decisions that impact the rights or welfare of others.

Religious diversity presents an intractable problem for someone who holds that moral knowledge comes from religious teachings. In addition to the major religious traditions mentioned above, there are also sects within those traditions, and practitioners within the same sect who have different interpretations of religious texts and doctrines. Some religions and sects have been at war with each other for more than a thousand years. Although most of these religions, sects, and interpretative frameworks agree on some common moral principles, they also disagree on many topics, such as the morality of abortion, the death penalty, women's rights, and homosexuality. How are we to decide which religion to follow for moral guidance? Moreover, how are atheists supposed to acquire moral knowledge? Can they only know what is morally right by embracing a religion? We would seem to have very little hope of achieving any kind of consensus on moral issues if we hold that moral knowledge depends on religious teachings (Timmons 2012).

None of this need imply that religion is totally irrelevant to moral decision-making. People may still choose to rely on their religious views for personal guidance, motivation, and inspiration. Also, many religious and secular approaches to morality recognize some of the same duties, obligations, and values and can therefore reach some consensus on social/moral causes, such as preventing violence, human trafficking, child pornography, famine, poverty, or homelessness. However, it is not realistic to expect that we can or should appeal to particular religious texts, teachings, or doctrines to address public ethical or policy issues that affect people who hold diverse religious beliefs or have no religious beliefs at all. Thus, the divine command theory cannot provide us with a suitable (i.e., applicable) framework for dealing with ethical dilemmas in research involving human subjects.

3.5 Virtue Ethics

Virtue ethics, like the divine command theory, is a very old view, tracing its history to ancient Greece. Plato and his student Aristotle (384–322 BCE) held that morality consists in living a good life and being a morally good (or virtuous) person. In the *Nichomachean Ethics*, Aristotle (2003) developed an argument linking the good life to virtue that goes as follows:

1. Something is good insofar as it performs its essential function well. For example, a good flute makes good music when played.

2. A good human being, therefore, is a person who performs the essential human function well.
3. Rational activity of the soul (or reasoning) is the essential human function. People have many functions that they share with other living things, such as reproduction, growth, and movement, but only human beings have the ability to reason.
4. Human goodness (or virtue) can therefore be equated with rational activity of the soul.

According to Aristotle, rational activity of the soul consists in practicing and developing various good character traits (or virtues), such as courage, moderation, kindness, benevolence, honesty, loyalty, integrity, compassion, and justice. Aristotle held that virtues fall in between two extreme forms of conduct. For example, too little courage is cowardice, which is a vice; conversely, too much courage is rashness, which is also a vice. Proper courage falls in between cowardice and rashness. Aristotle held that we develop virtue by imitating virtuous conduct we observe in other people: to become a good man (or woman) you must do what the good man does. Aristotle also included practical wisdom among the virtues. Practical wisdom consists in knowing how to act in particular situations. We can use our practical wisdom to determine what a virtuous person would do in various situations (Aristotle 2003).

Although virtue ethics was popular in ancient Greece, it fell out of favor among philosophers during the Enlightenment period until it was revived by Foot (1978), MacIntyre (1984) and others, who were dissatisfied with dominant approaches to moral theory (i.e., Kantianism and utilitarianism, discussed below) (Timmons 2012). Several authors have applied virtue theory to questions of professional ethics, including medical and research ethics (Pellegrino 1992; Pellegrino and Thomasma 1993; MacFarlane 2008; Pennock 2019; Poomkudy and Shah 2024).

Part of the appeal of virtue ethics is that it provides guidance on how one ought to live one's life and the kind of person one ought to be. Most people will readily accept the idea that we ought to strive be kind, just, benevolent, honest, and so on. All of the other theories discussed in this chapter also imply that virtue is an important moral value worth pursuing (Timmons 2012). Virtue ethics is a teleological theory because it holds that morality consists in striving toward the goal of virtue.

The main problem with virtue ethics is that it lacks an effective decision-making procedure for dealing with ethical dilemmas in which virtues pull in different directions. For example, suppose that you are the director of a transplant unit at a local hospital network. A heart has become available (Resnik 2012b; Timmons 2012). Three people on the waiting list for a new heart are immunologically compatible with the heart. All three people have similar prognoses if they receive the heart. There are some other differences, however. The first person is a 32-year-old woman with no family who has been waiting longer for the heart than the others, the second is a 38-year-old man who has a wife and two young children who depend on him, and the third is 55-year-old man who owns a small welding company but has no wife or children. Who should receive heart? The virtue theorist has no readily apparent way of dealing with this dilemma, other than to advise us to use our practical wisdom to discover what the virtuous person would do. However, it is not obvious how a virtuous person would decide what to do in this situation. He or she could act according to virtue of benevolence and give heart to the man with the family, since this would probably do the most good. Alternatively, he or she could act according to the virtue of justice and give the heart to the person who has been waiting the longest ("first come, first served"). It is possible that different people pursuing virtue would make very different choices (Timmons 2012).

This example also illustrates another problem with virtue ethics: it has difficulty dealing with complex questions of social justice or public policy, which often arise in biomedicine and biomedical research. (See Box 3.1 for discussion of justice.) It is not at all clear, for example, what it would mean to practice the virtue of justice when allocating organs to patients. Would justice require that the person waiting the longest receive a chance at the organ first, or should people waiting in line enter a lottery? Does justice require that we give an organ to the person who is the most meritorious? When it comes to public policy, the answers we are looking for must be couched in terms of rules. For example, if the hospital wants to develop transplantation policies it will need to consider some rules for transplantation. Since virtue ethics focuses on moral virtues, as opposed to moral rules, its ability to provide policy guidance is somewhat limited. Virtue ethics seems to be most applicable to the realm of personal relationships.

Another criticism is that virtue ethics construes moral conduct as a type of self-improvement, i.e., we should act morally so we can be better people (Athanassoulis 2015). Without a doubt, improving ourselves helps us to get along with others, but one might argue that morality consists in more than in improving oneself: it also includes being committed to the wellbeing of other people and doing the right thing for the right reason. As we shall see below, Kantians take this approach to morality. Even those who do not adopt the Kantian view might wonder whether there is more to morality than practicing virtue (Timmons 2012).

Although virtue ethics provides us with some insight into what it means to live a morally good life and offers useful guidance for how we should handle personal relationships, it has difficulty dealing with the ethical dilemmas that arise in complex and novel situations involving many people, which often occur in research with human subjects. Virtue ethics may compliment other approaches to ethical conduct, but it does not provide us with the decision-making framework we are seeking.

Box 3.1: Justice

Philosophers have debated about justice since antiquity. Plato argued that well-ordered societies exhibit justice. A well-ordered society is one in which the guardians (i.e., wise, philosopher-kings) rule the other social classes, which include the producers (laborers and merchants) and the auxiliaries (soldiers and police). Plato opposed democracy because he believed it led to chaos and mob rule (Plato 2007). Aristotle, like Plato, was also skeptical of democracy and believed that society should be ruled by benevolent kings and that social and economic goods should be distributed on the basis of merit.

(continued)

Box 3.1 (continued)

Issues related to justice or fairness often arise in human research ethics. For example, inclusion and exclusion criteria for research participation raise issues (discussed in Chap. 8) related to the justice or fairness of selecting subjects for studies (MacKay and Saylor 2020). As discussed in Chap. 2, the authors of the *Belmont Report* argued that justice was a fundamental principle of research ethics which requires that the benefits and burdens of research be distributed fairly (National Commission 1979). The authors distinguished between five ways of distributing benefits and burdens including on the basis of equality, need, contribution, merit, and desert, but they have little to say about justice beyond this. Moral and political theorists have distinguished between the following types of issues related to justice (Rawls 1999, 2001; Sandel 2010; Miller 2021):

- **Distributive justice**: justice pertaining to the distribution of basic goods in society, such as income, wealth, opportunities, and health;
- **Retributive justice**: justice regarding punishment for crimes;
- **Restorative justice**: justice in making amends for wrongs;
- **Procedural justice**: fairness of the procedures used to make decisions pertaining to distributive, retributive, or restorative justice;
- **Formal justice**: consistency in the application of principles of justice, i.e., similar cases should be treated similarly;
- **National justice**: justice among citizens of a nation or state;
- **International justice**: justice among nations or states;
- **Global justice**: justice among all the people in the world, regardless of nationality;
- **Intergenerational justice**: justice between generations.

Some of the influential theories of distributive justice include:

- **Egalitarianism**: basic goods should be distributed on the basis of equality, such as equality of resources or opportunities;
- **Libertarianism**: basic goods should be distributed according to fair procedures that protect individual rights, promote individual responsibility, and reward merit;
- **Utilitarianism**: basic goods should be distributed in a way that maximizes society's overall good or utility.

Questions concerning distributive justice are highly complex, and it may be the case the no single theory of justice can adequately account for all of our intuitions concerning justice, and, therefore, different approaches may apply to different situations.

For example, let's consider economic justice, as illustrated by three different societies in Fig. 3.2. In society A, wealth is distributed equally between

(continued)

Box 3.1 (continued)

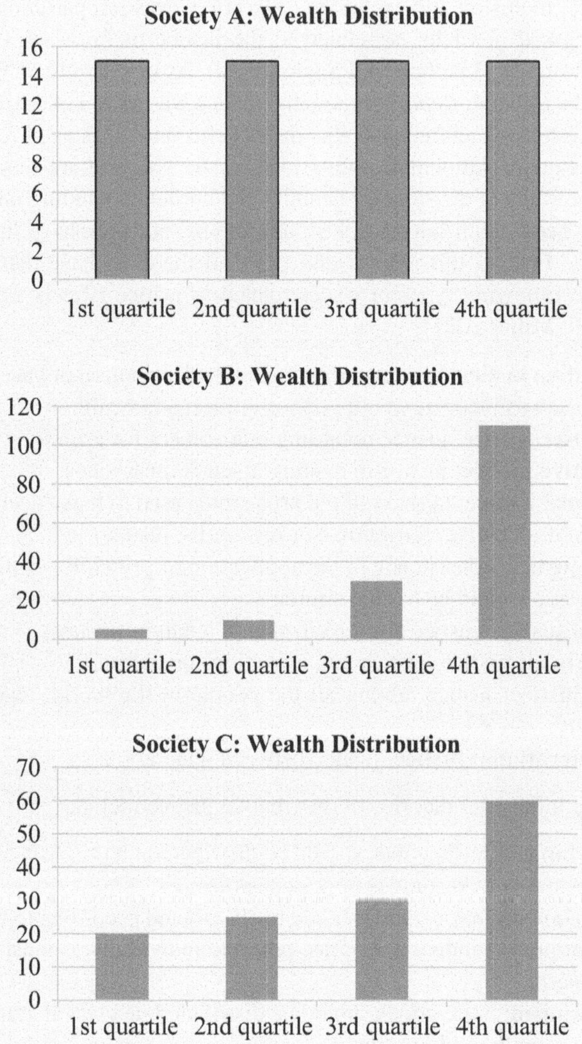

Fig. 3.2 Wealth distribution in three societies

four quartiles and the total wealth is 60. In society B, wealth is distributed unequally, and the total wealth is 150. In society C, wealth is distributed unequally but the worst-off members in society C have more wealth (30) than they do in society A (15) or B (20), although the total wealth in C (140) is less

(continued)

Box 3.1 (continued)

than society B. Which distribution of wealth is the most just? Egalitarians that emphasize equality of outcomes (for example, Marxists) would say that society A is the most just because wealth is distributed equally. Utilitarians would say that society B is the most just because it has the most total wealth. Libertarians say that whatever wealth distribution has arisen by a fair process which honors individual rights, promotes individual responsibility and rewards merit is the most just. For libertarians, huge differences in the distribution of basic goods are acceptable if they result from a fair process (Nozick 1974).

A highly influential moral and political theorist, John Rawls (1921–2002) defended an approach to distributive justice that reconciles conflicts between these three different approaches to justice. Rawls (1971, 2001) defended two principles of distributive justice: (1) rights and liberties should be distributed equally, and (2) differences in social and economic goods are acceptable if these differences benefit the worst-off members of society and there is equality of opportunity in society. According to Rawls, if there is equality of opportunity and equal protection of rights and liberties in society C, then society C is the most just because the worst-off members of society have more wealth the worst-off in A or B. Rawls used a thought experiment known as the Original Position to justify these principles. In the Original Position, rational individuals (or contractors) are choosing principles that should govern society. To ensure that their decision-making is fair and impartial, the contractors are behind a Veil of Ignorance, which means that they do not know who they are in that society. That is, they do not know their race, ethnicity, gender, socioeconomic status, age, and so on. Critics of Rawls have argued that contractors would choose other principles, such as principles that maximize utility (Freeman 2023).

A utilitarian approach to justice may make the most sense during public health emergencies, in which medical resources are scarce. For example, suppose that an explosion has occurred in a chemical factory and hundreds of people have been injured. The emergency department at the local hospital is overwhelmed and cannot provide treatment to everyone who needs it. To deal with the situation, it would seem wise to follow a policy, known as triage, in which health care providers distinguish between: (a) patients who are so sick or injured that they will probably die soon, even if they receive treatment; (b) patients who are healthy enough to do without treatment for a while; and (c) patients who are significantly sick or injured and can benefit from immediate treatment. In triage, one treats patients from group (c) first because this will make the best use of limited medical resources. Patients from group (b) are treated later, and those from group (a) may receive only comfort/supportive care (Kipnis 2002).

(continued)

Box 3.1 (continued)

A libertarian approach may make sense for distributing goods in situations where considerations of merit are paramount. For example, most people would agree that class grades should be distributed based on performance on tests, exams, or assignments, rather than on the basis of equality (give everyone the same grade) or need (give the highest grades to those who need them the most).

In most situations involving matters of justice, treating people as having equal moral value will be an overarching consideration, even when other principles are at play. For example, organ allocation networks use medical criteria (such as prognosis and immune-compatibility) and not social or economic criteria (such as person's occupation or health insurance), combined with some form randomness, such as length of time on the waiting list (Veatch and Ross 2015).

3.6 Natural Law Theory

Natural law theorists equate moral values with things that are naturally good, such as life or health and conceive of morality as stemming from human nature. St. Thomas Aquinas (1225–1274) held it is part of human nature to need or want four goods: life, procreation, knowledge, and social relationships. Other goods, such as health, food, shelter, and so on, are valuable as far as they help us attain these basic goods (Aquinas 1988). Things that oppose natural goods, such as disease, ignorance, famine, and suffering are natural evils. The natural law theory is a teleological approach to morality because it holds that our moral obligations are based on the goals of promoting natural goods and avoiding natural evils. Our duty—the moral law—is based on natural law. For example, we have a duty not to commit murder because human life is naturally good. We have an obligation to have children because procreation is naturally good. We should not lie, cheat, steal, betray our friends, or commit adultery, because these acts undermine social relationships (Aquinas 1988; Timmons 2012).

A key problem for the natural law theory, or any moral theory for that matter, is how to deal with moral conflicts. Suppose that someone is threatening me with a knife. Is it justifiable for me to kill in self-defense? My life is a natural good, but should I take another life (which is also a natural good) to save my own? The principle of double-effect is a key pillar of the natural law theory that provides an answer to conflicts involving good and bad effects. According to the principle, it is morally acceptable to perform an action with bad effects if: (a) the action itself is not wrong; (b) the bad effect is not intended; and (c) the bad effect is proportionate to the good effect. A bad effect would be unintended if it is not sought in itself or as a means to the good effect (Timmons 2012). For example, killing in self-defense morally acceptable, provided that the act (i.e., saving one's life) is not wrong, the killing is not intended, and the killing is proportionate (e.g., no more violence is used than is necessary to save one's life).

While the principle of double-effect provides the natural law theory with a way of dealing with moral conflicts, it has some weaknesses. Chief among these is that it may not be clear what makes a bad effect proportionate to the good outcome (Timmons 2012). To determine proportionality, one must engage in moral priority setting: what is more important: to promote the good or avoid the bad? The natural law theory does not have any simple answers to this question. For example, suppose that a man is terminally ill and suffering a great deal. To ease the patient's suffering, the doctors prescribe him heavy doses of opioid medications which relieve his suffering but hasten his death. Would this act of "terminal sedation" be unethical? The answer depends, in part, on how quickly the doctors hasten the patient's death. If the patient has only a week or so to live, and the doctors shorten his life by one day, most people would agree that this would be acceptable. But suppose the patient has 6 months to live and they shorten his life by 5 months. Most people would probably view this as murder or unjustified euthanasia. Therefore, to decide upon the morally correct course of action, one must consider two competing values: promoting life and relieving suffering (Billings 2011). If one applied the doctrine of double-effect to the HIV prevention trials discussed in Chap. 1, one would need to determine whether the bad effects (i.e., not treating HIV patients in the placebo group) would be proportionate to the good (i.e., developing a cheaper means of preventing perinatal HIV transmission).

Part of the appeal of the natural law theory is that it attempts to provide an objective and universal foundation for morality in human nature (Gert 2004). Unfortunately, what makes the theory attractive to empirically-inclined philosophers and ethicists may also be its downfall. While psychology, biology, sociology, anthropology, and other sciences can provide us with some insights into morality (Greene 2013), we cannot easily derive normative, moral conclusions from descriptions of human nature (Timmons 2012). The mere fact that something is natural does not make it good or right. For example, aggression is part of human nature, but acknowledging this fact does not tell us whether any particular aggressive action is right. Aggression may be justified to defend oneself but not to steal someone's property. We need a moral standard independent of nature to tell us when aggression is and is not justified. Also, the fact that something is unnatural does not make it wrong. It is not wrong, one might argue, for a fertile couple to decide not to have children, even though this choice would go against their natural capacities and inclinations. Looking at the bigger picture, if we accepted a strongly naturalistic approach to ethics, then would have to regard much of our technology and medicine as immoral because it changes or enhances natural human functions and capacities. While we can draw upon nature to provide us with a source of value—for example most people would agree that life and health are valuable—we should not use nature as the sole arbiter of morality.

3.7 Utilitarianism

English philosophers Jeremy Bentham (1748–1832) and John Stuart Mill (1806–1873) proposed an influential theory known as utilitarianism. According to Mill, utilitarianism is:

> The creed which accepts the foundation of morals or "utility" or "the greatest happiness principle" holds that actions are right in proportion as they tend to promote happiness; wrong as they tend to promote the reverse of happiness (Mill 1979:7).

On this view, morally right conduct consists in promoting the greatest net good for the greatest number of people, where the good is equated with happiness (Mill 1978). Mill argued that everything we regard as good is either part of happiness (e.g., knowledge or virtue) or is a means to obtaining happiness (e.g., health, food). Mill did not equate happiness with mere physical pleasure and distinguished between different types of happiness. Happiness could include pleasures of the intellect, such as reading books, writing poetry, listening to or playing music, and so on (Mill 1979). Modern utilitarians, such as Singer (1972, 1979), Brink (1989), Brandt (1998) and Hooker (2000), do not equate the good with happiness, because happiness can be difficult to define. Some equate the good with advancement of welfare, fulfillment of interests, or satisfaction of preferences. Others accept a plurality of goods, so that promoting the good consists in promoting many different things that we would want or need, such as well-being, health, wealth, social relationships, and so on. Utilitarianism can be characterized as a teleological theory because it holds that morality is goal-directed (Timmons 2012).

Mill held that the principle of utility is a basic moral axiom which is not capable of proof in the ordinary sense of the term. That is, anyone who understands the principle should accept it without the need for additional justification. Nevertheless, Mill offered a justification of the principle:

> No reason can be given why the general happiness is desirable, except that each person, so far as he believes it to be attainable, desires his own happiness. This, however, being a fact, we have not only all the proof which the case admits of, but all which is possible to require, that happiness is a good, that each person's happiness is good for that person, and that the general happiness, therefore, a good to the aggregate of persons (Mill 1979:34).

There are some problems with this "proof." For example, Mill moves from the descriptive fact that we desire our own happiness to the normative conclusion that happiness is desirable. But there might be many things which people desire that we would not regard as desirable (i.e., worth desiring or pursuing). For example, some people desire to injure themselves or others, to eat food or drink alcohol excessively, and so on. Additionally, even if we accept the notion that each person's happiness is good, it does not follow that the aggregate happiness is good. Indeed, some would argue that this form or reasoning commits the fallacy of composition: one cannot infer properties of a whole thing from properties of the parts (Timmons 2012). For example, from the premise that all members of society are human beings we cannot conclude that society is a human being. Other philosophers have offered proofs of the principle of utility similar to Mill's, which we do not need to dwell on too long here (see Brandt 1998).

Fig. 3.3 The trolley problem (Used with permission from Peter B. Resnik)

Instead of critiquing arguments for a principle of utility, we will focus our criticism on the principle itself and some of the troubling implications of accepting it as a guide to ethical decision-making. Before turning to these critiques, we should discuss the theory's strengths, which are significant. First, the principle of utility captures an important intuition that most of us probably share, i.e., that promoting good outcomes does matter in ethics (Timmons 2012). For example, most of us would probably agree that, other things being equal, saving 100 lives is better than saving 10, or that killing 100 people is worse than killing 10. Second, utilitarianism, unlike some of the other theories we have examined, offers a clear decision-making procedure. In deciding what one should do, one should consider the likely consequences of the different options and choose the option which is likely to yield the greatest balance of good/bad consequences for the most people. Third, utilitarianism is a highly general theory because it applies to many different ethical problems, including questions involving personal relationships, social justice, and public policy.

The main problem with utilitarianism is that it does not give sufficient consideration to the welfare or rights of the individual (Timmons 2012). Consider a famous ethics thought experiment, originally developed by Foot (1978) and Thomson (1985), known as the trolley problem.[7] (See Fig. 3.3.)

Suppose there is a runaway trolley heading down the tracks. If it continues on its course, the trolley will kill five people who have been tied up and placed on the tracks by a wicked person. There is not enough time to untie the people. However, you can pull a lever which will divert the trolley onto a sidetrack on which one person is tied down. Utilitarianism would tell you to pull the lever, which nets +4 lives (1 killed, 5 saved), as opposed to doing nothing, which nets −4 lives (5 killed, 1 saved). Psychological studies indicate that about 80% of respondents say that they would pull the lever (Greene 2013). But let's change the example a little bit. Suppose there is no sidetrack, but you can push a large person off a footbridge onto the tracks, which will stop the trolley. There is no time to ask the person for their consent. In this scenario, about 80% of people say they would not push the person onto the tracks (Greene 2013). We seem to have a strong moral intuition, which runs counter to utilitarian thinking, that there should be some limits on what we may to do to individuals to promote the common good. The ends do not always justify the means. Natural law theorists could say, for example, that pushing the person off the footbridge would be wrong because the bad effect would be a means of obtaining the good effect.

As noted in Chaps. 1 and 2, this type of conflict, i.e., the good/rights of the individual vs. the good of society, often arises in research with human subjects. According to utilitarianism, we should resolve these dilemmas in favor of the good of society, which is promoted by advancing scientific knowledge. The same logic which implies we should kill an innocent person to save five would also imply that we should withhold medical treatment from research subjects in placebo control groups to save many lives, even when a treatment is available and they may be significantly harmed by not receiving treatment. It is acceptable for some people to suffer or die as a result of research participation so that others may benefit. However, many people would object to this utilitarian thinking on the grounds that human research subjects should be protected from unnecessary harm or exploitation (Jonas 1985).

Another critique is that utilitarianism does not provide a satisfactory account of social justice because it focuses on aggregate utility rather than the distribution of utility (Greene 2013). For example, consider the distribution of wealth in three societies (see Box 3.1 and Fig. 3.2). Utilitarians would hold that society B is the most just because it has the greatest total wealth, but other theorists would claim that society B is unjust because wealth is distributed very equally (Rawls 1971). Utilitarianism also has disturbing implications for criminal justice. Suppose an unidentified man of race A has murdered a man of race B in a town with a history of violent conflict between races A and B. If the police do not arrest a man from race

[7] There are also many variations of this thought experiment. In one variation, the five people are all 90 years old, whereas the one person is only 12; in another variation, the five are strangers, whereas the one is your spouse or child; and in another variation, you can flip a switch that will make a robot push the large person onto the tracks (see Greene 2013).

A soon and charge him with a crime, people from race B are likely to start a riot, which would probably cause serious injuries, property damage, and possibly death. To prevent these harmful consequences, the police decide to arrest an innocent man from race A and charge him with a crime. The prosecutor agrees to put him on trial and seek a conviction if necessary. While a utilitarian would endorse this plan to arrest an innocent person to prevent a riot, most people would view it as unjust on the grounds that we should punish people only if we have strong evidence that they have committed a crime. Innocent people should not be punished to calm and angry mob or for any other reason (Greene 2013).

A third critique is that utilitarianism does not give sufficient weight to self-interest or self-care. For example, suppose I have $250 each month leftover after paying my bills. I can spend that money on myself, or I can send the money to a charitable organization that helps people who are starving. If I follow utilitarianism, I should give that money to charity, since this would probably produce more net good. Though the money would produce some utility for me, since my basic needs are already met, it would produce more utility by giving it those whose needs are not met. But it does not end there. Utilitarianism would also seem to instruct me to continue giving away my income and wealth until I am as deprived as those I am helping (Singer 1972). Taking utilitarianism seriously seems to demand a great deal from us—some would say too much (Greene 2013).

Another way of putting this point is to say that utilitarianism does not recognize the distinction between what is morally **obligatory** and what is morally **supererogatory**. An act is obligatory if it morally required; an act is supererogatory if it goes above and beyond the call of duty. While we might praise you as a moral saint if you give all of your money to help others, we would not say that you would be acting immorally if you failed to do this. Ordinary people are not required to be moral saints. Utilitarianism seems to require that we should all act like moral saints, which seems to place unrealistic and excessive demands on people (Timmons 2012).

Utilitarians have developed responses to these objections, most of which involve distinguishing between **act utilitarianism** (AU) and **rule utilitarianism** (RU). According to AU, one should perform the act that maximizes utility, whereas under RU one should follow a rule that maximizes utility (Timmons 2012). Rule-utilitarians argue that counter-intuitive implications of utilitarianism will not occur if one understands the theory as applying to rules instead of actions (Brandt 1998; Hooker 2000; Greene 2013). For example, utilitarianism need not imply that we should push the large person onto the railroad tracks to save five lives because the adverse social consequences of following a rule that required such conduct would outweigh the good results. If we lived in a society where people were always ready to trade one life for five, we would be in constant fear that we might be killed to save someone's life. This fear would undermine human relationships and cause a great deal of harm. Rule-utilitarians could argue that utilitarianism can handle distributive justice issues by accounting for the negative impacts of inequitable distributions of utility. Societies with huge disparities in wealth are likely to have many socioeconomic problems, such as poverty, disease, homelessness, drug abuse, and crime, which negatively impact overall utility. A rule-utilitarian could also argue that utility would not be maximized if police and prosecutors followed a rule that allowed them to accuse innocent people of crimes to quell social unrest, since the adverse effects of such as policy (e.g., increased distrust of the police and fear of the legal system) would outweigh any gains. Finally, rule-utilitarians could argue that utilitarianism can give adequate weight to self-interest and self-care. Society would not be better-off, a rule-utilitarian could argue, if everyone followed a rule which required them to give away virtually income and wealth to help people who are worse-off, since people would lack motivation to work hard (or at all) if this were the moral rule.

While the RU response to these objections to utilitarianism seems reasonable, it may not satisfy the critics of the theory because a great deal hinges on the relationship between the rules and utility (Timmons 2012). RU might recommend different rules, depending on the type of society one lives in and consequences of adopting the rules. For example, people who live in a collectivist society may be better able to accept a rule that mandates taking an innocent life to save five. It could turn out that the negative consequences of following such a rule would not outweigh the positive ones in that society. Critics would argue that individual rights/welfare should be protected, regardless of the likely consequences of various rules. Likewise, the adverse consequences of inequitable distributions of utility might not materialize in societies in which people are accustomed to large differences in income, wealth, legal rights, and so on, such as societies based on caste systems. Rule-utilitarians could respond to these counterarguments by claiming that the rules entailed by RU are rules for ideal societies, not actual ones (Brandt 1998; Greene 2013). For example, a rule that requires the killing of one innocent person to save five would not promote utility in an ideal society. But this reply leaves one wondering how rule-utilitarians would decide which rules would promote net utility in an ideal society, and how one should characterize an ideal society.

Rule utilitarians must also develop a procedure for adjudicating conflicts among rules when these arise. For example, suppose that RU includes a rule that protects a parent's right to make decisions for their child but also a rule allowing the state to mandate vaccinations for children. The rule utilitarian needs a way of deciding which rule should take priority if a parent refuses to allow her child to be vaccinated. The rule utilitarian could argue that all conflicts among rules should be settled by appealing to the principle of utility. While this strategy seems reasonable, it would seem to defeat the entire point of having rules, since utility is the ultimate arbiter when moral conflicts arise. However, the rule utilitarian could argue that rules still have a purpose, since they govern our behavior when we do not face moral dilemmas.

To summarize this section, RU shows some promise as a theory for making ethical decisions concerning research with human subjects. The theory provides a clear decision-making procedure which applies to a variety of complex ethical issues. However, RU has some problems dealing with our moral intuitions concerning the protection of the rights and welfare of individuals, respecting self-interest and self-care, and promoting fair distributions of goods.

3.8 Kantianism

Kant constructed a deontological moral theory which continues to have considerable influence. Even philosophers who do not consider themselves Kantians embrace some of his ideas about morality. Kant's key insight is that morality is about following rules for conduct that, we, as rational beings, make for ourselves. Actions are right or wrong not because of the consequences they produce, the virtues they display, or the feelings they manifest, but because of their connection to the moral law (Timmons 2012).

As we have seen above, different theories have different accounts of what is good or morally valuable. Kant held that the only thing which is a good without qualification is a good will. Having a good will is more fundamental than moral virtue, because moral virtues may be used to accomplish immoral goals. For example, a person with a bad will could draw upon their courage to commit many crimes. Likewise, wealth, power, health and other things thought to be good can be used by someone with a bad will to accomplish bad things. Even happiness is not good without qualification, because the happiness of a bad-willed person would be a bad thing. To even be worthy of happiness, one must have a good will (Kant 1981).

A person with a good will is a someone who is motivated to act not by selfish reasons, emotions, or ingrained character traits, but by the desire to do their moral duty for duty's sake. One's duty, according to Kant, is based on a general moral principle known as the **Categorical Imperative** (CI). The CI implies various categorical rules (Kant 1981). Categorical rules (or imperatives) are different from hypothetical ones in that they are not directed toward particular ends or goals. The rule (or maxim), "In order to be a good person, I will try to make others happy" would be a **hypothetical imperative** because it has the goal of developing moral virtue in one's self, whereas the rule "I will help others who are in need" is a categorical imperative because it is not directed toward any particular goal.

Kant developed several different formulations of the CI, each with slightly different proofs. The proofs are complex and we will not discuss them here (see Hill 1992).

The first formulation of the CI is: "Act only on that maxim whereby you can will at the same time it should become a **universal law** (Kant 1981:30)." The key insight behind this version of the CI is that morality is about following universal laws for all rational agents. A rule that applied only to a group of people, community, society, or religious sect would not be a moral rule. Although this version of the CI sounds like the Golden Rule, it is different. Indeed, Kant was critical of the Golden Rule because it depends on the agent's point of view, whereas the CI does not. For example, a masochist who inflicts pain on others would be satisfying the Golden Rule but not the CI because the rule "inflict pain on others" could not become a universal rule for all rational agents.

The second formulation of the CI is: "Act in such a way that you treat humanity, whether in your own person or in the person of another, always at the same time as an end and never simply as a means (Kant 1981:36)." The key insight behind this version of the CI is that we should not use rational agents (i.e., humanity) as mere objects or things to achieve certain goals, because rational agents have intrinsic dignity or worth which cannot be traded for or replaced by something else. Deception, fraud, coercion, manipulation, theft, exploitation, murder, rape, and many other actions widely regarded as immoral involve violations of **human dignity**. Incidentally, this version of the CI would prohibit us from pushing someone in front of a runaway trolley to save five lives (Greene 2013).

According to the third formulation of the CI, one ought to act according only according to maxims that "harmonize with a possible kingdom of ends" (Kant 1981:42). What Kant means by the Kingdom of Ends is similar to the idea of a **social contract** discussed by Thomas Hobbes (1588–1679), John Locke (1632–1704), Jean Jacques Rousseau (1712–1778) and more recently by Rawls (1971) and Robert Nozick (1938–2002) (Korsgaard 1996). A Kingdom of Ends is a hypothetical situation in which rational, autonomous agents formulate rules that they agree to live by in society. True freedom (or autonomy), for Kant, consists in living according to principles that one chooses as a rational agent. This a very different from modern notions of autonomy, which equate it with the ability to make free (e.g., uncoerced) and informed choices in accordance with one's values and preferences (see discussion in Chap. 5).

Kant distinguished between several different types of duties that are based on the CI, including duties to oneself vs. duties to others, and perfect vs. imperfect duties. Duties to oneself include the duty to promote one's own well-being and develop one's talents or skills. Duties to others include duties not to harm others and promote their well-being. One should also treat oneself and other people with respect, i.e., not as a mere end.

A perfect duty is absolute rule with no exceptions. A perfect duty has no exceptions because the negation of the duty could not become a universal law. For example, Kant held that we have a perfect duty to others not to make promise we do not intend to keep because the maxim "I will make a promise I do not intend to keep" could not become a universal law, since if everyone followed this rule, we would not trust each other's promises and the institution of promise-keeping would collapse. We have a perfect duty to our self not to commit suicide, because suicide could not become a general rule for all rational agents, since society would cease to exist if suicide became a rule (Kant 1981). **An imperfect duty**, by contrast, may have exceptions because the negation of the duty could become a universal law. For example, Kant held that we have an imperfect duty to help others. The maxim "I will not help others" could become a universal law because a society could exist where people do not help each other. However, moral agents would recognize that they would sometimes need help, so they would not want it to become a universal law. Since the duty to help others is imperfect, we are not obligated to help someone whenever we have the opportunity to do so, because we may have other duties which conflict with this duty, such as an imperfect duty to promote our own well-being. Kantianism does not imply that we should devote so much of our time, wealth, and energy toward helping others that we reduce ourselves to poverty or injure our own health. In deciding whether to help others, we must consider our competing obligations.

Kantianism has considerable strengths that we should mention before turning to critiques of theory. First, it accounts for our strong intuition that we should not use people as mere things to promote certain ends—even noble ones (Hill Jr. 1992). Most of us would not push someone in front of a runaway trolley to save five lives. The second version of Kant's CI provides us with a way of articulating this intuition. Second, Kantianism emphasizes the idea that morality is about following universal rules (Korsgaard 1996). As we saw in the previous section, rule-utilitarians used this insight to defend utilitarianism against various objections. Rules and principles also play an important role in other moral theories (Timmons 2012). Third, Kant's theory includes duties to oneself and duties to others, so it does not have the counter-intuitive implication that we should sacrifice our own well-being to help others (Timmons 2012). Fourth, Kantian theory provides a framework for thinking about justice. Rawls' defense of the principles of distributive justice was strongly influenced by Kant's conception of a kingdom of ends. Kant also held that punishments should be proportional to the crimes committed and that criminal procedures should respect human dignity (Hill Jr. 1992). Fifth, the CI implies a general decision procedure for making complex ethical decisions. To resolve a conflict, one can formulate a maxim for a proposed choice and test it against the CI. If the maxim is not consistent with the CI, then one should not make that choice and one should seek another option.

One of the main critiques of Kant's theory is that it appears to imply a kind of a kind of **moral absolutism** with counterintuitive results. Kant maintained that we have a perfect duty not to lie to others, since lying could not become a universal rule for all rational agents. Lying—even for a good reason—could not become a universal rule because if everyone did this no one could be trusted to tell the truth. Also, lying to someone fails to respect their humanity (Varden 2010). But what about lying for a very good reason, such as to prevent someone from being killed? Suppose that a would-be murderer comes to our door, and they are looking for a person they intend to kill, who happens to be hiding in our house. They ask us if we know where that person is. Should we tell them the truth or a lie? Kant's view implies that we should not lie to the murderer, since we have a perfect duty not lie but only an imperfect duty to help others. Kant, himself, discusses this very example and argues that we should not lie to the would-be murderer (Varden 2010). Most people, however, would probably say that it we should lie to a would-be murder to save someone's life. Defenders of Kantianism have argued that his theory need not have the counterintuitive implication that we should tell the truth to the murderer (Varden 2010). If we formulate a specific maxim tailored to this type of situation, e.g., "I will lie to a would-be murder who is planning to kill someone," then this maxim could become a general law, since adopting this maxim as a rule for conduct would not destroy the practice of truth-telling, because we would only rarely be in a situation in which we lie to a would-be murderer to save someone's life.

This reply to the objection that Kant's view is excessively absolutist brings to light another problem with the theory, namely, that a great deal depends on how we formulate maxims we test against the CI (O'Neill 1989; Timmons 2012). Highly general maxims, e.g., "never lie," or "never break a promise," may sometimes have counterintuitive implications, whereas more specific maxims may not have such implications. So how should we decide which maxim to test when considering our moral duties? Testing only highly specific maxims might allow us to avoid the counterintuitive implications of Kantianism, but this strategy would undermine the very purpose of the CI, which is to provide us with categorical moral rules. The more specific a maxim becomes, the more likely it is to be hypothetical or goal-directed. Indeed, in his defenses of the first formulation of the CI, Kant specifically says that the principle should be devoid of any references to motives or goals (Kant 1981).

Another problem with Kant's theory is that it does not provide us with adequate guidance for resolving conflicts of duties. The theory can deal with conflicts of perfect and imperfect duties because perfect duties take precedence (Timmons 2012). However, conflicts between imperfect duties can pose problems. Suppose I have $250 leftover after paying my bills. I could give the money to a charity, or I could use the money to buy myself some exercise equipment. Since duties to help others or oneself are both imperfect, I need a way of deciding which duty should have moral priority. Kantians could argue that I can decide what to do by formulating different maxims and testing them against the CI, but it is not clear which maxims I should use and why. A utilitarian could solve this dilemma by considering which option is likely to produce more overall good, but that strategy is not available to Kant.

Consider the issue, discussed in Chap. 1, of including placebo control groups in clinical trials when there is known effective treatment for the disease. If the clinical trial involves treating people as mere means for benefitting others, then Kantianism implies that the trial would be unethical. A clinical trial which enrolled human subjects without their consent would be treating the subjects as mere means and would be unethical, on a Kantian view. However, the dilemmas we are interested in addressing do not involve violations of informed consent. The subjects have presumably consented to being randomly assigned to receive the treatment under investigation or a placebo. The dilemmas the generate controversy involve conflicts between imperfect duties: the duty to help subjects with the disease in the placebo group versus the duty to help perhaps many more people by developing scientific knowledge concerning the treatment under investigation. Kantianism would also appear to have no readily available answer on how to resolve this dilemma other than to instruct us to consider different maxims and test them against the CI.

To summarize this section, Kantianism, like RU, shows some promise as a theory for making ethical decisions concerning research with human subjects. The theory provides a decision-making procedure which can deal with a variety of complex ethical issues and accounts for our intuition that we should not treat people as mere means to promote certain goals. It also instructs us to think about moral choices in terms of applicable rules or principles. However, Kantianism, like the other theories we have considered, has some significant weaknesses. First, it seems to imply a moral absolutism that conflicts with commonsense; and second, it can be difficult to interpret and apply. Thus, although Kantianism is by far the most sophisticated moral theory we have considered in this chapter, like the other theories, it has difficulty resolving moral conflicts.

3.9 Natural Rights Theories

We have mentioned rights at various points in this book without exploring this concept in any detail. We will now consider moral theories which focus on rights. Before we do, we need to say a few words about moral rights. A right is an entitlement an individual has to be treated by other people or the government in a certain way. For example, if I have a right to life, then other people have a duty not to kill me. Rights have at least two functions: (a) to protect an individual's interests and (b) to give an individual a sphere of influence. For example, my property rights pertaining to my house protect my interests in that house and give me dominion over the house (Wenar 2020).

Legal rights, as established in federal or state constitutions and other documents, are different from moral rights because legal rights are enforced by the coercive power of the state, whereas moral rights often are not. Also, one may appeal to moral rights to defend or critique legal rights. For example, one might argue that a woman should have a legal right to abortion based on her moral right to control her body. Very often states also enforce widely recognized moral rights, such as the right to life, but the legal right to life is still distinct from the moral right to life.

Most theorists distinguish between positive and negative rights. Negative rights are rights to be left alone: one can honor a negative right simply by not interfering with or violating that right. For example, the right to life, understood negatively, is simply a right not to be killed. Positive rights, in contrast, are rights to be provided with a good or service. To honor a positive right, one must do something to or for someone (Wenar 2020). For example, many state governments recognize that all school-age children have a right to an education. To implement this right, state and local governments must build schools, hire teachers, establish curricula, and develop educational standards. All of this requires considerable funding, which the state acquires through taxation. Most natural rights theorists, such as Nozick (1974) restrict rights to negative rights, since they view the role of government as limited to protecting rights rather than promoting them.

Rights are sometimes described as "trump cards" because they take precedence over other sorts of claims when conflicts occur (Dworkin 1984). For example, if I have a right to free speech, I could assert that this right takes precedence over society's interest in restricting my speech to prevent harm. However, "trump" is probably too strong a word to describe how rights function in argumentation, since most theorists agree that rights can sometimes be restricted when they conflict with other rights or with important social goals (Sinnott-Armstrong 1996; Wenar 2020). A better way of viewing the issue is to say that rights establish a burden of proof: for example, if I have a right to free speech, then the government must produce a compelling argument to justify restricting my speech. The importance of the right at issue determines the level of the burden of proof. For example, the U.S. courts apply a burden of proof known as strict scrutiny to government limitations on some of the rights protected by the Constitution, such as freedom of speech or religion. Under the strict scrutiny test, the government must establish that it has a compelling public interest in restricting the right and that it will use the least burdensome means to restrict it (Barron and Dienes 2013).

Several of the theories we have examined thus far can support moral rights. For Kantians, moral rules stemming from the CI imply moral rights. For example, if I have a duty not lie to other people, then they have a right not to be lied to by me. Also, the respect for humanity version of the CI can be interpreted as establishing a general right to respectful treatment (Hill Jr. 1992). Rule utilitarians can support individual rights insofar as rules which honor rights promote utility. For example, in *On Liberty* Mill (1978) defended many of the freedoms recognized by natural rights theorists on the grounds that allowing people to make their own choices generally produces more net utility than restricting their freedom, even when freedom is restricting their freedom is for their own good. However, as we have seen, utilitarian protections for rights could easily erode if utility is better served by restricting rights (Timmons 2012).

Natural rights theorists do not justify rights by appealing to moral duties or utility, since they regard human rights as foundational. Natural rights theorists claim that morality is based some fundamental rights shared by all human beings, such as rights to life, liberty and property. The right to liberty typically includes the right to freedom of thought, speech, action, association, religion, and peaceful assembly (Nozick 1974). The natural rights approach takes its inspiration from seventeenth century philosopher John Locke's political theory. Locke (1980) argued that all human beings are endowed with certain natural rights by their Creator. While modern natural rights theorists, such as Nozick (1974) and Thomson (1992), do not claim that rights come from God, they do treat rights as foundational axioms of moral theory. Natural rights theories are usually classified as deontological because they hold that morality is not goal-directed.

Locke used the social contract idea to give an account of the function of government and its authority to restrict rights. Nozick (1974) also appeals to this idea in his moral and political philosophy. According to Locke, during a hypothetical time before the existence of civil society known as the state of nature, people have the authority to enforce their own natural rights. For example, in the state of nature if someone steals my property, I have the authority to retrieve my property and punish the person who took it. As Hobbes (2006) observed, life in the state of nature can be perilous and arduous (or "nasty, brutish and short"), since we may easily fall prey to theft, rape, murder, and other violations of our rights, and we may need to devote considerable resources toward protecting our rights. People recognize that it is in their self-interest to form a government to protect their rights and to benefit from social cooperation (Locke 1980; Nozick 1974; Hobbes 2006). Thus, the sole function of government, under this theory, is to protect our natural rights (Locke 1980; Nozick 1974). The founding fathers of the U.S. government took inspiration from Locke's political writings. The second paragraph of the Declaration of Independence (1776) reads:

> We hold these truths to be self-evident, that all men are created equal, that they are endowed by their Creator with certain unalienable Rights, that among these are Life, Liberty and the pursuit of Happiness.–That to secure these rights, Governments are instituted among Men, deriving their just powers from the consent of the governed…

Natural rights theories appeal to our strongly held intuition that all human beings have some basic rights that should be protected. Some philosophers have challenged this idea, arguing that rights must be justified and cannot simply be assumed as morally foundational (Fagan 2015). I will not dwell on that issue, since the other moral theories we have reviewed also make some basic assumptions which could be questioned. For example, utilitarians treat the principle of utility as an axiom of moral theory, and Kantians regard the CI as foundational. All moral theories, it would seem, must make some basic assumptions that are accepted without further argument (Timmons 2012). My critiques will focus instead on the difficulties with applying the natural rights view to ethical issues and dilemmas.

All of the theories we have examined must confront the problem of how to resolve moral conflicts. Natural rights theorists must provide a way to resolve conflicts involving rights, which could include conflicts between different rights-holders or conflicts between rights and public interests (Sinnott-Armstrong 1996). For example, suppose that I own some farmland and I want to start raising pigs. My neighbors do not want me to raise pigs because they are concerned about the stench from the animals and possible adverse health impacts (such as increased risks of respiratory illnesses). This dispute would involve a conflict between my right to use my property and my neighbors' rights not to be harmed (by possible adverse the health impacts) or offended (by the smell). For another example, suppose that the state determines that all children should be vaccinated against an infectious disease to protect the public's health. Some parents do not want their children to be vaccinated, however, because they are concerned about possible adverse health effects of the vaccine. This dispute would involve a conflict between parental rights concerning child-rearing and the state's interest in promoting children's health.

Natural rights theorists have proposed different strategies for handling conflicts involving rights. One strategy is to develop a list of exceptions for a right (Oberdiek 2008). For example, the definition of the right to free speech could include various exceptions that specify when speech can be restricted; for example, to prevent violence, libel, or violations of copyrights. One problem with this approach is that the list of possible exceptions may become limitless, because new exceptions might need to be included when the situation demands it. Also, this approach could leave rights so riddled with exceptions that they would become meaningless (Thomson 1992). Rather than creating a list of exceptions, some natural rights theorists prefer to develop tests for balancing the rights and interests at issue, acknowledging that the burden of proof still falls on those who propose to restrict rights (Thomson 1992). This procedure for dealing with conflicts involving rights would involve the kind of priority-setting addressed elsewhere in this chapter.

Some have criticized rights-based approaches on the grounds that they interfere with productive dialogue on moral issues. When someone asserts that they have a right at stake in a moral issue, this often functions to stop debate rather than encouraging the kind of deliberation necessary for a fair and effective resolution of the issue (Glendon 1991). Rights can stop debate because, as we noted earlier, they function like moral trump cards. When someone claims that they have a right at stake, the burden of proof then falls on the party to give a convincing reason for restricting that right, which may result in a stand-off, since each side may assert that their argument is stronger or that the other side has the burden of proof.

Another objection to rights-based approaches is that they ignore our moral duties and responsibilities (Glendon 1991). These approaches tend to skew the moral discussion in the direction of what other people (or society) owe to the individual as opposed to what the individual owes to others. This distortion is especially pronounced in approaches which defend only negative rights, because no one has a duty to do anything for anyone else on this type of view (Callahan 1981). Of course, all approaches, including the natural rights approach, imply that we may acquire obligations to help others as a result informal agreements, legal contracts, or relationships that we voluntarily undertake. For example, a physician who agrees to care for a patient has a moral obligation to help the patient as a result of this relationship. But these obligations are different from obligations to help strangers. Other moral theories that we have considered, such as Kantianism and utilitarianism, hold that we have substantial moral obligations to benefit other people with whom we do not have an agreement, contract, or relationship. As we saw in the discussion of the controversial HIV prevention trials in Chap. 1, the extent of investigators' obligations to benefit research subjects can be an important issue in placebo-controlled studies.

To summarize this section, the natural rights theory provides a way of articulating our strongly held moral intuition that all human beings should be free to pursue their life plans as long as they don't harm others. However, it has difficulties handling the moral conflicts that arise when rights collide with each other or with public goods. Though some theorists have proposed strategies for dealing with these conflicts, these ultimately involve the kind of priority-setting that occurs in other theories. Fair and effective priority-setting may be more difficult for the rights-based approach than other theories, because rights-claims tend to skew the debate in the direction of protecting individual interests.

3.10 Feminist Ethics

Although feminism is frequently associated with the women's rights movement of the 1960s, feminist writers and activists, such as Mary Astell (1666–1731), Mary Wollstonecraft (1759–1797), Harriet Taylor Mill (1807–1858), and Catherine Beecher (1800–1878), defended women's rights to vote, to own property, and to be educated, and critiqued patriarchal social and political systems (Norlock 2019) hundreds of years ago. Harriet Taylor Mill was the wife of John Stuart Mill (discussed above). John Stuart Mill also advocated for women's rights, and it is likely that Harriet Taylor Mill, who was herself an outstanding philosopher, strongly influenced her husband's moral and political writings. In the twentieth century, writers and activists, such as Charlotte Perkins-Gilman (1860–1935), Simone de Beauvoir (1908–1986), and Gloria Steinem championed legal and economic rights for women (Norlock 2019).

In the early 1980s, psychologist Carol Gilligan (1982) pioneered an innovative approach to moral theory based on evidence concerning women's experiences of moral judgment, decision-making, and emotion. Gilligan's key insight is that women have different experiences of morality than men and tend to focus on caring relationships, rather than on abstract notions of duty, justice, or rights. Gilligan's work helped to establish a feminist approach to ethics known as the ethics of care (Baier 1985, 1986; Sander-Staudt 2019). Philosophers have debated about whether the notion of care can be understood in terms of traditional moral concepts like virtue, duty, or justice (Sander-Staudt 2019). Timmons (2012), for example, classifies that ethics of care as a form of virtue ethics that focuses on the virtue of being a caring person.

Other important feminist moral philosophers, such as Nel Noddings, Annette Baier, and Allisson Jaggar, have explored feminist themes, such as the role of gender in society and influence of male privileges (Norlock 2019). Some philosophers, known as ecofeminists, combine feminist ethics with concerns related to protection of the environment and stewardship of natural resources (Gaard and Gruen 1993).

While I think that feminist moral philosophers have made and continue to make important contributions to ethical theory, this approach, like all the others we have examined in this chapter, also has difficulties with resolving difficult ethical dilemmas, such as the Trolley Problem (discussed above) or including placebo control groups in clinical trials (discussed in Chap. 1). How would a person who practices care ethics deal with these dilemmas? How would their answers differ from those given by virtue ethicists, Kantians, or utilitarians? That being said, I do think that care ethics' emphasis on relationships is a crucial into insight into the nature of morality and I will discuss it in more depth in Chap. 4, when I examine the role of trust in ethics (Baier 1986; Peter and Morgan 2001).

3.11 Principlism

The final type of theory we will consider has been dubbed "principlism" because it holds that morality is founded on several (perhaps many) independent moral principles, each of which have intuitive appeal. Some of these principles may be teleological while others are deontological (Timmons 2012). Scottish philosopher W.D. Ross (1877–1971) pioneered this approach but several other prominent theorists, including Rawls, Tom Beauchamp, and James Childress have also defended principle-based theories (Rawls 1971; Beauchamp and Childress 2012; Shamoo and Resnik 2022). Also, as discussed in Chap. 2, the *Belmont Report* articulates three independent principles, i.e., respect for persons, beneficence, and justice. Principlists also hold that there is no meta-principle for ranking the principles, though Beauchamp and Childress (2012) suggest that autonomy should generally be given more weight than other principles.

Ross (1930) argued that morality is based on seven principles:

Duties of fidelity: keeping promises, telling the truth.
Duties of reparation: amending past wrongs one has committed.
Duties of gratitude: expressing gratitude for gifts or help that one has received.
Duties of justice: treating people fairly.
Duties of non-maleficence: not harming others.

Duties of beneficence: helping others.
Duties of self-improvement: improving one's character, talents, intelligence, or abilities.

Ross' principles are based on insights from Kantianism, virtue ethics, and utilitarianism. The principles of fidelity, reparation, and justice reflect a Kantian approach to morality, gratitude and self-improvement align with virtue ethics, and beneficence embodies utilitarian thinking. Other theorists have defended principles that deal with obligations not covered by Ross' view, such duties to non-human animals and the environment (Resnik 2012a).

Ross' principles imply *prima facie* duties or obligations. The duties are *prima facie* insofar as we should obey them unless there are no other conflicting duties. When *prima facie* duties conflict with each other we must use our moral judgment to decide which duty should take priority (Ross 1930). For example, suppose that I have made a promise to meet a friend for dinner. I have a duty to keep this promise. However, while driving my car on the way to dinner I encounter a pregnant woman who is walking by the side of the road. She signals for me to stop. She has gone into labor and needs a ride to a hospital. Let's assume there is not enough time to call an ambulance for the woman or call my friend. I also have a *prima facie* duty to help this woman. What should I do? According to Ross, I should reach a moral decision by carefully considering my duties, options, and the relevant facts (Beauchamp and Childress 2012). If the woman is in dire need of help, then the correct choice may be to help her and break my promise to my friend. However, if she is not in great need, e.g., if she is not in labor and is just taking a walk and wanted a ride to her friend's house, then maybe I shouldn't break my promise. If other options are available, this might also affect my decision. For example, if I have access to a phone, I may be able to call one of her friends or family members or an ambulance.

One of the advantages of principlism is that it instructs to set priorities when principles conflict by carefully considering the facts and circumstances relevant to our decision as well as the available options (Beauchamp and Childress 2012; Richardson 2018). To apply the principles to any particular choice, we must attend to the details of the situation, and the devil, as they say, it in the details.

This decision-making method may be especially helpful in thinking about moral dilemmas involving research with human subjects. For example, one might argue that the acceptability of placebo control groups in clinical trials depends on several key facts, such as (a) the availability of a standard (proven) therapy for the condition, (b) the adverse health impacts of receiving a placebo instead of a standard therapy, and the (c) the subjects' understanding of randomization, placebos, and other concepts related to the trial, (d) the importance of the knowledge expected to be gained from the study, and (e) the strength of the placebo effect (Miller and Brody 2002).

An interesting implication of principlism is that there may be more than one morally acceptable solution to an ethical dilemma, because there may different ways of setting priorities. As a result, different people may arrive at different solutions to a moral dilemma because they emphasize different moral principles. As we saw in Chap. 1, well-informed, thoughtful investigators and ethicists took different positions on several ethical controversies concerning research with human subjects.

Those who think that a moral theory should deliver one and only one solution to a moral dilemma regard the possibility of multiple solutions as a weakness of principlism. Others, however, see this implication as a strength since it allows the theory to acknowledge and explain moral disagreement (Timmons 2012; Richardson 2018).

Another criticism of principlism is it reduces morality to situational ethics because the correct moral decision depends on the facts and circumstances pertaining to a decision. For example, Kantians would say that you should never push an innocent person in front of trolley to save other people. It does not matter whether there are 5, 50, or 5000 lives at stake because you should not treat people as a mere means to achieving certain ends. A principlist, however, would say that if enough lives are at stake, then pushing someone in front of the trolley would be justified because the facts should impact our moral assessment. Some critics argue that this implication undercuts the very point of having moral principles. However, principlists can reply that moral principles still play an important role in ethics, since most of the time we should follow the principles. It is only when conflicts arise that we must make decisions based on facts inherent in the situation (Timmons 2012).

A related criticism is that principlism is unsystematic because it lacks a unifying principle or concept, such as the CI, utility, natural goods, natural rights, or virtue. The theory is a hodge-podge of different views (Gert 2004). While this charge has some merit, I would not say that the theory lacks any unity. What unifies the theory is the recognition that we have some basic moral intuitions concerning different rules, which sometimes conflict. Principlists see unity in this diversity, whereas other theorists may find this diversity unacceptable.

A final criticism of principlism is that it is no better than other theories at dealing with ethical conflicts, because solving these dilemmas still depends on exercising moral judgment. Whether one is Kantian, utilitarian, natural law theorist, natural rights theorists, virtue ethicist, or care ethicist, ethical decision-making still depends very much on how one sets priorities in a situation. While this charge has some merit, I would not say that principlism is no better than other theories at dealing with moral dilemmas. Part of the attraction of principlism is that it synthesizes insights from other approaches. The principlist can use ideas from different approaches to make ethical decisions. These ideas may give the principlist more conceptual resources to work with when thinking about ethical problems.

To summarize this section, the main appeal of principlism is that it recognizes that we have strong intuitions concerning some conceptually distinct moral principles. We can use different moral principles to guide our everyday conduct and make ethical decisions. Another advantage of principlism is a pragmatic theory because it includes a workable decision-procedure which applies to a variety of complex issues. While principlism still has some difficulties when it comes to dealing with ethical dilemmas, other theories also have this problem. Moreover, principlism may be better than other theories at addressing ethical dilemmas because it has more conceptual resources to draw upon.

3.12 Conclusion: Toward a Decision-Making Framework

In this chapter we have examined seven different moral theories which could provide a workable decision-making framework for ethical dilemmas involving research with human subjects. We have show that each of these theories have strengths and weaknesses (Timmons 2012). The theories emphasize different aspects of morality, i.e., following moral rules, promoting moral goods (or values), respecting human rights, or developing moral virtues or caring relationships. The theories also try to deal with common problems involving moral conflict, e.g., the individual vs. society, fair distribution of benefits and harms, and helping others vs. taking care of one's interests. Recalling our discussion of reflective equilibrium earlier, we may be at a point in time where we are still going back and forth between intuitions and theories (or principles) and we have not yet reached a point where they are in perfect agreement.

Given all of the above, I hold that principlism is the most reasonable approach to moral theory. Although the view also some weaknesses, it is better able to accommodate conflicting intuitions concerning moral principles and values than the other approaches and has a workable decision procedure (Timmons 2012). The view I will to ethical issues discussed in this book includes four, independent moral principles:

Respect for human dignity: Treat all human beings[8] as having inherent moral worth; do not restrict an autonomous person's decisions or actions with a sound justification.
Non-maleficence: Avoid causing harm to oneself or others.
Beneficence: Promote the welfare of oneself and others.
Justice: Distribute benefits and harms fairly; use fair procedures to make distributive decisions.

A few comments about these principles are in order. First, the four principles described above are nearly identical Beauchamp and Childress' (2012) principles. One key difference is that Beauchamp and Childress have a principle of respect for autonomy, while I have a principle of respect for human dignity and autonomy. I include respect for dignity in this principle because I view dignity as broader than autonomy. Respect for autonomy applies to people as far as they have the ability to make rational decisions. A key part of respecting a person's dignity is respecting their right to make autonomous decisions when they have the capacity to do so. But many people who lack the capacity to make autonomous decisions still have human dignity which should be respected. For example, we should treat neonates, infants, young children, and mentally disabled adults, and even people in persistent vegetative states as having inherent moral worth even if they lack decision-making

[8] By "human being" I mean not only members of the human species but also other beings that have the qualities that adult members of our species possess, such as consciousness, personality, reasoning, emotion, intelligence, and judgment. In theory, members of other terrestrial species, such as dolphins or chimpanzees; alien life forms; or even artificially intelligent machines, could posses these qualities. Thus, while I use the term "human being" in this book, what I have in mind is close to Kant's notion of humanity or personhood.

capacity. Dignity has to do with our status as human beings, not our status as moral agents.

The principles are also similar to three described in the *Belmont Report*, except respect for dignity replaces for respect for persons, and beneficence and non-maleficence replace beneficence, which are combined in the *Belmont Report*. While some may argue that my approach is not original because I am borrowing heavily from other writers, the uniqueness of my view will become apparent to readers in Chap. 4, when I discuss the role of trust in research ethics.

Second, one can apply these general principles to particular cases by means of a process known as specification (Richardson 2000). Specification involves developing subsidiary rules from general principles. The rules can be more easily applied to cases that the principles. For example, the principles of respect for dignity and non-maleficence justify rules for informed consent in research (Brock 2008). Informed consent respects dignity by giving subjects the right to decide whether to participate in research and it protects subjects from harm by allowing them to refuse to participate in research they judge to be too risky. The requirement to obtain consent justifies other subsidiary rules, such as rules pertaining to the disclosure of information or the documentation of consent.

Third, like Beauchamp and Childress, I view these principles as *prima facie* rules that may conflict in particular cases. Principlism, like other moral theories, must still deal with the problem of priority-setting when conflicts arise. Beauchamp and Childress do not include a meta-rule for prioritizing the principles, nor do I. Like Beauchamp and Childress, I hold that we might emphasize different principles in different circumstances. Though respect for dignity should usually be given the highest priority, there are situations (such as Mill's example of a person about to cross a damaged bridge, discussed at the end of Chap. 2) when we are justified in restricting a person's autonomous decisions or actions for their own good. In Chap. 4, I will propose that a fifth principle, the principle of trust, can help investigators, oversight committees, institutions, and funding agencies set priorities in research with human subjects when conflicts arise. As I shall explain later, one can decide what to do by appealing to goal of promoting trust in research with human subjects. Although this fifth principle is not a meta-rule for prioritizing principles, it can sometimes tip the scales in favor of one option or another when have difficulty choosing among conflicting obligations.

References

Afifi, R.Y. 2007. Biomedical research ethics: An Islamic view part II. *International Journal of Surgery* 5 (6): 381–383.

Aquinas, S.T. 1988 [1265–1275]. *On politics and ethics*. Trans. P.E. Sigmund. New York: W.W. Norton.

Aristotle. 2003 [350 BCE]. Nichomachean ethics. Ed. H. Tredennick and Trans. J.A. Thompson. New York: Penguin Books.

Athanassoulis, N. 2015. *Virtue ethics. Internet Encyclopedia of Philosophy*. Available at: http://www.iep.utm.edu/virtue/. Accessed: January 23, 2025.

Ayer, A.J. 1952. *Language, truth, and logic*. 2nd ed. New York: Dover.

Baghramian, M. and J.A. Carter. 2020. Relativism. In *Stanford Encyclopedia of Philosophy*. Available at: https://plato.stanford.edu/entries/relativism/. Accessed 23 May 2024.

Baier, A. 1985. What do women want in a moral theory? *Nous* 19: 53–65.

———. 1986. Trust and antitrust. *Ethics* 96: 231–260.

Barron, J.A., and C.T. Dienes. 2013. *Constitutional law*. 8th ed. Saint Paul: West Publishing.

Beauchamp, T.L., and J.F. Childress. 2012. *Principles of Biomedical Ethics*. 7th ed. New York: Oxford University Press.

Benedict, R. 1946. *Patterns of culture*. New York: Penguin Books.

Billings, J.A. 2011. Double effect: A useful rule that alone cannot justify hastening death. *Journal of Medical Ethics* 37 (7): 437–440.

Brandon, R.N. 1995. *Adaptation and environment*. Princeton: Princeton University Press.

Brandt, R. 1998. *A theory of the good and the right*. revised ed. New York: Prometheus Books.

Brink, D.O. 1989. *Moral realism and the foundations of ethics*. Cambridge: Cambridge University Press.

Brock, D.W. 2008. Philosophical justifications of informed consent in research. In *The Oxford handbook of clinical research ethics*, ed. E.J. Emanuel, C. Grady, R.A. Crouch, R.K. Lie, F.G. Miller, and D. Wendler, 606–612. New York: Oxford University Press.

Callahan, D. 1981. Minimalist ethics. *Hastings Center Report* 11 (5): 19–25.

Daniels, N. 1996. *Justice and justification: Reflective equilibrium in theory and practice*. New York: Cambridge University Press.

David, M. 2015. The correspondence theory of truth. In *Stanford encyclopedia of philosophy*. Available at: https://plato.stanford.edu/entries/truth-correspondence/. Accessed 27 Sept 2024.

de Waal, F. 2009. *Primates and philosophers: How morality evolved*. Princeton: Princeton University Press.

Declaration of Independence. 1776. Available at: http://www.archives.gov/exhibits/charters/declaration_transcript.html. Accessed 8 June 2024.

Dennett, D.C. 1995. *Darwin's dangerous idea: Evolution and the meaning of life*. Vol. 35, 34–40. New York: Simon and Schuster.

Driver, J. 2022. Moral theory. In *Stanford encyclopedia of philosophy*. Available at: https://plato.stanford.edu/entries/moral-theory/. Accessed 17 May 2022.

Dworkin, R. 1984. Rights as trumps. In *Theories of rights*, ed. J. Waldron, 153–167. Oxford: Oxford University Press.

Fagan, A. 2015. Human rights. In *Internet encyclopedia of philosophy*. Available at: http://www.iep.utm.edu/hum-rts/ Accessed 8 June 2024.

Foot, F. 1978. *Virtues and vice*. Oxford: Blackwell.

Freeman, S. 2023. Original position. In *Stanford encyclopedia of philosophy*. Available at: https://plato.stanford.edu/entries/original-position/. Accessed 27 May 2024.

Gaard, G., and L. Gruen. 1993. Ecofeminism: Toward global justice and planetary health. *Society and Nature* 2: 1–35.

Gert, B. 2004. *Common morality: Deciding what to do*. New York: Oxford University Press.

Gilligan, C. 1982. *In a different voice*. Cambridge, MA: Harvard University Press.

Glendon, M.A. 1991. *Rights talk: The impoverishment of political discourse*. New York: Free Press.

Gowans, C. 2021. Moral relativism. In *Stanford encyclopedia of philosophy*. Available at: https://plato.stanford.edu/entries/moral-relativism/. Accessed 23 May 2024.

Greene, J. 2013. *Moral tribes: Emotion, reason, and the gap between us and them*. New York: Penguin Press.

Haidt, J. 2013. *The righteous mind*. New York: Vintage.

Harman, G. 1977. *The nature of morality*. New York: Oxford University Press.

Hill, T.E., Jr. 1992. *Dignity and practical reason in Kant's moral theory*. Ithaca: Cornell University Press.

Hobbes, T. 2006. [1651] *Leviathan*. New York: Dover.

Hooker, B. 2000. *Ideal code, real world: A rule-consequentialist theory of morality*. Oxford: Oxford University Press.

Hume, D. 1896 [1739]. *A treatise of human nature*. Oxford: Clarendon Press.
Jonas, H. 1985. The imperative of responsibility. In *In search of an ethics for the technological age*. Chicago: University of Chicago Press.
Kant, I. 1981 [1785]. Groundwork for the metaphysics of morals. Trans. J.W. Ellington. Indianapolis: Hackett.
Kipnis, K. 2002. Triage and ethics. *Virtual Mentor* 4 (1): 19–21.
Kitcher, P. 1995. *The advancement of science*. New York: Oxford University Press.
Korsgaard, C. 1996. *Creating the kingdom of ends*. Cambridge: Cambridge University Press.
Locke, J. 1980 [1689]. *Second treatise of government*. Indianapolis: Hackett.
MacFarlane, B. 2008. *Researching with integrity: The ethics of academic inquiry*. New York: Routledge.
MacIntyre, A. 1984. *After virtue*. South Bend: University of Notre Dame Press.
MacKay, D.K.W., and Saylor. 2020. Four faces of fair subject selection. *American Journal of Bioethics* 20 (2): 5–19.
Macklin, R. 1999. *Against relativism: Cultural diversity and the search for ethical universals in medicine*. New York: Oxford University Press.
McCarthy, J. 2021. U.S. approval of interracial marriage at a new high of 94%. Gallup, 10 September. Available at: https://news.gallup.com/poll/354638/approval-interracial-marriage-new-high.aspx#:~:text=WASHINGTON%2C%20D.C.%20%2D%2D%20Ninety%2Dfour,asked%20the%20question%20in%201958. Accessed 18 May 2024.
McDonald, J. 2018. Solving a medical mystery with oral traditions. JSTOR Daily, 14 March. Available at: https://daily.jstor.org/solving-a-medical-mystery-with-oral-traditions/. Accessed 27 Sept 2024.
Mill, J.S. 1978 [1859]. *On liberty*. Rapaport E ed. Indianapolis: Hackett.
———. 1979 [1861]. *Utilitarianism*, ed. G. Sher. Indianapolis: Hackett.
Miller, D. 2021. Justice. In *Stanford encyclopedia of philosophy*. Available at: https://plato.stanford.edu/entries/justice/. Accessed 26 May 2024.
Miller, F.G., and H. Brody. 2002. What makes placebo-controlled trials unethical? *American Journal of Bioethics* 2 (2): 3–9.
Moore, G.E. 1903. *Principia ethica*. Cambridge: Cambridge University Press.
National Commission for the Protection of Human Subjects of Biomedical or Behavioral Research. 1979. *The Belmont Report: Ethical Principles and Guidelines for the Protection of Human Subjects of Research*. Washington, DC: Department of Health, Education, and Welfare.
Norlock K. 2019. Feminist ethics. In *Stanford encyclopedia of philosophy*. Available at: https://plato.stanford.edu/entries/feminism-ethics/. Accessed 8 June 2024.
Nozick, R. 1974. *Anarchy, state, utopia*. New York: Basic Books.
O'Neill, O. 1989. *Constructions of reason*. Cambridge: Cambridge University Press.
Oberdiek, J. 2008. Specifying rights out of necessity. *Oxford Journal of Legal Studies* 28 (1): 127–146.
Peck, J.R., and D. Waxman. 2018. What is adaptation and how should it be measured? *Journal of Theoretical Biology* 447: 190–198.
Pellegrino, E.D. 1992. Character and ethical conduct of research. *Accountability in Research* 2: 1–11.
Pellegrino, E.D., and D. Thomasma. 1993. *The virtues in medical practice*. New York: Oxford University Press.
Pennock, R. 2019. *An instinct for truth: Curiosity and the moral character of science*. Cambridge, MA: MIT Press.
Peter, E., and K.P. Morgan. 2001. Explorations of a trust approach for nursing ethics. *Nursing Inquiry* 8 (1): 3–10.
Plato. 2002 [395–399 BCE]. *Five dialogues: Euthyphro, apology, crito, meno, phaedo*, ed. J.M. Cooper. Trans. J.M.A. Grube. Indianapolis: Hackett.
———. 2007 [375 BCE]. *Republic*. Trans. D. Lee. New York: Penguin.
Pojman, L.P. 2005. *Ethics: Discovering right and wrong*. 5th ed. Belmont: Wadsworth.

Poomkudy, J.T., and S. Shah. 2024. Virtue ethics and the unsettled ethical questions in controlled human infection studies. *Bioethics* 38 (8): 692–701.

Quine, W.V. 1951. Two dogmas of empiricism. *Philosophical Review* 60 (1): 20–43.

Rawls, J. 1971. *A theory of justice*. Cambridge, MA: Harvard University Press.

———. 1999. *Law of peoples*. Cambridge, MA: Harvard University Press.

———. 2001. *Justice as fairness: A restatement*. Cambridge, MA: Harvard University Press.

Resnik, D.B. 2012a. *Environmental health ethics*. Cambridge: Cambridge University Press.

———. 2012b. Ethical virtues in scientific research. *Accountability in Research* 19 (6): 329–343.

Richardson, H.S. 2000. Specifying, balancing, and interpreting bioethical principles. *Journal of Medicine and Philosophy* 25 (3): 285–307.

———. 2018. Moral reasoning. In *Stanford encyclopedia of philosophy*. Available at: https://plato.stanford.edu/entries/reasoning-moral/. Accessed 18 May 2024.

Rosenberg, A. 1985. *The structure of biological science*. Cambridge: Cambridge University Press.

———. 2016. *Philosophy of social science*. 5th ed. New York: Routledge.

Ross, W.D. 1930. *The right and the good*. Oxford: Oxford University Press.

Sandel, M.J. 2010. *Justice: What's the right thing to do?* New York: Farrar, Straus and Giroux.

Sander-Staudt M. 2019. Care ethics. In *Internet encyclopedia of philosophy*. Available at: https://www.iep.utm.edu/care-eth/. Accessed 15 Oct 2019.

Sayre-McCord, G. 1996. Coherentist epistemology and moral theory. In *Moral knowledge?* ed. W. Sinnott-Armstrong and M. Timmons, 137–189. New York: Oxford University Press.

———. 2015, Moral realism. In *Stanford encyclopedia of philosophy*. Available at: https://plato.stanford.edu/entries/moral-realism/#2. Accessed 17 May 2024.

Shamoo, A.E., and D.B. Resnik. 2022. *Responsible conduct of research*. 4th ed. New York: Oxford University Press.

Singer, P. 1972. Famine, affluence, and morality. *Philosophy and Public Affairs* 1 (1): 229–243.

———. 1979. *Practical ethics*. Cambridge: Cambridge University Press.

Sinnott-Armstrong, W. 1996. Moral dilemmas and rights. In *Moral dilemmas and moral theory*, ed. H.E. Mason, 48–65. New York: Oxford University Press.

Thomson, J.J. 1985. The trolley problem. *The Yale Law Journal* 94 (6): 1395–1415.

———. 1992. *The realm of rights*. Cambridge, MA: Harvard University Press.

Timmons, M. 2012. *Moral theory: An introduction*. 2nd ed. Lanham: Rowman and Littlefield.

Tramel, P. 2024. Moral epistemology. In *Internet encyclopedia of philosophy*. Available at: http://www.iep.utm.edu/mor-epis/. Accessed 17 May 2024.

United Nations. 1948. Universal declaration of human rights. Available at: http://www.un.org/en/universal-declaration-human-rights/. Accessed 19 May 2024.

Varden, H. 2010. Kant and lying to the murderer at the door…one more time: Kant's legal philosophy and lies to murderers and Nazis. *Journal of Social Philosophy* 41 (4): 403–421.

Veatch, R.M., and L.F. Ross. 2015. *Transplantation ethics*. 2nd ed. Washington, DC: Georgetown University Press.

Wenar, L. 2020. Rights. In *Stanford encyclopedia of philosophy*. Available at: http://plato.stanford.edu/entries/rights/#5.2. Accessed 1 June 2024.

Chapter 4
The Trust Approach to Research with Human Subjects

Abstract The chapter describes and defends a trust-based approach to the ethics of research with human subjects. The key premise of the framework is that research is founded on trust: trust between participants and investigators and institutions; between communities and investigators and institutions; among investigators, staff, and trainees; and between the public and the research enterprise. The chapter argues that the goal of promoting trust augments other moral principles, and that reflecting on the nature and importance of trust can often help one decide how resolve ethical and policy dilemmas in research with human subjects. Although promoting trust does not solve all ethical dilemmas in research, it can be an essential tool in helping us to develop workable solutions. The chapter also explains what trust is, distinguishes between trust and verification, discusses alternative views about trust in research, and discusses methods for promoting trust in research with human subjects, such as community and public engagement.

Keywords Trust · Human subjects · Investigators · Communities · Public · Verification

The previous two chapters have provided some historical and philosophical background to set the stage for the discussions that will take place in this chapter. In Chap. 2, I examined the history of the ethics of research with human subjects and argued that existing regulations and guidelines were adopted to restore and maintain public trust and prevent ethical abuses from occurring again. However, since ethical issues concerning research with human subjects are likely to continue arising, we need a decision-making framework to help guide policy formation and establish priorities. In Chap. 3, I examined some influential moral theories to determine whether they could supply such a framework. I argued that the theories have strengths and weaknesses but that no single theory satisfactorily answers all objections and answers all critics, and that the most reasonable position is to adopt some form of principlism. I proposed that four moral principles, similar to those defended by Beauchamp and Childress (2012), can guide ethical decision-making and policy

D. B. Resnik, *The Ethics of Research with Human Subjects*, The International Library of Bioethics 111, https://doi.org/10.1007/978-3-031-82757-0_4

formation. I noted, however, that this approach still faces the issue of priority-setting when moral conflicts arise. I suggested that a fifth principle, the principle of trust, can help researchers, oversight committees, institutions, and funding agencies set priorities when conflicts arise. In this chapter I will elaborate on and defend an approach to research with human subject based on the notion of trust.

At the outset, I will note that many articles have discussed the importance of trust in research involving human subjects, but very few have tried to explain exactly what trust means; how trust is obtained, and maintained; and how reflecting on trust can help us addressing moral dilemmas in research.[1] Thus, while I acknowledge that the idea of thinking about trust in research with human subjects is not new, I hope that the framework and analysis presented in this chapter is.

4.1 What Is Trust?

To develop a trust-based approach to research ethics, I first need to explain what is meant by trust. Scholars from philosophy, psychology, sociology, economics, and many other disciplines have examined the meaning of trust and proposed different definitions of this concept (Baier 1986; Fukuyama 1995; Govier 1997; Blomqvist 1997; O'Neill 2002; Potter 2002; Hardin 2001). Webster's Dictionary (2024) defines trust as "assured reliance on the character, ability, strength, or truth of someone or something." Clearly, **reliance** is an important feature of trust. To trust someone is to depend on that person behaving according to certain expectations so one can plan accordingly. When I trust other drivers at a traffic light, I am expecting that they will follow the traffic laws, which allows me to drive safely. When I purchase collision insurance from an automobile insurer and sign a contract with them, I can rely on them to pay for the costs of an accident, after I meet my deductible.

However, trust in the sense that is important for understanding ethical obligations and issues, is based on more than mere reliance; it is based on a **relationship** another person or group of people (Baier 1986; Hardin 1996). I may rely on a clock to tell time, but I do not have a relationship with the clock. However, when I rely on my friend to be on time, it is because we have a relationship that involves the expectation of being on time. A trusting-relationship is made possible by the assumption of trust, which grows over time as members of the relationship demonstrate that they can be relied upon. For example, to play together well, members of a basketball team must learn to trust each other. This trust grows over time as they get to know each, they demonstrate concern for each other, and as they succeed and fail together (Smith 2004). Societies are webs of relationships among spouses, friends, family members, coworkers, teammates, doctors and patients, attorneys and clients, and teachers and students. Indeed, social and political scientists have characterized trust

[1] Some notable articles that analyze the notion of trust in research with human subjects include Richardson and Belsky (2004), Kerasidou (2017), and Foláyan and Haire (2023). These will be discussed below.

as the cement that holds society together (Fukuyama 1995; Govier 1997). It also worth noting that while the concept of trust can be applied to communities, organizations, businesses, institutions, nations, or any subject matter involving human relationships.

Of course, not all relationships involve trust: for example, I may have a relationship with a bully (albeit a bad one), and I may even rely on them to acting hostilely toward me, but that does not mean I trust them. Trust also has moral dimensions (Hardin 2001; O'Neill 2002). An important moral feature of trust is that it involves **vulnerability** (Baier 1986; Hall et al. 2001; Kerasidou 2017). When you trust a friend with a secret, you take a risk that they could tell it to other people and hurt you. You make yourself vulnerable to being harmed or taken advantage of by them. You also are vulnerable when you trust a psychotherapist to not disclose information shared in confidence, a surgeon to perform an appendectomy correctly, a pilot to fly an airplane, and a bank to keep your money. Indeed, almost all of us depend on family members, friends, professionals, businesses, and organizations that could cause us serious harm if they violated our trust.

Vulnerability implies another feature of trust: mutual expectations of **moral integrity** (Baier 1986; Kerasidou 2017). For example, when I trust someone whom I have a close relationship with, such as my spouse, parent, child, or best friend, I (hopefully) can trust that they will act morally, i.e., that they will help me, not harm me, not lie to me, and so on.[2] Likewise, when I trust a dentist to treat a cavity, I am trusting them to act ethically, competently, and professionally. I am trusting them to act in my best interests and to not take advantage of me. If I discover that my dentist has violated this trust, I may decide to look for a new dentist (and I may file a lawsuit against them).

Because trust involves integrity, it has close connections to moral and legal concepts, such **truthfulness** (trusting people to tell the truth), **promise-keeping** (trusting people to keep their promises), **fidelity** (trusting people to be faithful or loyal), and **confidentiality** (trusting people to keep information confidential) (Hall et al. 2001). Trust often implies ethical and legal duties since the entrusted person may have an obligation to do what is expected of them. For example, the trust you place in a surgeon to operate on your shoulder obligates the surgeon to inform you about the surgery and perform it correctly. In the law, fiduciary relationships involve a strong degree of trust between the trusting and entrusted party. The entrusted party (i.e., the fiduciary) has legal obligations to act in the best interests of trusting party (Black's Law Dictionary 2001). Some examples of fiduciary relationships include doctor-patient, lawyer-client, trustee-beneficiary, and guardian-ward.

Recalling the discussion of moral relativism from Chap. 2, there are good reasons for thinking that trust is relative to particular ethical or legal framework in which a relationship takes places, because this framework will help establish mutual expectations. For example, in many Islamic and African societies, medical

---

[2] I recognize, of course, that this is not always the case and that many people, unfortunately, are involved in abusive relationships.

decision-making is more communitarian than individualistic and may necessitate including the husband in informed consent discussions concerning his wife's treatment (Princewill et al. 2017; Padela 2018). Therefore, promoting trust in that context may involve including family members in decision-making who might not be included in similar situations in non-Islamic societies. There is a tension, therefore, between promoting trust, which seems to be partly a function of local circumstances and acting according to universal moral rules (Plöckinger and Auga 2022).

While I appreciate the inherent tension between trust and universal morality, I would argue that acting morally involves much more than maintaining someone's trust. When Henrietta Lacks underwent surgery at Johns Hopkins hospital in 1951 (see discussion in Chap. 1), it was standard medical practice to use tissues leftover from surgical procedures without consent. It seems reasonable, to conclude, therefore, that Ms. Lacks' doctors did not violate her trust because, at that time, she would not have expected to consent to the use of her tissue, nor would the medical community have expected the doctors to try to obtain her consent. Their actions were consistent with the ethical and legal standards that prevailed at that time and place. Nevertheless, one might argue that her doctors acted unethically because they did not respect her autonomy and took advantage of her (Skloot 2010). I will discuss this issue again below.

Except for the innocent way that a young child trusts a parent, people usually trust others based on evidence of their **trustworthiness**, which is a type of moral virtue (Hardin 1996; Potter 2002). Evidence may come from a variety of sources, such as personal acquaintance with a person or group, or knowledge of a person's or group's reputation, experience, or competence (Tullberg 2008). For example, you might trust a friend with a secret only after learning more about that person's integrity and discretion. You might trust a mechanic to work on your car upon recommendation from a friend. You might trust a dentist to work on your teeth because you believe that the dental profession is generally competent and skilled. A key function of a professional ethics code is to foster the public's trust in members of the profession. Since trust can be difficult to restore when it has been violated, codes and guidelines often emphasize the importance of maintaining trust (Bayles 1988; Walker 2006).

Sometimes trust is contrasted with **verification** (Russell 2015; Graham et al. 2023). The idea is that if you trust someone, you don't need to verify whether they are behaving as expected and if you try to verify that they are behaving as expected, then you don't really trust them. You can trust or verify but not both.

I don't think of trust as incompatible with verification. Instead, human relationships form a continuum from trust-based to verification-based. (See Fig. 4.1). On one end of the spectrum are relationships, such as between spouses or very close friends or family members, which are almost entirely trust-based, and on the other end are relationships, such as negotiations between warring nations, which are almost entirely verification-based. In between are relationships that involve trust and verification, such as contractual (e.g., business or commercial) and professional relationships. For example, a trust my bank to keep my money safe, but I also like

Trust	Trust and Verification	Verification
Spouses Close friends or family members	Contractual and professional relationships	Warring nations Competing companies

Fig. 4.1 Trust vs. verification

to be able look at my bank statement. In many situations involving trust, the Russian proverb "доверяй, но проверяй" (doveryay, no proveryay) applies (Reagan 1987).

As we shall see below, research with human subjects involves many different relationships between diverse peoples and organizations. While trust is the glue that holds the whole system together, verification is needed to address human failings, such as bias, error, and misconduct; to restore trust once it has been broken; and to manage conflicts within the system, such as advancing knowledge vs. protecting people and making a profit vs. advancing knowledge and protecting people (Kerasidou 2017; Foláyan and Haire 2023).

4.2 Trust in Research Involving Research with Human Subjects

There is growing body of empirical evidence on the importance of trust for ethical and high-quality research with human subjects (see Box 4.1 for some key studies). To understand why trust is so important, let's consider all the different types of relationships involved in human subjects research (see Fig. 4.2).

First and foremost, there is the relationship between the researcher and the human research subject. Subjects must be able to trust researchers to act professionally, competently, and ethically (Earl and Penney 2001; Yarborough and Sharp 2002; Faden et al. 2005; de Melo-Martín and Ho 2008; Resnik 2011, 2021; Kerasidou 2017). The trust that subjects place in researchers encompasses the entire range of research conduct, including recruitment and informed consent; executing research procedures and methods; minimizing risk; safety monitoring; analyzing, interpreting and publishing data; returning research results; protecting biological samples and private information; using and sharing samples and data; reporting adverse events and unanticipated problems; and complying with regulations, guidelines, and policies. Trust in the investigative team is a key factor subjects' decision to participate in research (Tindana et al. 2012; Simon and Mosavel 2010; Slegers et al. 2015).

In the Havasupai case discussed in Chap. 2, members of the tribe said they would not participate in studies conducted by Arizona State University as a result of the breach of trust which occurred when researchers used their blood samples for

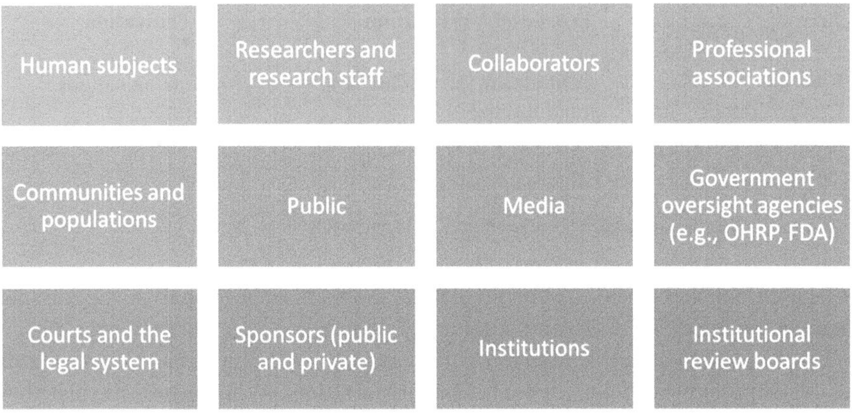

Fig. 4.2 The network of trust in research with human subjects

purposes the tribe said they had not approved (Mello and Wolf 2010). Many different studies have documented distrust of researchers and research institutions as an important barrier to clinical trial recruitment, especially among minority populations (Mouton et al. 1997; Corbie-Smith et al. 1999; Sengupta et al. 2000; Corbie-Smith et al. 2002; Ford et al. 2008). Studies have shown that African Americans' awareness of unethical behavior by researchers conducting the Tuskegee syphilis study have negatively impacted their willingness to participate in research (Shavers et al. 2000; Freimuth et al. 2001; Earl and Penney 2001).

Although most studies of trust in human subjects research emphasize that importance of subjects trusting researchers (and rightly so), researchers must also be able to trust subjects tell the truth about medical or other conditions that could disqualify them from participating in a study; comply with study requirements (such as taking medications as directed); and report symptoms and problems they experience during their participation; and honestly answer questions on surveys and during interviews (Resnik and Ness 2012). There is evidence that research subjects sometimes lie to researchers to qualify for studies or to avoid being dropped from a study due to non-compliance (Devine et al. 2013). Deception by research subjects can undermine the integrity of the study and place subjects at risk (Resnik and McCann 2015).

Another important relationship is between research subjects and research institutions, such as universities, hospitals, or academic medical centers. Institutions are responsible for overseeing researchers and ensuring that they uphold their professional and ethical obligations (Kass et al. 1996; Hall et al. 2001; Peckman 2006; de Melo-Martín and Ho 2008; Slegers et al. 2015). In the Havasupai case (discussed in Chap. 1), members of the tribe lost their trust in the researchers and Arizona State University as a result of misunderstandings about how their blood samples and data could be used. In the Jesse Gelsinger case (discussed in Chap. 2), Mr. Gelsinger questioned the university's trustworthiness because he was concerned that its financial interests in the research and lax oversight of the study led to the death of his son (Gelsinger and Shamoo 2008). In the John Moore case (discussed in Chap. 2), Moore sued the UCLA and the research team because they had profited from his tissues without informing him of their financial interests or plans to do so. Since institutions designate institutional review boards (IRBs) to oversee research with human subjects, trust in the institution extends to the IRB (Peckman 2006).

Communities and populations involved in research have relationships with researchers and institutions. Communities can provide important support for human subjects research: for example, community organizations (such as local health departments or clinics) can help with publicity and recruitment and may provide financial and other resources to researchers; and community members may enroll in studies, help with study design or survey development, or serve on committees that oversee research, such as IRBs (O'Fallon and Dearry 2002; Horowitz et al. 2002; Holzer et al. 2014). If researchers do not have support from communities and populations, they may have difficulty recruiting subjects or implementing study procedures (Ford et al. 2008; Tindana et al. 2012). The Havasupai case provides an instructive example of the importance of the community's trust, since the breach of trust in the case negatively impacted not only individuals but also the entire tribe (Mello and Wolf 2010).

Researchers have various relationships with other researchers founded on a trust (Whitbeck 1995). Research teams may include a leader (or principal researcher), post-doctoral trainees, students, staff scientists, staff clinicians, technicians, and research nurses. Teams may also collaborate with other teams at the same institution of different institutions. A multi-center clinical trial, for example, may involve hundreds of researchers at over a dozen institutions. Members of these teams must be able to trust each other to behave professionally, competently, and ethically (Shamoo and Resnik 2022). Researchers must trust members of their own research team and collaborators to comply with regulations and protocol requirements; collect, report, and analyze data accurately; implement study procedures and tests competently; report adverse events and other problems; and treat other members of the team with respect. Since researchers rely on the published literature to inform their own research, they must be able to trust that other members of the research community will not publish data known to be biased, misleading or fraudulent (Shamoo and Resnik 2022). The research misconduct cases discussed in Chap. 2 involving Roger Poisson, Eric Poehlman, Andrew Wakefield, and Woo Suk Hwang constituted serious betrayals of the trust of members of research teams as well the scientific community. The financial conflicts of interest and other problems related to industry bias discussed in Chap. 2 also had a negative impact on researchers' trust of each other. I will discuss research integrity issues in more depth in Chap. 9.

IRBs also have relationships with researchers and must encourage them to comply with ethical and legal requirements for conducting research with human subjects (Peckman 2006; McMurphy et al. 2013; Klitzman 2015). Although IRBs have the authority to monitor or audit research, they do not have the resources to observe research around-the-clock. Even if resources were available for intensive oversight of research, IRBs should refrain from doing this, as it would undermine researcher morale, interfere with academic freedom and study procedures, and transform the research environment into a police state. After the IRB reviews and approves a study, it must trust the researcher to follow the protocol, seek IRB approval for proposed changes to research, and report unanticipated problems, non-compliance, and adverse events. Trust between researchers and IRBs also goes the other way: Researchers must be able to trust IRBs to review their research fairly and competently (McMurphy et al. 2013; Klitzman 2015). They should expect IRBs to provide them with useful criticism and feedback concerning their research and not delay review unnecessarily. Researchers should also expect that IRB members will have a working knowledge of regulations and ethical guidelines, and that they will consult with external scientists or clinicians if they are reviewing a study which is beyond the scope of their collective knowledge or expertise. Since institutions designate IRBs to oversee research, institutions must also be able to trust IRBs.

The public must be able to trust the research enterprise (Kass et al. 1996; Kelch 2002; Resnik 2010, 2011; Oreskes 2019). The public expects: researchers to behave professionally, competently, and ethically; institutions to manage resources carefully and oversee researchers properly; government sponsors to fund worth-while projects; and the research enterprise to maintain the highest ethical and scientific standards. All of the cases involving abuses of human subjects or ethically questionable behaviors discussed in the Chap. 2 had a negative impact on the public's confidence in researchers, institutions, and the research enterprise. Lack of public trust can negatively impact science by eroding research funding, since citizens and politicians may be unwilling to allocate government money to pay for research they regard as untrustworthy (Resnik 2011). Although government support for scientific research is not as strong as it once was as a result of budgetary constraints, it still accounts for about one third of all research funding (Shamoo and Resnik 2022). Researchers need to maintain the public's trust to continue receiving generous funding support. Unfortunately, polls indicate the public's trust (at least the American public's trust) in scientific research is waning (Kennedy et al. 2022). While the increased politicization of science, especially during the COVID-19 pandemic, has probably played a major role in eroding the public's trust in the research enterprise, unethical behavior by scientists, academic institutions, and government science agencies has probably also played an important role (Paul 2023; Sutter 2024).

Box 4.1: Some Important Studies on Trust in Research Involving Human Subjects

Kass et al. (1996).
The authors report the results of interviews with patient-subjects conducted under the auspices of the Advisory Committee on Human Radiation Experiments. They found that patient-subjects often trust their physicians to guide them through decisions on research participation.

Corbie-Smith et al. (1999).
The authors conducted five focus groups with 33 African-American adults at urban public hospital concerning their attitudes toward medical research. Most participants indicated mistrust of doctors, scientists, and government, and many were concerned about unethical conduct of clinicians and investigators when poor or minority patients are involved in research. Participants also cited examples of exploitation to support their mistrust of the medical establishment. Although must participants knew about the unethical conduct of researchers involved in the Tuskegee Syphilis Study, they were all misinformed about the historical facts of the study. Few participants understood that informed consent is supposed to protect their autonomy and most regarded it as legal protection for physicians. The authors concluded that African-American distrust of the medical community is a prominent barrier to participation in clinical research.

Hall et al. (2001).
The authors reviewed and synthesized the theoretical, empirical, and public policy literature on trust in physicians and in medical institutions and developed a formal definition and conceptual model of trust, with implications for ethics, law, and public policy.

Molyneux et al. (2005).
The authors reported data on community views regarding the informed consent processes carried out by a large research center in Kenya. The authors found that community members regard trust as central to informed consent but that they questioned whether informed consent is relevant and practical. Community members also discussed how to balance trust and healthy mistrust and strengthen communication about medical and medical research.

McDonald et al. (2008).
The authors conducted in-depth interviews with research subjects about trust in research. Subjects indicated that trust in research cannot be assumed and that it is a dynamic concept, characterized by reciprocity and negotiation, which can be built and easily broken. Subjects were unsure about who, when, what, and how much to trust in research.

Moreno et al. (2009).
The authors described lessons learned from building research partnerships between an academic medical center and the community in Los Angeles County. Key lessons included the importance of clear and concrete community benefits, partnering with diverse organizations, political advocacy, supporting an academic–community champion, and long-term commitment and involvement.

(continued)

Box 4.1 (continued)

Yarborough et al. (2009).
The authors and their colleagues convened a group of national leaders from sectors outside of academic science and health care to discuss public trust. Strategies identified for promoting the public's trust were fostering different types of relationships and developing accountability practices beyond those required by law. The authors present some recommendations for building and restoring trust in biomedical research.

Guillemin et al. (2016).
The authors reported the results of interviews with 8 indigenous and 28 non-indigenous research participants in Victoria, Australia and found stark differences between these groups in terms of why and how they decided to participate in research. Indigenous participants took part in research primarily to benefit their communities not themselves. They also were cautious and deliberate in their decision-making. In contrast, non-Indigenous participants took researchers' claims at face value and deciding to participate was straightforward. They were also more likely to be motivated by personal interests.

Guillemin et al. (2018).
The authors interviewed 19 researchers about how they understand and practice trust in research. The authors found that researchers' understanding of trust is conceptual and concrete and is related to other similar concepts.

Smirnoff et al. (2018).
The authors conducted a survey of 355 New York City residents of their attitudes towards research conducted in emergency situations. A solid majority (85.3%) had a favorable opinion of emergency research but many also expressed mistrust.
Generally favorable attitudes towards research were expressed by a majority (85.3%), but many respondents expressed mistrust. Factors related to trust/mistrust included age, disability status, race/ethnicity, language preference, prior research experience and educational level.

Pulle et al. (2024).
The authors surveyed 296 participants from rural Uganda to determine whether there are differences in trust in medical research between community members vs. hospitalized people. Hospitalized people were significantly more trusting of medical research than community members. Previous research participation and willingness to participate in future research were positively associated with trust.

4.3 Trust as a Foundation for the Ethics of Research with Human Subjects: Some Alternative Views

Clearly, trust is an important value which applies to many aspects of research with human subjects. Before I articulate my framework for thinking about the ethics of trust in research with human subjects more fully, I will discuss and critique some other approaches to this topic.

Over three decades ago, Robert Veatch (1987) developed a theory of human research ethics based on the idea of the patient as a partner in research. Veatch observed that there is a mutual-dependence in clinical research with human subjects: subjects need researchers to gain access to experimental treatments which may benefit them, and researchers need human subjects in order to obtain data and samples they use in their research. Researchers depend on human subjects not only to enroll in studies but also to cooperate in study-related activities and procedures. Clinical research could not move forward without the participation of human subjects. Veatch argued that thinking of the relationship between researchers and human subjects as a partnership would provide a moral foundation for research with human subjects. Veatch applied this idea to a variety of topics in research with human subjects, including study design, informed consent, risk minimization, and research involving vulnerable subjects (Veatch 1987). Although Veatch did not discuss the concept of trust in any detail, it is clear that his proposal implies that researchers and human subjects must trust each other, since trust is essential to partnerships.[3]

While I think the partnership model is intriguing, it has a couple of limitations. First, while the patient as partner idea accounts for many of the types of trust which are important in research with human subjects, it does not address trust between human subjects and research institutions, communities and researchers/institutions; trust among researchers; or trust between the public at large and the research enterprise. Veatch's proposal was on the right track, but it did not go far enough.

Second, the patient as partnership idea is not completely apt because research subjects often do not function as full partners. Partnership implies equality of responsibility. For example, two business partners are both responsible for handling the affairs of the business, e.g., keeping records, paying debts, filing taxes, complying with regulations, and so on. One partner may delegate some responsibilities to the other, but they are both ultimately responsible. In most studies involving human subjects, researchers bear the brunt of the legal, ethical, and professional responsibilities associated with the study. Researchers are responsible for complying with IRB-approved protocol and regulations, performing study procedures, collecting data, etc. To be sure, research subjects also have some responsibilities, but not nearly as many as researchers in most cases (Resnik and Ness 2012).

While most types of research do not conform to the partnership model, some, such as community-engaged research (CER) and citizen science, come close. CER[4] is an approach to research that aims to produce knowledge that addresses the needs and concerns of the community and substantially involves community members in various aspects of research, including development of study aims and objectives, protocol and survey design, data collection, data interpretation, recruitment, and

[3] Rebecca Dresser (2016) also uses the partnership idea to describe the relationship between researchers and human subjects. However, Dresser invokes the idea of partnership primarily to highlight the lack of attention to the subject's perspective in debates about clinical research ethics and policy and not to develop a philosophical account of trust in human participant research.

[4] This is also sometimes referred to as Community-Based Participatory Research (CBPR) or community-based research (CBR).

consent (O'Fallon and Dearry 2002; Horowitz et al. 2002; Simon and Mosavel 2010; Holzer et al. 2014). Community members may participate as human subjects, research staff, or advisory board members. CER has become an influential approach in scientific disciplines that study communities and populations, such epidemiology, and public and environmental health. In citizen science, laypeople assist scientists with collecting specimens and data and are full-fledged members of the research teams. In some cases, laypeople take a leadership role in designing the research and obtaining funding for it and partner with scientists to help them with various aspects of the project (Smith et al. 2019; Resnik 2019).

In 1996, Nancy Kass and colleagues on the Advisory Committee on Human Radiation Experiments published an article in the *Hastings Center Report* titled "Trust: The fragile foundation of contemporary biomedical research." The article reported the results of interviews with over 1900 patients concerning their views about research participation (Kass et al. 1996). Many of the interviewees stressed the importance of the trust they placed in researchers, institutions, and the research enterprise. Kass and her colleagues drew some implications of their interviews for research practice and policy. For example, they argued that clinical researchers should be mindful of the tremendous influence they have over patients who are considering enrollment in research. They urged clinical researchers to describe the benefits and risks of research honestly and realistically and avoid biasing the discussion in favor of enrollment, especially when the benefits are small but the risks are significant. They also urged IRBs to take seriously their responsibilities to protect research subjects from harm and exploitation (Kass et al. 1996). Although the article had a major role in encouraging bioethicist to think more deeply about trust in human subject research, it did not include any substantive philosophical discussion that could serve as a foundation for research with human subjects.

In 2004, Henry Richardson and Leah Belsky published a highly influential paper which argued that the notion of partial entrustment provides a moral foundation for the ancillary care responsibilities for clinical researchers (Richardson and Belsky 2004). Richardson (2012) later published a book which expanded upon the ideas developed in that paper. Ancillary care is medical care provided to human subjects in a research study beyond what is required to (a) address the study's research questions, (b) protect subjects from harm related to their participation in the study, or (c) redress harms that have occurred as a result of participation (Belsky and Richardson 2004). Ancillary care ranges from simple and inexpensive interventions, such as treating mild skin infections, to complex and expensive interventions, such as performing dental work or providing treatment for cancer. Ancillary-care may include testing, diagnosis, treatment, or prevention (Belsky and Richardson 2004). Providing ancillary care can be a significant moral concern when human subjects (such as people living in developing nations) lack access to care outside of their participation in research (Dickert and Wendler 2009).

Federal research regulations and ethics codes provide little guidance concerning ancillary care. The Common Rule does not address ancillary care, but its requirement that the risks of research should be reasonable in relation to expected benefits could be interpreted as providing some support for providing ancillary care, because ancillary care is a potential benefit to the research subject (Dickert and Wendler 2009). However, one might challenge this interpretation on the grounds that the benefits mentioned in the Common Rule are benefits related to research participation or the products of research (knowledge), and ancillary care goes beyond these. The Helsinki Declaration also does not mention ancillary care, but its requirement that researchers and host country governments make provisions to ensure that all participants have post-trial access to interventions identified as beneficial in a clinical trial could be interpreted as supporting ancillary care after a study is completed. However, one could argue that post-trial access to treatments is not ancillary care because ancillary care occurs while a study is still underway. The Council for International Organizations of Medical Sciences (2002, 2016) guidelines state that sponsors are not required to provide health care beyond what is required to conduct research but that providing such care would be morally praiseworthy.

Ancillary care is ethically controversial because it often involves a conflict between promoting the welfare of human subjects and developing scientific knowledge. While ancillary care can provide significant benefits for human subjects, it can also increase the costs of conducting research and possibly interfere with a study's scientific objectives (Resnik 2009). To provide ancillary care, researchers may need to pay for additional medications, laboratory tests, medical devices, or clinical consultations. To cover these costs, they may need to divert money from other study-related activities, which could hamper their efforts to achieve their scientific objectives (Richardson and Belsky 2004). In some cases, providing ancillary care may compromise the validity of the study by biasing health outcomes under investigation. For example, suppose that researchers are conducting a long-term observational study about the effects of aging on sexual function. If the researchers provide subjects with medications to treat sexual dysfunction this would benefit subjects but also potentially bias the outcome of the study.

Richardson and Belsky (2004) argue for providing research subjects with ancillary care in some circumstances based on two considerations. First, they argue that all moral agents, including researchers, have a general duty to rescue those in need of help. For example, if you see someone drowning in a pool you should do something to help them, such as calling a lifeguard or throwing them a life-preserver. The duty of rescue has broad support from moral theories, such as Kantianism or utilitarianism, natural law theory, and virtue ethics. The principle of beneficence also supports a duty to help others (National Commission 1979).

Second, Richardson and Belsky (2004) argue that researchers also have special duties to benefit research subjects based on the trust subjects justifiably place in them.[5] They call their view the "partial entrustment" approach to distinguish it from the full entrustment view implied by conceiving of the researcher-subject relationship as fiduciary in nature, and the no entrustment view implied by viewing the relationship as purely contractual. According to the fiduciary approach, the

[5] Richardson and Belsky's view is different than mine in they do not equate trust with a psychological attitude or state. The fact that one relies on someone else does not create a trust relationship. For there to be an entrustment, one must have a good reason for relying on the entrusted party.

researcher-subject relationship is similar to the physician-patient relationship in that researchers have strong moral duties to promote the best interests of research subjects because patients entrust them with their health (Miller and Weijer 2003, 2006). The fiduciary view implies that it is unethical to include placebo control groups in clinical trials if an effective therapy is available, because withholding treatment from subjects in the placebo group would violate researchers' fiduciary obligations to provide subjects with optimal medical care (Miller and Weijer 2003, 2006). According to the contractual view, the researcher-subject relationship is defined by the terms the parties reach during the consent process (Morreim 2005). If a subject consents to participate in a study without the expectation of any extra benefits (such as ancillary care), then the researcher is not obligated to provide those benefits to the subject. The contractual view implies that using placebo control groups in clinical trials is ethical, even when an effective therapy exists, provided that researchers obtain consent from the subjects and minimize harm to them (Miller and Brody 2002, 2003).

Richardson and Belsky (2004) argue that subjects' trust in researchers lies somewhere in between no entrustment and full entrustment. The researcher-subject relationship is more than merely contractual but less than fiduciary. Researchers have some obligations to promote the interests of research subjects, but these obligations are not as extensive as the obligations that physicians have toward their patients. The scope of researchers' ancillary care obligations depends on two factors: discretion and vulnerability (Richardson and Belsky 2004). Discretion, which refers to the extent of the researcher's ability to impact the subject's well-being, is a function of the researchers' knowledge, skills, and available resources. Vulnerability, which refers to the subject's susceptibility to harm or exploitation, is a function of factors such as maturity, mental health, institutionalization, and socioeconomic circumstances. Subjects who do not have access to health care outside of a study may become dependent on clinical researchers for their health care.

To illustrate their approach, Richardson and Belsky consider a hypothetical example of a pediatric study in a region of Africa where malaria is endemic, with 10% of the population infected. They argue that the researchers have an obligation to diagnose and treat the children for malaria because (1) the malaria parasite is easily detectable by urinalysis, (2) treating malaria is inexpensive and won't interfere with achieving the goals of the study, (3) the researchers have the expertise and ability to treat malaria, and (4) the children will probably not receive treatment outside of their participation in the study (Richardson and Belsky 2004). They argue, however, that if the costs of diagnosing and treating malaria were much higher, then the researchers would not have an obligation to provide this ancillary care, because this could compromise their ability to carry out the research.

While Richardson and Belsky's partial entrustment idea represents a valuable contribution to the literature on the ethics of research with human subjects, it does not provide us with an entirely satisfactory account of the nature of trust in research with human subjects. The first problem is that Richardson and Belsky do not clearly explain what it means to partially trust someone. Does this mean that you trust them to some degree, ranging from almost no trust to almost complete trust? For example, you might partially trust an acquaintance to keep a secret but fully trust a close friend. Or does partial entrustment mean that you trust someone completely, but only with respect to specific activities? For example, you might trust a real estate agent to give you advice about the market value of a home you intend to buy but not about the tax implications of your purchase. The metaphor Richardson and Belsky use to capture their notion of partial entrustment, i.e., the notion of a bailment, does not accurately represent what they seem to convey when they claim that partial entrustment implies ancillary care responsibilities, since a bailor has an obligation only to take care of your property, not to improve it (Black's Law Dictionary 2001).[6] For example, a restaurant that offers valet parking is under no obligation to repair existing damage to your vehicle; it only has an obligation to repair damage that occurs while it is taking care of your vehicle (Resnik 2009).

A second problem with Richardson and Belsky's view is that it does not consider variation in the extent of trust involved between subjects and researchers. In some situations, the trust between the subject and researcher may be limited. A human subject who completes an anonymous survey, for example, may rightfully expect that the researcher will maintain confidentiality, but he or she would not be entitled to anything more than this, such as ancillary care. A human subject who donates a blood sample for a biobank may only rightfully expect that the researcher maintains confidentiality and performs the blood draw competently, not that the researcher will treat his or her existing medical problems. In some other situations, the extent of the trust may be very extensive; for example, when a researcher is also the subject's personal physician or when the subject is a child or a mentally disabled adult. The partial entrustment view makes the most sense in relationships that fall somewhere between these two extremes (Dickert and Wendler 2009).

A third problem with Richardson and Belsky's view is that it focuses on only one aspect of trust related to research with human subjects, namely, the trust subjects place in researchers. While this is certainly an important kind of trust, as discussed earlier, there are many other types of trust in research with human subjects, including trust between researchers and subjects, trust between subjects and research institutions, trust between communities and researchers, trust between communities and institutions, trust among researchers, and trust between the public and the research enterprise. A comprehensive account of trust in research with human subjects should address all these different types of trust (Peckman 2006).

In 2016, Sanchini et al. (2016) defended a trust-based approach to informed consent in biobanking. They argue that traditional informed consent is inadequate for donating samples to biobanks, because one may not know what type of research the samples will be used for, so consent cannot be fully informed. They argue that one way to get around this difficulty is to reconceive consent for biobanking as primarily about building trust rather than disclosing information.[7] While the authors proposed some interesting ideas about trust in research with human subjects, they did not

[6] Richardson (2012) drops the bailment metaphor.

[7] See also Hansson 2005.

articulate a comprehensive theory or approach. (In Chap. 5 I will discuss ethical issues concerning informed consent related to biological samples in more detail.)

In 2017, while the first edition of this book was just about to appear in print, Angeliki Kerasidou published an excellent analysis of trust in biomedical research that is very similar to my view. I was not aware of her article when I wrote the book and I believe that we independently came to very similar views. I wish to give her credit here. In her article, Kerasidou, reviewed some empirical studies on trust; discussed the relationship between trust and vulnerability, integrity, verification, and trustworthiness; and explained how the current regulatory scheme arose in response to the need to restore trust. She also discussed or mentioned several different parties involved in the web of trust in biomedical research, including subjects, researchers, institutions, professional organizations, and governments. Although Kerasidou did not view trust as incompatible with verification, she argued that institutions tend to focus on verification-related activities, such as ensuring compliance with regulations, and that more emphasis is needed on promoting trustworthiness:

> If researchers and research institutions want to restore trust in biomedical research, they should focus not only on promoting their reliability through regulatory and compliance systems, but also on promoting their trustworthiness. This can be achieved by fostering and encouraging ethical conduct and ethical research through educational programs, dedicated ethics teams and most importantly engagement with stakeholders to help promote the social value of research (Kerasidou 2017:49).

I completely agree with Kerasidou's assertions about the importance of promoting trustworthiness and her suggestions for doing this. In the discussion below, I will expand on her suggestions as well as developing some of my own.

4.4 Trust as a Foundation for the Ethics of Research with Human Subjects

Having discussed some alternative approaches to trust in research with human subjects, I would like to develop my own view in more depth. My position builds on the idea that promoting trust is an important goal in research with human subjects which should influence decision-making and policy. The principle of trust can be simply stated as: "Promote trust in research involving human subjects." Although some critics may claim that there is no need to defend trust as an ethical principle because it is so obvious, it is worth noting that the Nigeria's National Code of Health Research Ethics includes a principle of trust that implies obligations related to the researcher-subject relationship According to the Code:

> For research to be ethical, nothing must be done to undermine the **trust** relationship that is at the heart of the researcher(s)-participant(s) relationships. This requires that there is transparency in all matters relating to the research enterprise including clear description of goals, risks, benefits, alternatives to participation and voluntariness. It is also necessary to determine the social value of the research and engage in creative approaches for effective representations and involvement of researchers and communities in the entire enterprise.

Table 4.1 Strategies for promoting trust in research with human subjects

Strategy	Relationships
Researcher integrity and professionalism	Researchers, subjects, communities, public, professional associations
Informed consent	Researchers, subjects, communities
Community engagement	Researchers, subjects, communities, institutions
Public engagement	Researchers, institutions, government agencies, sponsors, professional associations, public
Transparent communication	Researchers, subjects, communities, public
IRB-researcher dialogue	IRBs, researchers
Management of conflicts of interest	Researchers, subjects, communities, public, IRBs, government agencies, sponsors, professional associations
Oversight/compliance	Institutions, researchers, government agencies, sponsors, IRBs

> Strategies for dynamic and reciprocal collaboration that leads to transformation of essential relationships based on reciprocity are also essential. This **trust** principle encourages the engagement of individual participants and communities, respects local socio- cultural values and encourages the provision of relevant and timely feedback to communities. (National Health Research Ethics Committee of Nigeria 2007; emphasis added).

Promoting trust may include restoring trust which has been damaged, maintaining trust, and nurturing trust (Kass et al. 1996; Walker 2006). As mentioned earlier, trust encompasses all aspects of research with human subjects, including relationships between subjects and researchers, subjects and institutions, communities and researchers, communities and institutions, and relationships among researchers (Peckman 2006; National Health Research Ethics Committee of Nigeria 2007). Promoting trust is an overarching goal of research, which applies to researcher-subject interactions, institutional oversight of research, interactions with communities, as well as storing, sharing, and analyzing samples and data, and funding and publishing research. There are a number of strategies for promoting trust in research with human subjects, which I briefly describe below (see Table 4.1).

4.5 Promoting Trust in Research with Human Subjects

The researcher's integrity is by far most important part of promoting trust in human subject research (or almost any type of research) because it impacts many different aspects of trust (Miller et al. 1998; Kerasidou 2017). Well-designed studies will not generate useful knowledge if they are not executed properly, and well-written consent forms will not help individuals make sound decisions if they are not administered properly.

Although there is little doubt that investigator integrity is essential for safeguarding trust in research, it is not entirely clear how best to ensure that investigators act with integrity. This question is as old as philosophy itself. Greek philosophers wondered whether morality can be taught because some had argued that it cannot be taught because people are inherently good or evil. As we saw in Chap. 3, Aristotle argued that people can learn how to be moral by observing moral behavior in their peers and imitating it. Many have argued that this strategy also applies to ethical behavior in science (Shamoo and Resnik 2022).

While most scholars agree that education and mentoring in research ethics play an important role in promoting integrity, evidence concerning the effectiveness of education and mentoring is inconclusive. Studies have shown that education/mentoring can improve ethical knowledge and reasoning but that it may have little effect on behavior (Anderson et al. 2007; Antes et al. 2010; Todd et al. 2017). In addition to education and mentoring, the culture within the research institute, institutional policies that support ethics, and ethical leadership, appears to play a key role in promoting investigator integrity (National Academies of Sciences, Engineering, and Medicine 2017). Clearly, more research is needed on how to promote investigator integrity.

Integrity is instrumental in ensuring that informed consent (discussed further in Chap. 5) is effective. During the consent process, researchers can assure subjects that they will treat them with respect, safeguard their confidential private information and biological samples, protect them from avoidable harm, and follow ethical and legal rules and guidelines. They can assure subjects that they are well-qualified, by education or experience, to conduct research according to scientific and professional standards. Researchers should view informed consent as a process for building trust, not just as the signing of a form (Berg et al. 2001). Researchers should provide relevant information to subjects and give them opportunity to ask questions and discuss their concerns. They should inform subjects who to contact if they are injured in research or if they want to withdraw from a study. If subjects are harmed in research, researchers should help them obtain medical care. Researchers should honestly and openly communicate with subjects during the entire course of research, including recruitment, enrollment, and follow-up. Communication may also include appointment reminders, newsletters, and thank you or follow-up letters, emails, texts, or phone calls. All forms of communication, including consent, should be transparent, i.e., information should be honestly and openly shared.

Engagement with the local community and/or the community being studied is an important strategy that institutions can use to promote their trust (Anderson and Solomon 2013; Council for International Organizations of Medical Sciences 2016). Institutions should conduct outreach activities that inform community members about the research conducted at the institution, opportunities to participate, and oversight mechanisms (Holzer et al. 2014; Bromley et al. 2015; National Academies of Sciences, Engineering, and Medicine 2016; World Health Organization 2020). Outreach activities may include educational forums, such as conferences, focus groups, or lectures open to the public; and sharing of information on websites and in newsletters, press releases, and interviews with the local media. Engagement should also involve the solicitation of information from the community. Community members should have the opportunity to expresses their needs, interests, and concerns related to research to institutional officials, researchers, and the IRB. Institutional officials should ensure that IRBs and other oversight committees have adequate representation from the community and that community members on the IRB are meaningfully involved in decision-making (Klitzman 2012). Researchers should also consider involving community members in the planning and implementation of studies, as community members may provide valuable insights into recruitment, consent, and research design. Researchers and institutional officials should be accountable to the community, i.e., they should take responsibility for problems that occur in research and should be prepared to answer questions from community members.

Public engagement is important for promoting the public's trust in the research enterprise (Wynne 2006; Holzer et al. 2014; National Academies of Sciences, Engineering, and Medicine 2016). Many of the public engagement activities described for promoting the community's trust can also promote the trust of the greater public. For example, researchers can provide the public with information about their research via media interviews, blog postings, website development, and other forms of communication. Researchers can learn about the public's needs and concerns by surveying or interviewing members of the public or soliciting public input at open meetings or focus groups (National Academies of Sciences, Engineering, and Medicine 2016).

Researchers can also promote public trust by taking steps to ensure that research is conducted according to scientific and ethical standards, and responding quickly and appropriately when problems arise. Researchers, institutional officials, and government funding agencies should be accountable to the public. For example, researchers should not avoid answering difficult ethical questions about their research and should be able to publicly defend it. Government officials should also answer questions about research and justify funding decisions. Funding agencies should solicit the advice of the public concerning research priorities and should fund research that the public considers to be socially valuable (Resnik 2001).

Transparent communication is important for building the public's trust (National Health Research Ethics Committee of Nigeria 2007; Elliott and Resnik 2019). Information about the types of studies being conducted, how they are being conducted, where they are being conducted, and why they are being conducted, should be readily available to the public. Research results and findings, consent forms, and protocols should be made available to the public.

Constructive dialogue can help build trust between researchers and IRBs. To promote constructive dialogue, IRBs should invite researchers and research staff to attend IRB meetings, and IRB chairs should encourage investigators to talk to them about human research issues, problems, or concerns (McMurphy et al. 2013). IRBs should also periodically survey researchers concerning their perceptions of the IRB review process (Keith-Spiegel et al. 2006). IRB decision-making should be transparent and accountable to researchers (McMurphy et al. 2013). For example, IRBs should clearly explain their decisions to investigators. If a researcher disagrees with a decision, the IRB should inform them that they can ask the IRB to reconsider its decision. IRB SOPs should address researcher-IRB interactions and communications (Klitzman 2015).

Proper management of conflicts of interest is also important for promoting the trust of subjects, communities, and the public (Shamoo and Resnik 2022). Jesse Gelsinger's tragic death and the subsequent adverse public reaction and the John Moore tissue patenting case (both discussed in Chap. 2) illustrate the importance of effective COI management. Although laypeople may not have a good understanding of scientific methods and procedures, they generally do understand how money affects behavior and can become highly suspicious and distrustful when COIs are not properly disclosed or managed. (COIs will be discussed in more depth in Chap. 9).

Finally, oversight activities, which help to ensure compliance with regulations and policies, and foster public and community accountability, are also important for safeguarding trust (Foláyan and Haire 2023). Some of these include developing policies and procedures for the responsible conduct of research; providing education and training in research regulations and policies; creating systems for reporting and investigating non-compliance (and other problems with research); protecting people who report non-compliance (often referred to as whistleblowers); offering resources for obtaining advice concerning ethics and compliance issues; seeking and maintaining accreditation[8]; and auditing of research records (Shamoo and Resnik 2022; National Institutes of Health 2023).

While IRBs play a key role in oversight of human subjects research, responsibility for oversight of research should extend beyond IRBs and include institutional officials, support staff, and researchers (Institute of Medicine 2002). Institutional officials should understand the importance of oversight of research with human subjects and be aware of relevant regulations, policies, and procedures. They should take a leadership role in research oversight and emphasize the importance of adhering to legal, ethical, and professional standards. Institutional officials should provide adequate resources to support the human research protection program (HRPP) and should also seek accreditation for the HRPP. Institutional officials should manage conflicts of interest wisely and ensure that research is protected from outside corrupting influences (Kelch 2002).

Recalling our earlier discussion of the distinction between trust and verification, one could make similar distinction between **ethics** and **compliance**, where ethics is the basis of trust and compliance is the basis of verification (Yarborough et al. 2009; Geller et al. 2010; National Academies of Sciences, Engineering, and Medicine 2017). Compliance sometimes is characterized as an ethically vapid activity that involves filling out forms, checking boxes, writing reports, and auditing records but does not involve genuine moral reflection or judgment. Probably most readers of this book are familiar with on-line educational modules that fulfill research ethics training requirements but involve very little in the way of meaningful learning (Hunt 2022).

[8]The Association for the Accreditation of Human Research Protection Programs (AAHRPP 2024) accredits research organizations. Currently, 249 organizations have AAHRPP accreditation. To obtain accreditation, an organization must comply with AAHRPP standards for the protection of human subjects. The standards are based on the federal regulations but go beyond them.

I agree that compliance is not the same as ethics. However, both are a necessary part of a culture that promotes research integrity (National Academies of Sciences, Engineering, and Medicine 2017; All European Academies 2023; Magalhães 2024). Integrity includes understanding and following the rules and making good decisions when the rules do not apply or require interpretation. Ethics education should therefore provide students with information about the rules and offer useful advice and practical experiences in ethical decision-making (Shamoo and Resnik 2022).

4.6 The Relationship Between Moral Principles and Research Regulations and Guidelines

The four principles discussed in Chap. 3 (i.e., respect for human dignity, non-maleficence, beneficence, and justice), together with the principle of trust discussed in this chapter, provide a moral foundation for regulations and guidelines for protecting human research subjects. The members of the National Commission who authored the *Belmont Report* were charged with developing ethical principles to provide a conceptual foundation for revisions of the federal research regulations (see discussion in Chap. 2). The four principles described in Chap. 3 are similar to the *Belmont* principles and can function in the same way. I will briefly describe how these five principles justify some common research regulations and guidelines (see Table 4.2). For further discussion and illustration see Emanuel et al. (2000).

As mentioned in Chap. 3, respect for human dignity justifies informed consent rules, because these rules enable autonomous individuals to decide whether to participate in research. Non-maleficence provides additional justification for these rules because consent allows individuals to avoid participating in studies that they judge to be excessively risky (Brock 2008). Informed consent requirements can promote trust by facilitating honest and open communication between researchers and subjects. Adherence to informed consent rules can help prevent the sense of betrayal and loss of trust that subjects may experience if they discover that they have not been fully informed about some key aspect of a study (as occurred in the Gelsinger case, discussed in Chap. 2) or that they have participated in research without giving their permission (as occurred in the Tuskegee study case, also discussed in Chap. 2).

Respect for dignity, non-maleficence, and trust also justify rules concerning the protection of confidentiality and privacy in research. Rules pertaining to privacy and confidentiality respect autonomy by giving subjects control over the use of their private information.[9] These rules also protect subjects from harm by preventing discrimination, stigma, embarrassment, identity theft, and other adverse consequences

[9] Chapter 6 will explore privacy and confidentiality issues in greater depth.

Table 4.2 Relationship between moral principles and ethical/legal requirements

Requirement	Justifying principle(s)
Informed consent	Respect for human dignity
	Non-maleficence
	Trust
Confidentiality/privacy protection	Respect for human dignity
	Non-maleficence
	Trust
Risk minimization	Non-maleficence
	Trust
Social value	Beneficence
	Trust
Reasonableness of risks	Non-maleficence
	Beneficence
	Trust
Sound scientific design	Beneficence
	Non-maleficence
Equitable subject selection	Non-maleficence
	Beneficence
	Justice
	Trust
Protection for vulnerable subjects	Respect for human dignity
	Non-maleficence
	Justice
	Trust
Independent review	Respect for dignity/autonomy
	Non-maleficence
	Beneficence
	Justice
	Trust

from unauthorized disclosure of private information.[10] Confidentiality protections can also promote trust by assuring subjects that their private information will be protected from deliberate or inadvertent disclosure. This assurance can make subjects comfortable disclosing private information to researchers that they might not disclose without confidentiality protections.

[10] In the U.S., various federal and state laws protect individuals from discrimination in employment and health insurance. Some of these include: the Americans with Disabilities Act, the Civil Rights Act, the Genetic Non-Discrimination Act, and the Health Insurance Portability and Accountability Act (HIPAA). However, these laws do not cover every type of discrimination that could occur, such as discrimination when applying for life insurance or a mortgage, and not all of them have been well-tested by the courts. Thus, discrimination may still occur in the U.S. and other countries, despite laws that prohibit it (Hudson 2007; Hodge and Gostin 2008).

The principle of non-maleficence justifies the requirement for minimizing risks to research subjects, since minimizing risks helps to prevent causing harm. The principle of trust also justifies this requirement since subjects reasonably expect that they will be protected from unnecessary harms related to their research participation. Under the category of risk minimization, I include requirements for data and safety monitoring in clinical trial and reporting unanticipated problems, non-compliance, and adverse events to the IRB.

The principles of beneficence and trust justify the requirement that research should be expected to yield valuable results for society. Beneficence justifies this requirement since this principle obligates researchers to conduct research that can promote the welfare of other people. Trust justifies this requirement since research subjects are likely to expect that their participation in research is necessary to obtain socially valuable knowledge (Emanuel et al. 2000).

The principles of beneficence, non-maleficence, and trust justify the requirement that risks are reasonable in relation to expected benefits. Risks are reasonable if they are justified in terms of expected benefits to the subjects or society via the knowledge gained. The principles of non-maleficence and benefice play a key role in justifying this requirement, since for risks to be reasonable, risks must be minimized and benefits maximized (Emanuel et al. 2000). Trust also helps to justify this requirement, since minimizing risks and maximizing benefits helps to assure subjects that they are participating in valuable research and not being unnecessarily exposed to risks.

The principles of beneficence, non-maleficence, and trust also justify the requirement that research be scientifically sound. If research is poorly designed, it may not yield any useful knowledge, i.e., social benefits. If research is not likely to yield useful knowledge, then the risks may not be reasonable in relation to expected benefits and subjects may be unnecessarily exposed to risks (Emanuel et al. 2000). Trust also plays an important role in justifying this requirement, since subjects are likely to expect that research is well-designed and that their participation is likely to help address important scientific questions.

The principles of non-maleficence, beneficence, justice, and trust justify the requirement that subject selection be equitable. Subject selection is equitable insofar as a study focuses an appropriate class of subjects for addressing a research question (Levine 1988; MacKay 2024). Risk minimization is related to subject selection, since subjects should be excluded to protect them or others from harm. For example, subjects who are not healthy should be excluded from a Phase I safety studies of new drugs to protect them from harm, and pregnant women may also be excluded from these same studies to protect their fetuses (Levine 1988).[11]

[11] The primary goals of Phase I studies are to assess drug safety and obtain information on dosing, toxicology, and pharmacokinetics. If a drug makes it past Phase I, the manufacturer can go ahead with a Phase II study. Phase II trials are larger studies (typically around 300 subjects) which enroll individuals with a disease or condition treated by the drug and may last several years. The primary goals of Phase II studies are to assess safety and efficacy. Phase III trials are similar to Phase II studies, except they are much larger (typically around 3000 subjects). Phase IV trials are also large

Beneficence is also related to subject selection, since choosing the appropriate class of subjects is important for ensuring that the research will yield valuable knowledge. For example, a Phase II study concerning the safety and efficacy of a drug to treat prostate cancer should include only men with prostate cancer, since including subjects who do not meet this criterion (e.g., women or men without prostate cancer) may compromise the usefulness of the data. Justice plays an important role in subject selection since studies should not unfairly include or exclude certain classes of subjects. For example, some argue it would be unfair to include mentally disabled subjects in a dangerous study unrelated to their disease or condition (National Commission 1979). It would also be unfair to exclude categories of subjects, such as members from certain racial or ethnic groups, from a study concerning a disease or condition which is prevalent in among members of these groups (Meltzer and Childress 2008; Mackay 2024). Finally, justice also plays an important role in justifying the requirement that subject selection is equitable since it helps to assure subjects that they will be protected from unnecessary harm and that their participation is likely to help produce knowledge useful to people with their disease or condition or society as a whole (MacKay 2024).

The principles of respect for human dignity, non-maleficence, justice, and trust justify the requirement to include additional protections for vulnerable subjects in research. A common procedure for protecting vulnerable subjects is to limit the risks they may be exposed to in research which is not likely to benefit them. The rationale for this requirement is that subjects who have a compromised ability to consent should be protected from harm, since they are not able to make sound decisions concerning their exposure to research risks (National Commission 1979). Subjects who have good decision-making abilities can decide to enroll in risky studies which are not likely to benefit them, such as Phase I drug trials. The principle of respect for human dignity comes into play when researchers include procedures for assessing the decision-making capacity of vulnerable subjects, such as mentally disabled individuals. If a mentally disabled individual has the capacity to consent to research, then they should be allowed to do so. If not, then a legally authorized representative, such as a guardian, may consent for them. Justice is also an important concern in additional protections for vulnerable subjects, since it would be unfair to include vulnerable subjects in research that is not relevant to their disease or condition. Finally, the principle of trust can help to justify additional protections for vulnerable subjects, since these safeguards can assure vulnerable individuals and their parents, guardians, or caregivers (as the case may be) that they will not be unfairly included in research.

All five principles justify the requirement for an independent board, such as an IRB, to review and oversee research. Researchers have an inherent conflict of

(typically thousands of subjects) and are conducted after the drug has been approved. Phase IV trials also collect data on safety and efficacy and may include classes of subjects not included in earlier studies. For example, if people 60 years or older were excluded from earlier studies to protect them from harm, they might be included in Phase IV trials if they data indicate that it is likely to be safe to include them (Shamoo and Resnik 2022).

interest when it comes to reviewing and overseeing their own research, since they may overlook problems that could interfere with study recruitment or research implementation (Emanuel et al. 2000). An IRB is in a better position to identify these problems and make recommendations to correct them. IRB review can help to assure that the research meets various ethical and legal requirements, such as risk minimization, sounds scientific design, informed consent, protection of confidentiality/privacy, and equitable subject selection. IRB review can also promote the trust of research subjects and community members by assuring them that an independent board is reviewing and overseeing research. For review and oversight to be independent, board members must not have any conflicts of interest that could compromise their judgment. For example, an IRB member should not review a study in which they have a financial interest. An IRB member should also not review a study they are significantly associated with as a principal researcher, associate researcher or collaborators. IRB members who have conflicts of interest should declare them to the board and recuse themselves from discussion and voting on matters in which they are conflicted (Wolf and Zandecki 2007).

4.7 The Role of Trust in Resolving Ethical Dilemmas in Research with Human Subjects

The principle of trust not only provides additional justification for regulations and guidelines that protect human subjects, but it also has applications for disputes involving the interpretation of regulations and guidelines or other ethical dilemmas involving competing values or principles. For example, consider the ethical dispute in the SUPPORT study (discussed in Chap. 1) concerning informing the parents about the risks of randomizing their newborns to different blood oxygenation levels (85–89% vs. 91–95%). Much of the debate focused on whether the regulations required the researchers to inform the subjects about the risks of randomization. Those who favored informing the participants asserted that these potential risks could be reasonably anticipated, whereas those who were opposed to informing them disputed this claim. One way of thinking about the controversy is to ask: how would informing parents about the potential risks of randomization impact their trust in the investigators or institutions? If informing the parents would probably promote trust, then researchers should do it, according to the view developed here. One reason why informing the parents could have promoted trust is that it could have helped to avoid that sense of betrayal that they might feel if they were to find out after consenting that they were not informed about the risks of randomization, an unfortunate situation which, in fact, happened. Informing the parents could have also promoted honest and open communication between the parents and the researchers concerning the risks of the study, the research design, and the importance of the research. Finally, informing them could have avoided the investigation by the Office and Human Research Protections and the ensuing adverse publicity, which may have negatively impacted the public's trust in this type of research and the scientific enterprise.

In the HIV prevention trials, also discussed in Chap. 1, there controversy focused on whether the studies should have included placebo control groups. Critics argued including placebo control groups was unethical because doing so denied participants in these groups an effective therapy. One could argue, however, that the researchers who designed these studies followed a process that fosters trust, because they consulted with public health officials and researchers in the host nations about the study design. Local populations did not raise significant concerns about including placebo control groups because they wanted access to an inexpensive, effective treatment as soon as possible to deal with crisis they faced (Mbidde 1998). Objections to the study design came largely from outside medical professionals and bioethicists who had little involvement with the affected populations and, therefore, knew very little about how to promote their trust.

For another example, consider the Facebook study, also discussed in Chap. 1. This study involved a dispute about the adequacy of informed consent. While some defended the study on the grounds that the user agreement constituted consent for the study, others said that this agreement did not satisfy the requirements of informed consent because it did not include enough information (Goel 2014). Viewing this dispute from the perspective of promoting trust, one can see that the study probably undermined Facebook users' trust in the organization and may have undermined the public's trust in social media research. To avoid breeding distrust, the researchers could have taken additional steps to ensure that Facebook users had more information about the study. For example, they could have notified all Facebook users about the study and allowed them to agree or not agree to participate. To minimize the potential for bias, the researchers could select a random sample of users who agreed to participate for inclusion. That way, the subjects would not know that their behavior was being studied. Although the federal research regulations did not apply to this study because it was conducted by a private organization, one could argue that Facebook should nevertheless have complied with these rules; not to meet legal mandates, but to promote trust in the organization and in social media research. As it turned out, the episode proved to be embarrassing for Facebook and the investigators (Goel 2014).

4.8 Objections to My View

Before concluding this chapter, I will consider several objections to my view. Addressing these objections is important for answering potential critics and explaining my view in more detail.

The first objection I will consider was raised by Susan Lederer (2019) in a review of the first edition of this book. The objection is that the trust principle may not be very helpful in resolving ethical dilemmas because trustworthiness depends on the expectations of the person, community, population, or public, which may vary a great deal. Thus, in complex ethical dilemmas, such as the SUPPORT study or the HIV prevention trials, the admonition to "promote trust" may be of little help, since different people may have different perspectives on trust, depending on their expectations.

I think this is an important objection to my view and I will try to answer it more fully in Chaps. 5, 6, 7, 8, and 9, when I apply the principle of trust to a variety of ethical and regulatory disputes in research with human subjects. I hope to show that the goal of promoting trust can provide us with some useful insights for developing rules or guidelines or making decisions in particular cases. While the principle cannot solve every complicated dilemma, it can be a very useful adjunct to the four other ethical principles.

The second objection is that whether a particular strategy, policy, or decision is likely to promote trust is an empirical question best answered by conducting research into the social psychology of trust. The points I make in this book are little more than philosophical speculation unless they are supported by evidence from empirical research on trust. A psychologist or sociologist is better qualified to write a book on trust in research with human subjects than a philosopher.

I agree that a great deal of what I have to say about trust is based on empirical assumptions concerning effective methods for promoting trust. To try to justify these assumptions, I have reviewed the empirical literature and included some key studies in Box 4.1. Some of the claims I make about trust are supported by empirical studies which I have referred to in this chapter (see, for example, Kass et al. 1996; Corbie-Smith et al. 2002; Yarborough et al. 2009; Guillemin et al. 2016; Smirnoff et al. 2018). Other claims I make concerning trust are based on our common knowledge of human relationships. For example, you do not need to conduct an extensive survey of human attitudes and beliefs to understand that lying and deception undermine trust. However, some claims I make are somewhat speculative (for example, the claim that better management of COIs can promote trust) and require additional empirical support. I acknowledge the empirical character of much of what I have to say about trust and encourage additional scientific research on the nature of trust in research involving human subjects.

Additionally, it is not a weakness of my view that it rests on empirical premises, since many of the normative claims in applied ethics depend on empirical claims or assumptions from disciplines such as psychology, sociology, or economics. In research ethics, one of the main arguments for limiting the amount of money that human subjects can receive for their participation is that excessive financial compensation can compromise informed consent by affecting subjects' judgments concerning risks and benefits. To evaluate this argument, it is necessary to obtain empirical data on the impact that financial incentives have on the decision to participate in research (Resnik and Ness 2015; Gelinas et al. 2018; see Chap. 5 for review of evidence). To develop well-supported arguments and positions which provide guidance for real-world problems, it is often necessary to draw upon and interpret relevant facts and scientific research. Indeed, many bioethicists have urged their peers to conduct empirical research on bioethics topics and to engage with the relevant scientific literature to enhance the practical value of bioethical reflection (Borry et al. 2005; Kon 2009).

The third objection is that trust approach implies moral relativism, which is an untenable position (see Chap. 3 for more on this point). The trust approach implies moral relativism because what makes someone or something trustworthy can vary a great deal across communities and cultures. For example, as mentioned previously, in some Islamic and African societies it is customary for a woman's spouse or older male relative to provide consent for her (Afifi 2007; Princewill et al. 2017). The trust framework I have developed would seem to imply that researchers conducting studies in these societies should honor their traditions concerning consent to promote trust. However, one might argue that the obligation to obtain consent from the research participant is a universal ethical requirement that researchers should follow, regardless of the customs, values, or expectations of the local community.

I have a couple of responses to this objection. First, I would argue that what makes research trustworthy does not vary across cultures and communities as much as this example suggests. Some values, such as minimizing risks and maximizing benefits, are widely held across different populations or communities.

Second, I would also point that none of the ethical principles I am defending are absolute: they are *prima facie* rules that may conflict with each other in particular cases. When conflicts arise, we must decide which principle should have priority. I have argued that we can appeal to the goal of promoting trust when dealing with conflicts among the other principles, such as non-maleficence vs. beneficence. We can do the same when trust conflicts with other principles. In the example discussed above, one needs to decide whether it is more important to promote trust by following the community's traditions concerning consent, or to respect the woman's dignity/autonomy by seeking consent from her, not from her spouse or older male relative. To establish priorities in this situation, one could appeal to another principle, such as non-maleficence. One could argue that obtaining consent only from the woman could harm her, since she might face repercussions from her family. A compromise position would be to obtain consent from the woman and ask her whether she would also like her spouse or older male relative to give permission for the research. The woman could designate her spouse or older male relative as the decision-maker, if she so chooses. Following this plan of action would protect the woman from harm, honor the local tradition, and respect the woman's decision-making since she could decide whether to include her family in the decision. The woman would also retain the right refuse to participate in the study.[12] (See Chap. 5 for further discussion of this issue.)

Although I think this is a reasonable response to the third objection, I am mindful of the issues it raises. To promote trust among researchers, subjects, institutions, and communities it is important to understand local traditions, customs, and values, so

[12] Mohammad Hosseini pointed out that this strategy may not work in many Muslim countries because of their adherence to Islamic laws related to medical decision-making. I recognize the crafting a workable compromise between local traditions and Western ideals of individual autonomy may be very difficult in many cases.

that these can be taken into account when designing research, recruiting subjects, and implementing the study. Community engagement can help researchers obtain information about the local research context. When honoring local traditions, customs, and values conflicts with research regulations or moral principles, researchers and IRBs must consider how best to promote trust while adhering to ethical and legal rules. Promoting trust is an important consideration, but not the only consideration, in conducting ethical research with human subjects.

A fourth objection, which is related to the third one, is that sometimes promoting trust will require researchers and sponsors to accede to unreasonable demands from subjects or communities. For example, suppose that members of a particular community expect medical researchers to provide subjects with a wide array of free medical and dental care in exchange for their participation. One might argue that this demand is unreasonable, since it would strain the research budget and compel researchers to exceed the scope of their expertise. The sponsor might not have sufficient resources to fund the study.

To answer this objection, we need to ask what it means for subjects or communities to make unreasonable demands. If the demands placed on researchers, sponsors, or institutions would make it practically impossible to conduct a study as a result of resource limitations or other constraints, then these demands would prevent the study from being conducted, which would deny the subjects and community important benefits. In deciding whether—or how—to conduct the study, the investigators, sponsor, and institution would need to balance two competing principles, beneficence and trust. They may be able to find a workable solution in which they provide subjects with some medical care that fits within the research budget even though do not meet all the community's demands (de Melo-Martín and Ho 2008).

A fifth objection is that trust is incompatible with verification and legal oversight/compliance. Since I have discussed this issue already, I will not reiterate my position here.

A sixth objection is that sometimes trust may be misplaced.[13] For example, the participants in the Tuskegee study (discussed in Chap. 2) probably trusted the doctors and medical staff to take good care of them, but their trust was misplaced because the researchers withheld important information from them, i.e., that they were in a research study and that they were being denied treatment for syphilis. Thus, we would not say that trust should be promoted in this case because it was based on deception.

In response to this objection, I would say that trust involves the expectation that the people you are trusting are behaving ethically. When trust is misplaced, the entrusted parties are not behaving ethically. Thus, the goal of promoting trust should include the proviso that trust is properly founded on expectations of ethical conduct. The fact that trust may sometimes be misplaced provides additional justification including verification and oversight in trust-based systems.

[13] I am grateful to Henry Richardson for pointing out this objection.

A final objection is that trust is not unique to research with human subjects, since it is also an important concern in medicine, government, business, and many other professions or endeavors. Hence, what I have to say about trust in research with human subjects is not particularly original or noteworthy.

I'm not sure this is really an objection to my view but more of an observation. I agree that much of what I have to say about trust in research with human subjects applies to other professions or endeavors. I could envision someone developing a trust-based approach to medical or government ethics, for example. However, I think there are two important differences between research with human subjects and other activities where trust is also paramount.

First, researchers and research institutions often have interests which are contrary to the interests of research subjects and communities. Researchers and institutions are interested in obtaining knowledge, whereas subjects and communities are mostly interested in deriving benefits from research participation and having their rights protected (Merritt 2005). Researchers and institutions may also have significant financial interests related to research (Shamoo and Resnik 2022). Because the interests of researchers, institutions, subjects, and communities do not always coincide, trust can become an important issue, since stakeholders in these relationships may be concerned that other stakeholders will act in ways that compromise their interests. For example, research subjects may be concerned that researchers' interests in obtaining knowledge will prevent the researchers from taking appropriate steps to protect subjects from harm. In contrast, researchers may be concerned that research subjects will not disclose information truthfully during enrollment because they are interested in receiving money for participation. Second, as documented in Chap. 2, the history of research with human subjects unfortunately includes many instances involving the betrayal of trust. Researchers and institutions have betrayed the trust of subjects, communities, and the public; and researchers have betrayed each other's trust. Regulations and ethical guidelines have been adopted, in large part, to restore and maintain trust. For these two reasons, trust is especially important in research with human subjects, perhaps even more so than in other human activities.

4.9 Conclusion

In this chapter I have argued that the goal of promoting trust serves as a principle for conducting ethical research with human subjects. I have explored the nature of trust in research with human subjects, discussed some strategies for promoting trust, considered some alternative approaches to trust, and addressed objections to my view. The principle of trust complements four other moral principles developed by Beauchamp and Childress (2012), i.e., respect for human dignity, non-maleficence, beneficence, and justice. One can appeal to these five principles to justify research regulations and guidelines to make decisions in particular cases. The principles are not absolute requirements, but *prima facie* rules which may conflict in some situations. When conflicts arise, one can consider the relevant facts and options to decide which one should take priority. We can use the principle of trust to resolve conflicts among the other four principles, but the principle of trust is not a meta-rule for decision-making. That is, the goal of promoting trust may sometimes conflict with the other principles, and we must decide whether it is more important to promote trust or to honor these other principles. In Chaps. 5, 6, 7, 8, and 9 I will apply my trust-based approach to a variety of topics in the ethics of research with human subjects.

References

Afifi, R.Y. 2007. Biomedical research ethics: An Islamic view part II. *International Journal of Surgery* 5 (6): 381–383.

All European Academies. 2023. The European code of conduct for research integrity. Available at: https://allea.org/code-of-conduct/. Accessed 20 June 2024.

Anderson, E.E., and S. Solomon. 2013. Community engagement: Critical to continued public trust in research. *American Journal of Bioethics* 13 (12): 44–46.

Anderson, M.S., A.S. Horn, K.R. Risbey, E.A. Ronning, R. De Vries, and B.C. Martinson. 2007. What do mentoring and training in the responsible conduct of research have to do with scientists' misbehavior? Findings from a National Survey of NIH-funded scientists. *Academic Medicine* 82 (9): 853–860.

Antes, A.L., X. Wang, M.D. Mumford, R.P. Brown, S. Connelly, and L.D. Devenport. 2010. Evaluating the effects that existing instruction on responsible conduct of research has on ethical decision making. *Academic Medicine* 85 (3): 519–526.

Association for the Accreditation of Human Research Protection Programs. 2024. About. https://www.aahrpp.org/

Baier, A. 1986. Trust and antitrust. *Ethics* 96: 231–260.

Bayles, M. 1988. *Professional ethics*. 2nd ed. Belmont: Wadsworth.

Beauchamp, T.L., and J.F. Childress. 2012. *Principles of biomedical ethics*. 7th ed. New York: Oxford University Press.

Belsky, L., and H.S. Richardson. 2004. Medical researchers' ancillary clinical care responsibilities. *British Medical Journal* 328 (7454): 1494–1496.

Berg, J.W., P.S. Appelbaum, L.S. Parker, and C.W. Lidz. 2001. *Informed consent: Legal theory and clinical practice*. 2nd ed. New York: Oxford University Press.

Black's Law Dictionary, 2nd Pocket Edition. 2001. St. Paul: West Publishing.

Blomqvist, K. 1997. The many faces of trust. *Scandinavian Journal of Management* 13: 271–286.

Borry, P., P. Schotsmans, and K. Dierickx. 2005. The birth of the empirical turn in bioethics. *Bioethics* 19 (1): 49–71.

Brock, D.W. 2008. Philosophical justifications of informed consent in research. In *The Oxford handbook of clinical research ethics*, ed. E.J. Emanuel, C. Grady, R.A. Crouch, R.K. Lie, F.G. Miller, and D. Wendler, 606–612. New York: Oxford University Press.

Bromley, E., L. Mikesell, F. Jones, and D. Khodyakov. 2015. From subject to participant: Ethics and the evolving role of community in health research. *American Journal of Public Health* 105 (5): 900–908.

Corbie-Smith, G., S.B. Thomas, M.V. Williams, and S. Moody-Ayers. 1999. Attitudes and beliefs of African Americans toward participation in medical research. *Journal of General Internal Medicine.* 14 (9): 537–546.

Corbie-Smith, G., S.B. Thomas, and D.M. St George. 2002. Distrust, race, and research. *Archives of Internal Medicine* 162 (21): 2458–2463.

Council for the International Organizations of Medical Sciences. 2002. *International ethical guidelines for biomedical research involving human subjects*. Geneva: Council for the International Organizations of Medical Sciences.

———. 2016. *International ethical guidelines for health-related research involving humans*. Geneva: Council for the International Organizations of Medical Sciences.

de Melo-Martín, I., and A. Ho. 2008. Beyond informed consent: The therapeutic misconception and trust. *Journal of Medical Ethics* 34 (3): 202–205.

Devine, E.G., M.E. Waters, M. Putnam, C. Surprise, K. O'Malley, C. Richambault, R.L. Fishman, C.M. Knapp, E.H. Patterson, O. Sarid-Segal, C. Streeter, L. Colanari, and D.A. Ciraulo. 2013. Concealment and fabrication by experienced research subjects. *Clinical Trials* 10 (6): 935–948.

Dickert, N.W., and D. Wendler. 2009. Ancillary care obligations of medical researchers. *Journal of the American Medical Association* 302 (4): 424–428.

Dresser, D. 2016. *Silent partners: Human subjects and research ethics*. New York: Oxford University Press.

Earl, C.E., and P.J. Penney. 2001. The significance of trust in the research consent process with African Americans. *Western Journal of Nursing Research* 23 (7): 753–762.

Elliott, K.C., and D.B. Resnik. 2019. Making open science work for science and society. *Environmental Health Perspectives* 127 (7): 75002.

Emanuel, E.J., D. Wendler, and C. Grady. 2000. What makes clinical research ethical? *Journal of the American Medical Association* 283 (20): 2701–2711.

Faden, R.R., A.C. Mastroianni, and J.P. Kahn. 2005. Beyond Belmont: Trust, openness, and the work of the advisory committee on human radiation experiments. In *Belmont revisited: Ethical principles for research with human subjects*, ed. J.F. Childress, E.M. Meslin, and H.T. Shapiro, 41–54. Washington, DC: Georgetown University Press.

Foláyan, M.O., and B. Haire. 2023. What's trust got to do with research: Why not accountability? *Frontiers in Research Metrics and Analysis* 8: 1237742.

Ford, J.G., M.W. Howerton, G.Y. Lai, T.L. Gary, S. Bolen, M.C. Gibbons, J. Tilburt, C. Baffi, T.P. Tanpitukpongse, R.F. Wilson, N.R. Powe, and E.B. Bass. 2008. Barriers to recruiting underrepresented populations to cancer clinical trials: A systematic review. *Cancer* 112 (2): 228–242.

Freimuth, V.S., S.C. Quinn, S.B. Thomas, G. Cole, E. Zook, and T. Duncan. 2001. African Americans' views on research and the Tuskegee Syphilis Study. *Social Science and Medicine* 52 (5): 797–808.

Fukuyama, F. 1995. *Trust: The social virtues and the creation of prosperity*. New York: Free Press.

Gelinas, L., E.A. Largent, G. Cohen, S. Kornetsky, B.E. Bierer, and H. Fernandez Lynch. 2018. A framework for ethical payment to research participants. *New England Journal of Medicine* 378 (8): 766–771.

Geller, G., A. Boyce, D.E. Ford, and J. Sugarman. 2010. Beyond "compliance": The role of institutional culture in promoting research integrity. *Academic Medicine* 85 (8): 1296–1302.

Gelsinger, P., and A.E. Shamoo. 2008. Eight years after Jesse's death: Are human research subjects any safer? *Hastings Center Report* 38 (2): 25–27.

Goel, V. 2014. As data overflows online, researchers grapple with ethics. *New York Times*, August 14, 2014:B1.

Govier, T. 1997. *Social trust and human communities*. Montreal: McGill-Queen's University.

Graham, M., R. Milne, P. Fitzsimmons, and M. Sheehan. 2023. Trust and the Goldacre review: Why trusted research environments are not about trust. *Journal of Medical Ethics* 49 (10): 670–673.

Guillemin, M., L. Gillam, E. Barnard, P. Stewart, H. Walker, and D. Rosenthal. 2016. "We're checking them out": Indigenous and non-indigenous research participants' accounts of deciding to be involved in research. *International Journal of Equity in Health* 15: 8.

Guillemin, M., E. Barnard, A. Allen, P. Stewart, H. Walker, D. Rosenthal, and L. Gillam. 2018. Do research participants trust researchers or their institution? *Journal of Empirical Research in Human Research Ethics* 13 (3): 285–294.

Hall, M.A., E. Dugan, B. Zheng, and A.K. Mishra. 2001. Trust in physicians and medical institutions: What is it, can it be measured, and does it matter? *Milbank Quarterly* 79 (4): 613–639.

Hansson, M.G. 2005. Building on relationships of trust in biobank research. *Journal of Medical Ethics* 31 (7): 415–418.

Hardin, R. 1996. Trustworthiness. *Ethics* 107 (1): 26–42.

———. 2001. Trust. In *Encyclopedia of ethics*, ed. L. Becker and C. Becker, 1728–1731. New York: Routledge.

Hodge, J.G., and L.O. Gostin. 2008. Confidentiality. In *The Oxford handbook of clinical research ethics*, ed. E.J. Emanuel, C. Grady, R.A. Crouch, R.K. Lie, F.G. Miller, and D. Wendler, 673–681. New York: Oxford University Press.

Holzer, J.K., L. Ellis, and M.W. Merritt. 2014. Why we need community engagement in medical research. *Journal of Investigative Medicine* 62 (6): 851–855.

Horowitz, C.R., M. Robinson, and S. Seifer. 2002. Community-based participatory research from the margin to the mainstream: Are researchers prepared? *Circulation* 119 (19): 2633–2642.

Hudson, K.L. 2007. Prohibiting genetic discrimination. *New England Journal of Medicine* 356 (20): 2021–2023.

Hunt C. 2022. Is it time to move away from tick-box ethics training? Forbes. 28 July. Available at: https://www.thomsonreuters.com/en-us/posts/news-and-media/tick-box-ethics-training/. Accessed 20 June 2024.

Institute of Medicine. 2002. *Integrity in scientific research: Creating an environment that promotes responsible conduct*. Washington, DC: The National Academies Press.

Kass, N.E., J. Sugarman, R. Faden, and M. Schoch-Spana. 1996. Trust: The fragile foundation of contemporary biomedical research. *The Hastings Center Report* 26 (5): 25–29.

Keith-Spiegel, P., G.P. Koocher, and B. Tabachnick. 2006. What scientists want from their research ethics committee. *Journal of Empirical Research on Human Research Ethics* 1 (1): 67–81.

Kelch, R.P. 2002. Maintaining the public trust in clinical research. *New England Journal of Medicine* 346 (4): 285–287.

Kennedy, B., Tyson, A., and Funk, C. 2022. Americans' trust in scientists, other groups declines. Pew Research Center, February 15. https://www.pewresearch.org/science/2022/02/15/americans-trust-in-scientists-other-groups-declines/

Kerasidou, A. 2017. Trust me, I'm a researcher!: The role of trust in biomedical research. *Medicine, Health Care and Philosophy* 20 (1): 43–50.

Klitzman, R.I. 2012. Institutional review board community members: Who are they, what do they do, and whom do they represent? *Academic Medicine* 87 (7): 975–981.

Klitzman, R.L. 2015. *The ethics police? The struggle to make human research safe*. New York: Oxford University Press.

Kon, A.A. 2009. The role of empirical research in bioethics. *American Journal of Bioethics* 9 (6–7): 59–65.

Lederer, S. 2019. Book review: The trust principle and research ethics. *Ethics and Human Behavior* 41 (3): 39–40.

Levine, R.J. 1988. *Ethics and the regulation of clinical research*. 2nd ed. New Haven: Yale University Press.

Mackay, D. 2024. Fair subject selection in clinical and social scientific research. In *The Oxford handbook of clinical research ethics*, ed. A.S. Iltis and D. MacKay, 236–252. New York: Oxford University Press.

Magalhães, S. 2024. Ethics and integrity in research: Why bridging the gap between ethics and integrity matters. *Journal of Academic* 22: 137–147.

Mbidde, E. 1998. Bioethics and local circumstances. *Science* 279 (5348): 155.

McDonald, M., A. Townsend, S.M. Cox, N.D. Paterson, and D. Lafrenière. 2008. Trust in health research relationships: Accounts of human subjects. *Journal of Empirical Research on Human Research Ethics* 3 (4): 35–47.

McMurphy, S., J. Lewis, and P. Boulos. 2013. Extending the olive branch: Enhancing communication and trust between research ethics committees and qualitative researchers. *Journal of Empirical Research on Human Research Ethics* 8 (4): 28–36.

Mello, M., and L. Wolf. 2010. The Havasupai Indian tribe case—Lessons for research involving stored biologic samples. *New England Journal of Medicine* 363 (3): 204–207.

Meltzer, L.A., and J.F. Childress. 2008. What is fair participant selection? In *The Oxford handbook of clinical research ethics*, ed. E.J. Emanuel, C. Grady, R.A. Crouch, R.K. Lie, F.G. Miller, and D. Wendler, 377–385. New York: Oxford University Press.

Merritt, M. 2005. Moral conflict in clinical trials. *Ethics* 115 (2): 306–330.

Miller, F.G., and H. Brody. 2002. What makes placebo-controlled trials unethical? *American Journal of Bioethics* 2 (2): 3–9.

———. 2003. A critique of clinical equipoise. Therapeutic misconception in the ethics of clinical trials. *Hastings Center Report* 33 (3): 9–28.

Miller, P.B., and C. Weijer. 2003. Rehabilitating equipoise. *Kennedy Institute of Ethics Journal* 13 (2): 93–118.

———. 2006. Fiduciary obligation in clinical research. *Journal of Law, Medicine & Ethics* 34 (2): 424–440.

Miller, F.G., D.L. Rosenstein, and E.G. DeRenzo. 1998. Professional integrity in clinical research. *Journal of the American Medical Association* 280 (16): 1449–1454.

Molyneux, C.S., N. Peshu, and K. Marsh. 2005. Trust and informed consent: Insights from community members on the Kenyan coast. *Social Science and Medicine* 61 (7): 1463–1473.

Moreno, G., M. A. Rodríguez, G. A. Lopez, M. A. Bholat and P. T. Dowling. 2009. Eight years of building community partnerships and trust: The UCLA family medicine community-based participatory research experience. *Academic Medicine* 84 (10): 1426–1433.

Morreim, E.H. 2005. The clinical researcher as fiduciary: Discarding a misguided idea. *Journal of Law, Medicine & Ethics* 33 (3): 586–598.

Mouton, C.P., S. Harris, S. Rovi, P. Solorzano, and M.S. Johnson. 1997. Barriers to black women's participation in cancer clinical trials. *Journal of the National Medical Association* 89 (11): 721–727.

National Academies of Sciences, Engineering, and Medicine. 2016. *Gene drives on the horizon: Advancing science, navigating uncertainty, and aligning research with public values.* Washington, DC: National Academies Press.

———. 2017. *Fostering integrity in research.* Washington, DC: National Academies Press.

National Commission for the Protection of Human Subjects of Biomedical or Behavioral Research. 1979. *The Belmont report: Ethical principles and guidelines for the protection of human subjects of research.* Washington, DC: Department of Health, Education, and Welfare.

National Health Research Ethics Committee of Nigeria. 2007. National code of health research ethics. Available at: http://www.nhrec.net/nhrec/NCHRE_10.pdf. Accessed 18 June 2024.

National Institutes of Health. 2023. Guidelines and policies for the conduct of research in the Intramural Program of the National Institutes of Health. Available at: https://oir.nih.gov/system/files/media/file/2023-11/guidelines-conduct_research.pdf. Accessed 20 June 2024.

O'Fallon, L.R., and A. Dearry. 2002. Community-based participatory research as a tool to advance environmental health sciences. *Environmental Health Perspectives* 110 (Suppl 2): 155–159.

O'Neill, O. 2002. *Autonomy and trust in bioethics.* Cambridge: Cambridge University Press.

Oreskes, N. 2019. *Why trust science?* Princeton: Princeton University Press.

Padela, A.I. 2018. Reflecting and adapting informed consent to fit within an Islamic moral landscape and in Muslim contexts. *Studia Bioethica* 11: 31–39.

Paul, D. 2023. *Deception: The great COVID cover up.* Washington, DC: Regnery.

Peckman, S. 2006. Chapter 1–4: A shared responsibility for protecting human subjects. In *Institutional review board management and function*, ed. R. Amdur and E. Bankert, 2nd ed., 16–20. Boston: Jones and Bartlett.

Plöckinger, U., and Auga. 2022. The "four principles" of Western medical bioethics and the bioethics of Shīʿī Islam in Iran—Is the claim of universality by both justified? *Religions* 13 (11): 1118.

Potter, N.N. 2002. *How can I be trusted? A virtue theory of trustworthiness*. Lanham: Rowman & Littlefield.

Princewill, C.W., A.S. Jegede, K. Nordström, B. Lanre-Abass, and B.S. Elger. 2017. Factors affecting women's autonomous decision making in research participation amongst Yoruba women of western Nigeria. *Developing World Bioethics* 17 (1): 40–49.

Pulle, J., S. Loue, G.N. Kiwanuka, and B. Francis. 2024. Trust in medical research: A comparative study among patients at a regional referral hospital and community members in Lira District, Northern Uganda. *Journal of Empirical Research in Human Research Ethics* 19 (1–2): 48–57.

Reagan, R. 1987. Remarks on Signing the Intermediate-Range Nuclear Forces Treaty, 8 December. Available at: https://web.archive.org/web/20150924111809/http://www.reagan.utexas.edu/archives/speeches/1987/120887c.htm. Accessed 10 June 2024.

Resnik, D.B. 2001. Setting biomedical research priorities: Justice, science, and public participation. *Kennedy Institute for Ethics Journal* 11 (2): 181–205.

———. 2009. The clinical researcher-subject relationship: A contextual approach. *Philosophy, Ethics, and Humanities in Medicine* 4: 16.

———. 2010. Public trust as a policy goal for research with human subjects. *American Journal Bioethics* 10 (6): 15–17.

———. 2011. Scientific research and the public trust. *Science and Engineering Ethics* 17 (3): 399–409.

———. 2019. Citizen scientists as human subjects: Ethical issues. *Citizen Science: Theory and Practice* 4 (1): 1–7.

———. 2021. Treat human subjects with more humanity. *American Scientist* 109: 232–237.

Resnik, D.B., and D.J. McCann. 2015. Deception by research participants. *New England Journal of Medicine* 373 (13): 1192–1933.

Resnik, D.B., and E. Ness. 2012. Participants' responsibilities in clinical research. *Journal of Medical Ethics* 38 (12): 746–750.

———. 2015. Bioethical issues in providing financial incentives to research participants. *Medicolegal and Bioethics* 5: 35–41.

Richardson, H.S. 2012. *Moral entanglements: The ancillary care obligations of medical researchers*. New York: Oxford University Press.

Richardson, H.S., and L. Belsky. 2004. The ancillary-care responsibilities of medical researchers. An ethical framework for thinking about the clinical care that researchers owe their subjects. *Hastings Center Report* 34 (1): 25–33.

Russell, N.S. 2015. The problem with a trust by verify approach. Psychology Today, 25 July. Available at: https://www.psychologytoday.com/us/blog/trust-the-new-workplace-currency/201507/the-problem-with-a-trust-but-verify-approach. Accessed 10 June 2024.

Sanchini, V., G. Bonizzi, D. Disalvatore, M. Monturano, S. Pece, G. Viale, P.P. Di Fiore, and G. Boniolo. 2016. A trust-based pact in research biobanks. From theory to practice. *Bioethics* 30 (4): 260–271.

Sengupta, S., R.P. Strauss, R. DeVellis, S.C. Quinn, B. DeVellis, and W.B. Ware. 2000. Factors affecting African-American participation in AIDS research. *Journal of Acquired Immune Deficiency Syndrome* 24 (3): 275–284.

Shamoo, A.E., and D.B. Resnik. 2022. *Responsible conduct of research*, 4th ed. New York, NY: Oxford University Press.

Shavers, V.L., C.F. Lynch, and L.F. Burmeister. 2000. Knowledge of the Tuskegee study and its impact on the willingness to participate in medical research studies. *Journal of National Medical Association* 92 (12): 563–572.

Simon, C., and M. Mosavel. 2010. Community members as recruiters of human subjects: Ethical considerations. *American Journal of Bioethics* 10 (3): 3–11.

Skloot, R. 2010. *The immortal life of Henrietta Lacks*. New York: Crown Publishing.

Slegers, C., D. Zion, D. Glass, H. Kelsall, L. Fritschi, N. Brown, and B. Loff. 2015. Why do people participate in epidemiological research? *Journal of Bioethical Inquiry* 12 (2): 227–337.

Smirnoff, M., I. Wilets, D.F. Ragin, R. Adams, J. Holohan, R. Rhodes, G. Winkel, E.M. Ricci, C. Clesca, and L.D. Richardson. 2018. A paradigm for understanding trust and mistrust in medical research: The community VOICES study. *AJOB Empirical Bioethics* 9 (1): 39–47.

Smith, D. 2004. *The Carolina way*. New York: Penguin.

Smith, E., J.C. Bélisle-Pipon, and D.B. Resnik. 2019. Patients as research partners; how to value their perceptions, contribution and labor? *Citizen Science* 4 (1). https://doi.org/10.5334/cstp.184.

Sutter, P.M. 2024. *Rescuing science: Restoring trust in an age of doubt*. New York: Routledge.

Tindana, P., S. Bull, L. Amenga-Etego, J. de Vries, R. Aborigo, K. Koram, D. Kwiatkowski, and M. Parker. 2012. Seeking consent to genetic and genomic research in a rural Ghanaian setting: A qualitative study of the MalariaGEN experience. *BMC Medical Ethics* 13: 15.

Todd, E.M., B.S. Torrence, L.L. Watts, T.J. Mulhearn, S. Connelly, and M.D. Mumford. 2017. Effective practices in the delivery of research ethics education: A qualitative review of instructional methods. *Accountability in Research* 24 (5): 297–321.

Tullberg, J. 2008. Trust—The importance of trustfulness versus trustworthiness. *The Journal of SocioEconomics* 37: 2059–2071.

Veatch, R.M. 1987. *The patient as partner: A theory of human experimentation ethics*. Bloomington: Indiana University Press.

Walker, M.U. 2006. *Moral repair: Reconstructing moral relations after wrongdoing*. Cambridge: Cambridge University Press.

Webster's Dictionary. 2024. Trust. Available at: https://www.merriam-webster.com/dictionary/trust. Accessed 8 June 2024.

Whitbeck, C. 1995. Truth and trustworthiness in research. *Science and Engineering Ethics* 1 (4): 403–416.

Wolf, L.E., and J. Zandecki. 2007. Conflicts of interest in research: How IRBs address their own conflicts. *IRB* 29 (1): 6–12.

World Health Organization. 2020. Community engagement. Available at: https://www.who.int/publications/i/item/9789240010529. Accessed 19 June 2024.

Wynne, B. 2006. Public engagement as a means of restoring public trust in science–hitting the notes, but missing the music? *Community Genetics* 9 (3): 211–220.

Yarborough, M., and R.R. Sharp. 2002. Restoring and preserving trust in biomedical research. *Academic Medicine* 77: 8–14.

Yarborough, M., K. Fryer-Edwards, G. Geller, and R.R. Sharp. 2009. Transforming the culture of biomedical research from compliance to trustworthiness: Insights from nonmedical sectors. *Academic Medicine* 84 (4): 472–477.

Chapter 5
Informed Consent

Abstract This chapter explores various issues related to informed consent in research including the elements of consent; the moral and legal foundations of consent; research without consent; medical records research; emergency research; consent by parties other than the subject; documentation of consent; general (or broad) consent; tiered consent; readability of consent forms; opt-in vs. opt-out consent; deception in research; payment for research participation; information disclosure standards; (mis) understanding of information disclosed during consent; the therapeutic misconception; the difference between decision-making capacity and legal competence; assent; coercion and undue influence; the relationship between consent and trust; the right to withdraw from research; and empirical studies of informed consent. Although the federal regulations provide considerable guidance pertaining to informed consent, ethical issues still arise due to controversies concerning the interpretation of the regulations, conflicts between ethical principles (e.g., respect for human dignity vs. beneficence) and the need to acknowledge local customs and traditions concerning consent. The chapter argues that trust plays a key role throughout the consent process and that reflecting on the importance of promoting trust can help us address ethical issues pertaining to consent.

Keywords Ethics · Human subjects · Informed consent · Trust · Information · Understanding · Law · Regulation · Deception · Coercion · Undue influence · Payments

In the first four chapters of this book, I examined the historical and philosophical foundations of the ethics of research with human subjects and developed a theoretical framework for decision-making. The key insight of this framework is that promoting trust is an important principle for making ethical decisions concerning research with human subjects. Beginning with this chapter, I will apply my framework to various topics in human research ethics and policy. The first topic I will address is informed consent. I have chosen this topic as a starting place for my application of the trust framework to research issues because (1) as we saw in Chap. 4, trust and consent are intimately connected; and (2) consent is a

D. B. Resnik, *The Ethics of Research with Human Subjects*, The International Library of Bioethics 111, https://doi.org/10.1007/978-3-031-82757-0_5

foundational concept in research ethics. Indeed, with a few exceptions, consent is a moral prerequisite for exposing people to significant risks in research (Miller and Brody 2002; Capron 2008; Wertheimer 2011).

5.1 Moral and Legal Foundations of Consent

Philosophers and political and legal theorists agree that informed consent obligations are based primarily on the idea that rational individuals should be allowed to make autonomous choices concerning their body, their property, and their lives (Faden et al. 1986; Berg et al. 2001; Manson and O'Neill 2007; Brock 2008; Beauchamp and Childress 2012; Courtland et al. 2022). As we saw in Chap. 3, the principle of autonomy is supported by several moral theories, including Kantianism, utilitarianism, and natural rights theory. Legally, rights to make autonomous choices are usually grounded on fundamental rights enumerated or implied in constitutions or other important legal documents (Berg et al. 2001; Capron 2008). For example, the US constitution (1789) guarantees many different rights, including rights to freedom of speech, religion, peaceful assembly, and the press; the right to due process under the law; the right against self-incrimination; and the right to vote. The United Nations (1948) Universal Declaration of Human Rights, which has been adopted by 193 nations, holds that all human beings have rights to life, liberty, legal due process, and freedom from discrimination, slavery, torture, and cruel or degrading punishment. Although autonomy is a widely recognized principle of ethics and law in most Western nations, it has less importance in some nations, such as China, Iran, Russia, Saudi Arabia, Singapore, Uganda, and Vietnam, which place less of an emphasis on individual rights and have more of a collectivist approach to ethics and law. As we have noted previously, ethical issues can arise in research when Western notions of autonomy clash with local traditions or laws (Resnik and Hecking 2022; Plöckinger and Auga 2022; Kudaibergenova 2024).

While there is widespread agreement among Western philosophers and political and legal theorists about the importance of informed consent as a means of protecting and promoting individual autonomy, some have argued that consent also serves other goals, such as establishing trust-based relationships between people, communicating information, protecting people from harm, and protecting institutions and researchers from legal liability (Manson and O'Neill 2007; Dickert et al. 2017; Millum and Bromwich 2021; Resnik 2009a, 2021). Some of these goals can reinforce each other. For example, relationship-building and information-sharing often go together. However, the goal of protecting institutions and researchers from legal liability may conflict with the other goals, since disclosure of information helps to reduce legal liability but too much information can overwhelm or confuse the participant and interfere with their decision-making (Flory and Emanuel 2004; Capron

2008; Bothun et al. 2021).[1] Although ethicists and researchers have repeatedly emphasized the importance of making consent forms readable, studies indicate that most forms are written at a 10th grade reading level or higher, even though the typical U.S. adult reads at only about a 7th grade level. Forms average about 12 single-spaced, typed pages in length, though many are much longer (Paasche-Orlow et al. 2003; Resnik et al. 2008; Pandiya 2010; Larson et al. 2015; Hadden et al. 2017; Bothun et al. 2021). A key challenge in writing and reviewing consent forms is how to balance the conflicting goals of protecting the institution and researcher from legal liability and building trust-based relationships.

5.2 Trust, Understanding, and Informed Consent

As argued in Chap. 4, informed consent is important for promoting trust between subjects and researchers, because it helps to subjects understand what they can expect when they participate in research and can assure them that they will be treated with respect and protected from unnecessary harm. During the consent process researchers describe not only the study's goals, methods, and procedures but also measures they will take to protect subjects' rights and welfare. Consent is a kind of mutual agreement between researchers and subjects in which both parties affirm their intentions to behave in particular ways (Faden et al. 1986; Berg et al. 2001; Menikoff 2006). The consent form provides a written record of this agreement, but the process of consent includes much more than reading and signing a form and encompasses other types of communication between researchers and subjects, such as advertisements, brochures, reminder emails and letters, newsletters, informal conversations, and so on (Berg et al. 2001). To help solidify trust, researchers should communicate honestly and openly with subjects during the consent process (Menikoff 2006; Sanchini et al. 2016). The consent process also presumes a certain amount of trust because prospective subjects are not likely to even have discussions with researchers about participating in studies unless they already have some trust in them (Corbie-Smith et al. 1999, 2002; Guillemin et al. 2016).

Trust plays a crucial role in helping subjects to understand information that is being shared with them (Resnik 2021). Studies have shown that subjects often do not fully understand crucial items of information disclosed in research, including randomization, placebo control groups, and the difference between research and therapy, but that understanding can be improved when researchers take ample time to answer subjects' questions and explain scientific terms to them (Flory and Emanuel 2004; see studies in Box 5.1). Also, as noted earlier, consent forms are often very long and contain technical and legal jargon even though the Common Rule requires that consent documents be written in "language understandable to the

[1] Assumption of risk is an affirmative defense to a negligence lawsuit. A risk if assumed if one is made aware of the risk and proceeds to take it anyway. See Berg et al. (2001).

subject or the legally authorized representative (45 CFR 46.116a3)." Because consent forms can be difficult to read and comprehend, subjects often rely on researchers (or research staff) to help them understand important information contained in these documents (Berg et al. 2001; Albala et al. 2010). For these oral communications to be meaningful, subjects must trust that researchers will not to hide or obfuscate important information (Menikoff 2006; Resnik 2021). One of the key problems with the informed consent process in the Gelsinger case (discussed in Chap. 2) is that the principal investigator and his staff members did not clearly explain the risks of the research or their financial interests to the research subject (Gelsinger and Shamoo 2008).

Subjects in clinical research often succumb to a misunderstanding, known as the **therapeutic misconception**, when they mistakenly believe that the primary goal of a study is to provide them with treatment rather than to gain scientific knowledge (Henderson et al. 2007; Lidz et al. 2015). Subjects who do not understand that the primary goal of a study is to develop scientific knowledge may overemphasize the benefits they may gain from participating in the study and deemphasize the risks. They may also fail to understand that they will be randomized to different treatment groups or that they may receive a placebo (Appelbaum and Lidz 2008). As a result, they may decide to participate in research without carefully considering the benefits, risks, and alternatives. The federal research regulations require researchers to inform subjects that the study they are participating in involves research (45 CFR 116a1). However, many subjects fail to appreciate the difference between research and therapy even when they are told that they are participating in research (Appelbaum and Lidz 2008).

Appelbaum and colleagues described the therapeutic misconception 40 years ago, and numerous studies have documented the phenomenon since then (Appelbaum et al. 1982, 1987; Henderson et al. 2007). This misunderstanding continues to persist even though clinical researchers have known for many years that they should try their best to dispel it (Appelbaum and Lidz 2008). There are a couple of possible explanations for why clinical research subjects often fail to appreciate the difference between research and therapy. First, subjects who a seriously ill may cling to whatever hope they have of finding a cure and doctors may be hesitant to undercut their hope, because they know that hope can be a powerful psychological healing force in medicine (Appelbaum et al. 1987; Glannon 2006). Second, it can be difficult for subjects who have been treated by doctors for many years to make the mental and emotional leap from therapy to research and they may therefore assume that the clinical researchers who are also treating them will always act in their best interests. Some patients have so much trust in their doctors that they will agree to participate in a clinical trial before reading the consent form (de Melo-Martín and Ho 2008).

Because the therapeutic misconception can compromise informed consent, clinical researchers should carefully explain the difference between research and therapy and identify procedures done only for research purposes and experimental treatments. They should also help their patients understand the risks and benefits of the research and forms of treatment which may be available outside of the study (Menikoff 2006; Appelbaum and Lidz 2008). Clinical researchers can help their patients avoid the therapeutic misconception without taking away their hope or betraying their trust.

5.3 The Elements of Consent

Legal theorists and medical ethicists have developed five different necessary conditions (or elements) for consent to be valid (Faden et al. 1986; Berg et al. 2001):

Competence In the law, competence refers to the legal capacity to make a decision. Adults are presumed to be competent unless they have been adjudicated incompetent by a court. When a court declares someone to be incompetent it will appoint a guardian to make decisions for that person (i.e., the ward). A court may grant a guardian full authority to make decisions for the ward, or it may limit the guardian's authority to making medical or financial decisions. Children are legally incompetent unless they have been emancipated under a legal statute or by a court. A judge may emancipate a minor if they determine that the minor is economically independent and emancipation is in the minor's best interests (e.g., emancipation is necessary to escape parental abuse). Marriage automatically emancipates minors in some U.S. states. The age of adulthood is 18 in most states (Berg et al. 2001).

Competence is not the same as decision-making capacity (DMC), which is an ethical and clinical concept. A person has DMC if they are capable of making a responsible choice, i.e., they are autonomous. A mentally ill or disabled adult could be legally competent but lack DMC, and an intelligent 17-year-old might have DMC but not be legally competent. Unlike competence, DMC comes in degrees (Buchanan and Brock 1990). For example, a person who is fully awake and alert may have near perfect DMC, but someone who is tired or ill may have compromised DMC. The degree of DMC required to make a decision is a function of the choice's complexity and level of risk. For a simple, low-risk choice, such as ordering a meal, a low degree of DMC may be sufficient, but for a complex, high-risk choice, such as deciding whether to accept or refuse life-saving medical care, a high degree of DMC would be required. To establish the degree of DMC needed to make a decision one must balance two competing ethical considerations: the obligation to respect the person's autonomy and the obligation to protect the person from harm. Factors that can impact DMC include age (or maturity), mental illness or disability, emotional distress, and the use of medications or recreational drugs. DMC can also be situational and task-specific (Buchanan and Brock 1990). For example, an elderly woman with mild dementia may be capable of making decisions when she is in her home but become confused when she is in the hospital, and a mentally disabled adult may be capable of mowing lawns but unable to manage their money.

The federal research regulations address the issue of competence by stating that consent must be obtained from the subject or the subject's legally authorized representative (LAR) (45 CFR 46.116). Different states and nations have different ways of determining who should be the LAR. In many jurisdictions, the following preferred ordering is used: guardian or parent (if the person is a minor), health care agent (i.e., someone who is named on a health power of attorney form), close family member (such as a sibling or adult child), other family member or close friend. The regulations also include additional protections for children (45 CFR 46, Subpart D) but not for adults with cognitive impairments that can impair decision-making.

Researchers who are enrolling prospective subjects are responsible for making sure that subjects are legally competent to decide whether to participate and have sufficient DMC for making this choice. They are also responsible for identifying an appropriate LAR if the person is not legally or ethically qualified to make a choice concerning participation and obtaining consent from the LAR. Issues concerning assessment of DMC and use of LARs will be discussed in more detail in Chap. 8.

Information The person providing consent must have enough information to make a responsible choice. As noted in Chap. 2, the federal research regulations include requirements concerning the types of information that researchers must disclose to subjects during consent, such as the nature of research (e.g., study aims, design, procedures, methods, length of participation), reasonably foreseeable risks, potential benefits, alternatives to participation, confidentiality/privacy protections, the right to withdraw, and whom to contact with questions (45 CFR 46.116).

The regulations also require that researchers disclose information that a reasonable person would want to have to make a participation decision (45 CFR 46.116c). However, the regulations do not define the "reasonable person" or indicate what types of information this person would want to have. Hence, even though the regulations include numerous informed consent requirements, there is still considerable room for interpretation, and therefore, disputes (Odwazny and Berkman 2017). For example, the controversy over the SUPPORT study (discussed in Chap. 1) revolved around whether the risk of randomization was a reasonably foreseeable risk that should be disclosed.

Historically, there have been three different legal/ethical standards for disclosing medical information during the consent process: the professional standard, the reasonable person standard (described above), and the subjective standard (Berg et al. 2001). According to the professional standard, a medical professional (such as a clinical investigator) should disclose the information that a reasonable member of the profession would disclose. Commentators have criticized the professional standard on the grounds that it may compromise the subject's autonomy, since professional norms for information disclosure often emphasize non-maleficence and beneficence over autonomy (Faden et al. 1986; Berg et al. 2001). For example, for many years professional norms permitted doctors not to inform their patients that they had terminal illnesses (Faden et al. 1986).

According to the subjective standard, the medical professional should disclose the information that their patient would want to know.[2] While the subjective

[2]The shared decision-making approach can be viewed using a subjective standard for information disclosure because a shared decision is one in which the physician and patient both work together to share information relevant to the decision. This approach is subjective because what is relevant to the decision is determined, in part, by the patient's preferences and values. See Whitney et al. (2004).

standard promotes patient autonomy, commentators have criticized it on the grounds that it places unreasonable demands on medical professionals, since it compels them to find out what a particular subject would want to know (Berg et al. 2001). To accomplish this task, one might need to engage in an extensive dialogue with prospective research participants to learn about their values, interests, goals, and background knowledge. For example, suppose that an 18-year-old male is considering volunteering for a study of fat metabolism that involves a muscle biopsy. The informed consent document discusses the risks of infection from the biopsy but has little to say about the resulting scar. The prospective research subject would like to know more about the biopsy scar because he is considering a career in modeling. To obtain this information the researcher (or research staff) would need to find out more this person's career plans. Researchers might not have enough time to ask these sorts of questions, and even if they do, it might not occur to them to ask them. Adopting the subjective standard as a matter of law could open researchers to lawsuits from subjects who decide, after the fact, that they would not have participated in a study had they known some idiosyncratic piece of information (King and Moulton 2006).

Given the problems with the other two standards, the reasonable person standard has emerged as the most popular approach to disclosing information in biomedical research and clinical practice. The 2018 revision of the Common Rule (45 CFR 46) reflects this consensus. However, the reasonable person standard also has some of its own problems, such as defining or interpreting the concept of "reasonable person" (Odwazny and Berkman 2017). There are two approaches to this issue (Miller and Perry 2012). According to the descriptive approach, the reasonable person is the statistically normal (or average) person within a population. Specifying the relevant population poses a significant problem for the descriptive approach, however, since the statistical norms concerning informational needs may vary considerably across populations. For example, as we saw in Chap. 2, members of the Havasupai tribe had some concerns about the use of their blood samples that were different from those of the average person living in, for example, Phoenix, AZ. If the "reasonable person" is a person who lives in Phoenix, there might not be obligation to inform them about how their blood is handled, whereas if they are a member of the Havasupai tribe, there would be. Problems with the descriptive approach have led some commentators to defend a normative approach, which equates the "reasonable person" with a hypothetical individual who makes appropriate judgments concerning risks and benefits. However, determining what makes a judgment "appropriate" is a key problem with the normative approach, since "appropriateness" is a normative concept that is based on prior acceptance of some moral theory or tradition, but, as we saw in Chap. 3, there is considerable disagreement about moral theories.

The goal of promoting trust can provide some useful guidance for disclosing information to research subjects. According to principle of trust defended in Chap. 4, disclosure should promote trust between the relevant parties, i.e., subjects, researchers, communities/populations, and institutions. The principle implies that researchers should adapt their disclosure plans to the needs of subjects, communities, or populations, because meeting these needs is likely to foster trust, whereas ignoring these needs could have the opposite effect. Researchers should attempt to avoid the sense of betrayal which may occur when research subjects discover that they did not receive some relevant information during recruitment and enrollment. For example, if it is likely that members of community have special concerns about the handling of blood samples, then researchers should share information designed to address these concerns during consent.

Thus, the principle of trust recommends disclosure standards that address the informational needs of a person or population. So why not simply adopt the subjective approach on the grounds that it promotes trust and respects autonomy? The problem with the option is that, as discussed above, it could place excessive demands on researchers and strain budgets and undermine the goal of conducting beneficial research. While researchers and staff should allocate sufficient time to the consent process to address subject-specific needs, they should not spend so much time on consent that they do not have enough time left for other research activities. Likewise, budgetary and resource constraints could place limits on community engagement activities. In sum, disclosure standards in informed consent in research with human subjects should carefully balance three principles: respect for human dignity, beneficence, and trust.

Before concluding this section, it will be useful to say a few words about the amount of information disclosed. As noted earlier, informed consent documents have steadily increased in length over the years (Menikoff 2006). This trend is probably due to several factors, including the increasing complexity of research, greater attention to the informational needs and concerns of human subjects, and pressures to protect the institution from legal liability. As also noted earlier, providing more information to a subject may not always improve the consent process, since the subject may not be able to understand and process the information. One of the revisions to the Common Rule in 2018 was adding two requirements designed to consolidate information and enhance understanding:

(i) Informed consent must begin with a concise and focused presentation of the key information that is most likely to assist a prospective subject or legally authorized representative in understanding the reasons why one might or might not want to participate in the research. This part of the informed consent must be organized and presented in a way that facilitates comprehension.
(ii) Informed consent as a whole must present information in sufficient detail relating to the research, and must be organized and presented in a way that does not merely provide lists of isolated facts, but rather facilitates the prospective subject's or legally authorized representative's understanding of the reasons why one might or might not want to participate (45 CFR 46.116c5),

Some writers (e.g., Flory and Emanuel 2004) have suggested that consent could be improved if subjects were provided with a substantially shorter form to read, with additional information contained in an appendix. Grady et al. (2017) conducted an

RCT to determine whether a short document would be more effective at promoting understanding than the standard, long document, but the study did not show that either one was better than the other.

Voluntariness The person providing consent must be able to make a free choice. The federal research regulations state that consent should take place under circumstances that "minimize the possibility of coercion or undue influence (45 CFR 46.116)." Coercion involves the use of force, intimidation, or threats to make someone comply with a demand or request. A threat could involve the prospect of physical, psychological, or economic harm, including the loss of a benefit that one is entitled to (Wertheimer and Miller 2008). For example, if a professor made participation in their experiment a requirement for one of their courses, this would be coercion, since students who refused to participate could suffer a grade penalty. Undue influence (or inducement) is exerting inappropriate influence over someone's decision-making to cause them to make a particular choice. For example, if a professor seeks to have sex with one of their students, this would be undue influence because the professor's position of power over the student would inappropriately influence the student's decision to engage in sexual activity. I will discuss coercion and undue influence in greater depth in the Consent by Parties other than the Subject, Right to Withdraw, Payment for Research Participation sections (below) and in Chap. 8, which deals with vulnerable subjects.

Comprehension The person providing consent must be able to comprehend or understand the information. The federal research regulations have very little to say about comprehension other than to state that information presented to the subject or LAR must be in language understandable to them (45 CFR 46.116). This phrase is usually interpreted as implying: (1) that consent shall take place in a language that the subject or representation is capable of understanding (e.g., if the subject only speaks Spanish, then the consent form should be written in Spanish and someone who speaks Spanish should conduct the consent); (2) technical terms from science, medicine, or law should be interpreted into lay-language (Berg et al. 2001). As noted earlier, one of the reasons why trust is so important in that subjects often have trouble understanding key information conveyed during consent and they must depend on researchers to help them understand. This dependency, as stressed in Chap. 4, makes subjects vulnerable to manipulation.

Authorization The person providing consent must agree to the interaction that requires consent, such as participating in research or receiving medical care. The person may communicate their agreement verbally, by signing a document, or by making some other sign of authorization (e.g., nodding their head).

Box 5.1: Some Key Studies on Subjects Understanding of Information Disclosed During Consent

Flory and Emanuel (2004)
The authors studied different methods for improving understanding of information conveyed during consent and found that most had only limited success. The most effective way of improving understanding was having a study team member or a neutral educator spend time with the participant to explain the study and answer questions.

Mandava et al. (2021)
The authors compared studies of informed consent in developed and developing countries and found that in both settings understanding of randomization and placebo controls were poorer than understanding of study design. In studies in developing countries, participants were less likely to understand that they could withdraw from a study without penalty.

Nishimura et al. (2013)
The authors reviewed 54 interventions to improve understanding of information conveyed during consent and found that multimedia interventions, enhanced consent forms, and extended discussions significantly improved understanding.

Montalvo and Larson (2014)
The authors conducted a systematic review of 27 articles to assess the quality of studies to improve understanding of information conveyed during consent. The authors concluded that to study informed consent rigorously researchers need to use validated tools and methods for assessing health literacy, reading level, and comprehension of research participants.

Tam et al. (2015)
The authors conducted a systematic review of 103 studies to estimate the proportion of participants in clinical trials who understand different components of informed consent. 75.8% of participants understood freedom to withdraw at any time, followed by nature of the study (74.7%), voluntary nature of participation (74.7%), potential benefits (74%), the study's purpose (69.6%), potential risks and side-effects (67%), confidentiality (66.2%), availability of alternative treatment if withdrawn (64.1%), knowing that treatments were being compared (62.9%), use of placebo (53%) and randomization (52.1%). Most participants (62.4%) understood that the purpose of the study was to develop knowledge not to optimize their treatment.

Thong et al. (2016)
The authors reviewed 31 studies of therapeutic misconception (TM) in psychiatric research populations and found that TM ranged from 12.5% to 86%. Factors contributing to TM included perceived medical roles of researchers, media influence, and the research setting.

Beskow et al. (2017)
The authors conducted a study on 3931 subjects and found that a simplified consent form was more effective at promoting understanding than more complex forms.

Pietrzykowski and Smilowska (2021)
The authors performed a systematic review of 14 studies assessing the understanding of information disclosed during consent. Over 50% of participants understood voluntary participation, blinding, and freedom to withdraw at any time, but less than 50% understood placebos, randomization, safety issues, risks, and side effects.

Wu et al. (2024)
The authors conducted a me a-analysis of 117 studies assessing the understanding of information disclosed during consent. 97.5% of participants understood confidentiality, followed by compensation (95.9%), the nature of the study (91.4%), knowing that treatments were being compared (68.1%), and voluntary participation (67.3%). Only 4.8% understood the concept of placebo and only 39.4% understood randomization.

5.4 Documentation on Consent

The federal research regulations require that consent be documented by means of an IRB-approved form which is signed by the subject or the subject's LAR (45 CFR 46.117a). The consent form may be a document that includes all the required information (i.e., the long form) or a short document stating that the required information has been presented orally to the subject (or their LAR). When the shorter method is used, the subject shall also receive a written summary of the long form and an oral presentation of the long form, which could be a translation into the subject's preferred language. A witness to the oral presentation must sign the short form and the summary. The subject LAR's shall receive a copy of the short form and summary (45 CFR 46.117b). The purpose of using the short form is to accommodate subjects who are illiterate or blind or are not fluent in the language used in research (e.g., English). If researchers expect to enroll subjects from a particular linguistic group (e.g., Spanish) they should translate the long form into their language. The short form should be used for unanticipated enrollment of non-English speakers.

The Common Rule allows the IRB to waive documentation requirements when (1) the consent document is the only written record linking the subject to the research and the only risk of the research is breach of confidentiality, (2) the research is no more than minimal risk and does not include any procedures for which consent is usually required outside or research participation, or (3) the research is no more than minimal risk and the subjects are from a cultural group or community in which signing a document is not the norm (45 CFR 46.117c). The FDA regulations do not include waivers for documentation of consent.

The main reasons for documenting consent are legal: a signed consent form is proof that the subject or LAR has been informed about the risks of research. However, documentation can also serve ethical goals. For example, documentation of consent can help to show respect for the subject's autonomy and promote trust, because the subject may want to have a written record that includes information about research procedures and methods, risks, benefits, confidentiality protections, and so on. Subjects who forget important information can refer to the consent document. It is especially useful for subjects to have a written record of whom to contact if they are injured in the study, have a question about the research, or want to withdraw (Resnik 2009b).

Documentation of consent may undermine trust in some situations, however. For example, if research is conducted in a community in which most people are illiterate and prefer oral communication, using a written form to document consent may generate suspicion and undermine the community's trust (Onvomaha et al. 2006). In some cultures, people may distrust written documents or see no need for them (Rashad et al. 2004; Resnik and Hecking 2020). In some cases, the informed consent document may be misleading or confusing due to difficulties with translating the document into the local written language (Onvomaha et al. 2006). Researchers should use culturally appropriate methods for documenting consent (Council for the International Organizations of Medical Sciences 2002, 2016; Liamputtong 2008;

Krogstad et al. 2010). In some cases, no documentation may be the best strategy; in others, use of the short form may be appropriate. Community engagement can help researchers learn about culturally appropriate methods of documenting consent (Seibert et al. 2002). The principle of trust would support a method of documentation that is most likely to promote the subject's, community's, or population's trust.

5.5 Consent by Parties Other Than the Subject

As noted in Chap. 2, research regulations and guidelines require researchers to obtain consent from the subject or from the subject's LAR if the subject is not capable of providing consent, due to immaturity or mental disability or illness (Berg et al. 2001).[3] However, in some communities or societies it is customary for someone other than the research subject to provide consent, even when the subject is capable of consenting. As mentioned in Chap. 4, in some Islamic and African cultures it is customary for a woman's spouse or older male relative to provide consent for medical procedures (Rashad et al. 2004; Afifi 2007; Princewill et al. 2017). Family involvement in the consent process is also common among Latino populations in the Americas (Resnik and Hecking 2020). In some African societies it is customary for tribal leaders to provide consent for members of the tribe (Onvomaha et al. 2006; Liamputtong 2008). While it is important for researchers to be sensitive to cultural traditions concerning consent (Seibert et al. 2002; Krogstad et al. 2010), allowing someone other than the competent, adult subject to provide consent contravenes research regulations and guidelines adopted by the U.S. and other nations, which emphasize protection of individual rights.

Consent by someone other than the competent, adult research subject raises at least two ethical concerns. The first is coercion: consenters might pressure prospective subjects to participate in research against their wishes or better judgment. Coercion could be explicit, e.g., the prospective subject faces a threat from the consenter, or implicit, e.g., the prospective subject complies with the decision of the consenter because they want to please the consenter or avoid making them angry. Conversely, the consenter might refuse to allow the prospective subject to participate in research, which could deny potential benefits to the prospective subject, such as participation in a clinical trial. Even if the prospective subject is not likely to derive medical benefits from participation, they might want to contribute to advancing the goals of the study. The second concern is privacy/confidentiality: the consenter might learn about private information disclosed during the enrollment process that the subject does not want to share with other people. For example, during the enrollment process, the researchers might ask subjects about drug or alcohol use, sexual abuse, trauma, sexually transmitted diseases, and so on. Even if the consenter

[3] In this section, I am not considering situations in which the subject is a child or is an adult who lacks DMC. These will be considered more fully in Chap. 8.

is not involved in the enrollment process, they would learn that the subject is participating in the study, which could also compromise the subject's privacy or confidentiality. For example, if the study is interviewing women who have a particular sexually transmitted disease, the consenter would learn that the subject has this disease.

Given the concerns about coercion and privacy/confidentiality, there are strong moral reasons for not allowing anyone besides the prospective subject (who is a competent adult) to provide consent for research participation. The principles of respect for human dignity and beneficence would recommend that no one else besides the subject provides consent. However, prohibiting outside parties from participating in the consent process might undermine the trust of the community and prospective research subjects (Krogstad et al. 2010). Also, as noted in Chap. 4, prohibiting outside parties from being involved in the consent process may result in harm to the subjects. For example, a husband might retaliate against his wife for enrolling in a study without his permission, or tribal leaders might punish members of the tribe who enroll in a study without their authorization.

One way of dealing with this ethical dilemma is to develop a compromise solution that respects cultural traditions and the prospective subject's autonomy (Krogstad et al. 2010). In Chap. 4 I argued that women's consent for research participation in some Islamic cultures could be obtained by either: (a) asking the woman to consent for research, (b) asking her if she also wants her spouse or older male relative to give permission, and (c) allowing the woman to designate her spouse or older male relative as the decision-maker, if she so chooses. Which of these options should be chosen depends on the facts pertaining to the situation, i.e., what appears to work best, given what one has learned about the community and population. All I would add to these recommendations is that to protect the woman's privacy and confidentiality, researchers should not include the spouse or male relative in the enrollment process, unless she asks that he be included. Researchers should also not inform the spouse or older male relative that the woman is participating in research if she does not want them to know.

For the situation in which consent is traditionally given by tribal leaders, I recommend that researchers consult with the leaders about the research and obtain their approval to recruit members of the tribe but not ask them to provide consent for specific participants (Weijer and Emanuel 2000; Dickert and Sugarman 2005). Members of the tribe would still be free to decide whether to participate in the study (Juengst 1998). To protect their privacy and confidentiality, tribal leaders would not be told who is or is not participating in the study. Onvomaha and colleagues implemented this type of methodology in a research project involving a tribal community in Northern Ghana (Onvomaha et al. 2006). The researchers consulted with tribal leaders and obtained their permission before recruiting research subjects.

These compromise solutions may not satisfy everyone. For example, husbands or male relatives in Islamic cultures may want to provide consent for their wives or female relatives, not just grant their permission once the woman has already consented. If a woman decides to participate in a study and does not want her husband or older male relative to know about it, she could face repercussions if they find out.

Likewise, tribal leaders might want to provide consent for tribal members, not just grant researchers permission to enroll members of the tribe in the study. Human rights advocates may object that these solutions to do not do enough to respect autonomy. A human rights advocate might argue that husbands or older male relatives should have no involvement in a woman's decision to participate in research, even if she asks them to be included. The fact that a compromise does not satisfy all parties is by no means a reason to reject it, since compromises involve some give-and-take. Unless we say that a particular moral principle must always trump other principles, then most of our answers to ethical dilemmas will be compromise solutions.

5.6 Research Without Consent

5.6.1 Research on De-identified Data and Biospecimens

Although informed consent is one of the most important principles for ethical research with human subjects, one might argue that sometimes it is acceptable to conduct research on human subjects without their consent (Miller 2008; Gelinas et al. 2016). As noted in Chap. 2, the Common Rule (45 CFR 46.116e-g) allows an IRB to waive or alter consent requirements for minimal risk research that could not be conducted without a waiver or alteration.[4] Some common situations in which an IRB might grant a waiver include:

- Pre-screening (for example, asking someone questions to see if they qualify for enrollment)
- Reviewing or analyzing medical records[5]
- Quality assurance/quality improvement (QA/QI) projects
- Secondary use of de-identified biospecimens from a study (typically an older one) in which the consent document did not discuss the use of biospecimens.
- Cluster randomized trials, in which an intervention is given to an entire group.[6]

Additionally, the Common Rule and the FDA regulations do not require researchers to obtain consent for conducting research on de-identified human biospecimens (e.g., tissue, blood, etc.) left over from medical procedures because their de-identified biospecimens are not considered to be human subjects (Bathe and McGuire 2009).

[4]The FDA has similar regulations at 21 CFR 50.22.

[5]The privacy rule of the Health Insurance Portability and Accountability Act (HIPAA) allows an IRB or privacy board to approve research involving review of medical records (National Institutes of Health 2004). HIPAA will be discussed in greater depth in Chap. 6.

[6]The Johns Hopkins IV catheter infection control study discussed in Chap. 1 was a cluster randomized trial.

The moral argument for allowing research without consent in situations like those mentioned above is based on the following premises: (1) the research poses no more than a minimal risk to the subjects as long as adequate protections are in place to safeguard the confidentiality and security of data and biospecimens; (2) the research is likely to produce knowledge that advances socially valuable goals, such as developing medical treatments or promoting public health; and (3) it would be highly impractical (and perhaps impossible) to do the research if consent needed to be obtained (Miller 2008; Bathe and McGuire 2009; Wendler 2021).

Objections to this argument are: (1) the risks of the research are small but not negligible, especially since re-identifying individuals in de-identified datasets is becoming increasingly easier to do; and (2) using data or biospecimens without consent is a significant violation of autonomy because people care about how their data and biospecimens are used (Ploug 2020).

Proponents of this research with consent argue that autonomy does not need to be honored in these situations because the risks to person are so small and want matters to most people are risks (Gelinas et al. 2016). Proponents also claim that a patients relinquish any interests they might have in their biospecimens when they allow them to be treated as medical waste. Opponents argue that autonomy involves more than control over risks but also control over what is done with one's data and biospecimens. Some people might not want their data or specimens used for certain purposes, such as human cloning experiments or training of artificial intelligence systems. Moreover, most people never have a chance to say whether they want their biospecimens to be treated as waste (Lewis et al. 2013; Morse et al. 2023).

How should this issue be resolved? I believe the principle of trust offers some useful guidance. To promote trust in biomedical research (and biomedicine more generally) it is important to understand what individuals, communities, and the public expects concerning control over their data and biospecimens. Two empirical studies have addressed this question. Page et al. (2016) surveyed 211 patients at a medical clinic in Calgary, Canada concerning their views the use of health information and biospecimens in research. Although most respondents were comfortable with using their health information (63.5%) or biospecimens (61.2%) for research without their consent, most of them also believed that informed consent should be sought (Page et al. 2016). Yamanaka et al. (2021) surveyed 361 patients at a hospital in Kyoto, Japan concerning their views about the of health information and biospecimens. Most of the respondents did not think it was acceptable to use biospecimens or health information in research without consent. 36% of respondents said that consent was unnecessary for the use of biospecimens in research and 32% said that consent was unnecessary for the use health information.

What these studies suggest is that **some form of consent** should be required for research with medical information or biospecimens leftover from clinical procedures because continuing to do this research without consent could undermine the public's trust in biomedical research. However, requiring a form of consent that meets all the regulatory requirements in 45 CFR 46.116a-c may not be necessary, especially since it is not clear that all of these requirements are needed to promote trust, and imposing these requirements would be a time-consuming and costly administrative burden on that could significantly impeded medical progress. A solution that solicits consent but also reduces burdens may be the best way forward. Under the approach proposed here,[7] patients would be informed when they receive treatment at a hospital or clinic and that their de-identified health information and biospecimens may be used in biomedical research.[8] They would also be informed that they can opt-out[9] of this system by signing a form stating that they do not want their information and data used in research. A potential drawback with this approach is that some patients may not consent to using their medical records or biospecimens in research, but enough probably will that the data will still be useful. Indeed, surveys have shown that most people are willing to consent to broad use of their data and specimens in research (Wendler 2006; Hill et al. 2013; Warner et al. 2018).

5.6.2 *Emergency Medical Research*

For hundreds of years, physicians have been testing experimental treatments and procedures on patients who are facing life-threatening medical emergencies. For example, Pasteur's administration of a rabies vaccine to Joseph Meister and Baby Fae's baboon heart transplant would qualify as emergency research (see Chap. 2). In 1995, the FDA adopted regulations for emergency use of a test article (which is not considered research but therapy) and for emergency research. The emergency research regulations allow IRBs to waive informed consent requirements for conducting emergency medical research under the following conditions (21 CFR 50.24)[10]:

(1) The human subjects are in a life-threatening situation, available treatments are unproven or unsatisfactory, and the collection of valid scientific evidence, which may include evidence obtained through randomized placebo-controlled investigations, is necessary to determine the safety and effectiveness of particular interventions.
(2) Obtaining informed consent is not feasible because:
    (i) The subjects will not be able to give their informed consent as a result of their medical condition;
    (ii) The intervention under investigation must be administered before consent from the subjects' legally authorized representatives is feasible; and.
    (iii) There is no reasonable way to identify prospectively the individuals likely to become eligible for participation in the clinical investigation.

[7] According to Wendler (2021), many hospitals are already implementing this approach.

[8] 45 CFR 46.116d includes requirements for broad consent for the use of data or biospecimens. Institutions follow these requirements even if they are not funded by DHHS or another federal agency.

[9] The ethics of opt-out consent is discussed in Sect. 5.7.

[10] The Common Rule does not include any provisions for emergency medical research.

(3) Participation in the research holds out the prospect of direct benefit to the subjects because:

   (i) Subjects are facing a life-threatening situation that necessitates intervention;
   (ii) Appropriate animal and other preclinical studies have been conducted, and the information derived from those studies and related evidence support the potential for the intervention to provide a direct benefit to the individual subjects; and.
   (iii) Risks associated with the investigation are reasonable in relation to what is known about the medical condition of the potential class of subjects, the risks and benefits of standard therapy, if any, and what is known about the risks and benefits of the proposed intervention or activity.

(4) The clinical investigation could not practicably be carried out without the waiver.
(5) The proposed investigational plan defines the length of the potential therapeutic window based on scientific evidence, and the investigator has committed to attempting to contact a legally authorized representative for each subject within that window of time and, if feasible, to asking the legally authorized representative contacted for consent within that window rather than proceeding without consent. The investigator will summarize efforts made to contact legally authorized representatives and make this information available to the IRB at the time of continuing review.
(6) The IRB has reviewed and approved informed consent procedures and an informed consent document consistent with § 50.25. These procedures and the informed consent document are to be used with subjects or their legally authorized representatives in situations where use of such procedures and documents is feasible. The IRB has reviewed and approved procedures and information to be used when providing an opportunity for a family member to object to a subject's participation in the clinical investigation consistent with paragraph (a)(7)(v) of this section.
(7) Additional protections of the rights and welfare of the subjects will be provided, including, at least:

   (i) Consultation (including, where appropriate, consultation carried out by the IRB) with representatives of the communities in which the clinical investigation will be conducted and from which the subjects will be drawn;
   (ii) Public disclosure to the communities in which the clinical investigation will be conducted and from which the subjects will be drawn, prior to initiation of the clinical investigation, of plans for the investigation and its risks and expected benefits;
   (iii) Public disclosure of sufficient information following completion of the clinical investigation to apprise the community and researchers of the study, including the demographic characteristics of the research population, and its results;
   (iv) Establishment of an independent data monitoring committee to exercise oversight of the clinical investigation; and

(v) If obtaining informed consent is not feasible and a legally authorized representative is not reasonably available, the investigator has committed, if feasible, to attempting to contact within the therapeutic window the subject's family member who is not a legally authorized representative, and asking whether he or she objects to the subject's participation in the clinical investigation. The investigator will summarize efforts made to contact family members and make this information available to the IRB at the time of continuing review.

An example of a type of study that was conducted in accordance with the FDA regulations was a clinical trial of Polyheme, an artificial blood substitute manufactured by Northfield Laboratories. The trial, which took place from 2003 to 2006 at 32 hospitals across the U.S., enrolled hundreds of trauma patients who were being transported to the hospital in ambulances, where there is a limited blood supply. Patients were enrolled if: (1) they were facing a medical emergency for which they needed a blood transfusion; (2) blood matching their blood-type was unavailable; (3) they were unable to consent, and (4) their LAR was unavailable. The investigators conducting the study advertised it in local papers and on radio and television and held meetings to inform community members about it. Patients could opt-out of the study by wearing wrist band indicating they did not want to be in the study (Dickert and Sugarman 2007).

The moral argument for allowing emergency research without consent is based on the following premises: (1) the research has the prospect of significantly benefitting the patient because they are facing a life-threatening condition for which there are no viable alternatives; (2) the research is socially valuable because it can generate knowledge that can be used to develop treatments or therapies; (3) consent is not possible because the patient is incapable of consenting due to the medical condition (for example, they are unconscious or delirious) and their LAR is not readily available; and (4) most people would probably consent to participation under these circumstances because they would want to have life-saving treatment (Karlawish 2008; Dickert and Sugarman 2007). For example, in medical law, the emergency exception allows a physician to provide emergency medical care to a patient when the patient is unable consent and their LAR is not available, based on the rationale that the physician is acting in the patient's best interests (Berg et al. 2001).

Objections to this argument include the following points: (1) to respect the patient's autonomy and dignity, participation in medical research should be based on the consent of the patient or their LAR; (2) community consultation, public education, and opt-out bracelets are no substitutes for informed consent; (3) emergency research is often not in the patient's best interests because there may be alternative treatments; and (4) emergency research without consent often results in disproportionate (and unfair) enrollment of socioeconomically disadvantaged or racially or ethnically minoritized participants, because they are less likely to have an LAR available (Karlawish 2008; Dickert and Sugarman 2007).

How should this debate be resolved? Again, I believe the principle of trust offers some practical solutions. Emergency medical research without consent is acceptable if steps are taken to foster the trust of the community and potential participants. To secure the community's trust, meaningful engagement with the community is essential (Food and Drug Administration 2013). Engagement could include various methods for educating community members about a proposed study, soliciting their input, and addressing their questions and concerning the study (Resnik 2018; Smischney et al. 2018; World Health Organization 2020; see discussion of community engagement in Chap. 4). Engagement should involve transparent sharing of information, respectful consideration of differing opinions, and inclusive and balanced dialogue among community members, researchers, and institutional officials. Engagement should occur in early in the process (e.g., before a protocol is submitted to the IRB) to ensure that community members have meaningful input and don't feel like the decision is a *fait accompli* (Food and Drug Administration 2013; Resnik 2018). To secure trust from potential participants, there should be an easy method of opting out, such as wearing a bracelet indicating that one has opted out of the study. To reduce the possibility of a mistake, opt-out decisions should be recorded. However, the subject (if conscious and coherent) or their LAR should retain the right to opt-in after they have opted out. Also, the opt-out option should be well-publicized in the local media and in various settings, such as hospitals, clinics, churches, schools, businesses, and government offices.

A fundamental issue related to community engagement is whether the community should have the right to veto a research proposal (Juengst 1998; Resnik 2019). This is a challenging issue that may involve a potential conflict between the interests of medical science (and possibly the larger public) and the local community. It also raises questions about how the community registers it decision. Can an elected official or tribal leader speak for the community? Does the community need to vote on a resolution opposing the research? The principle of trust would recommend that an IRB should not approve emergency medical research without consent when the community expresses strong opposition to the research by means of political process accepted by the community (such as voting or decision-making by tribal leaders). For example, in 2016, residents of Key Haven, Florida voted against a field trial of genetically modified mosquitoes in the community that had been approved by the Food and Drug Administration and the State of Florida. The community's wishes were respected, and the field trial did not take place. However, field trials of these mosquitoes did take place in other areas, such as Brazil and Malaysia, in which community support was strong (Resnik 2019).

5.7 Opt-Out Consent

Opt-out consent procedures, like those mentioned above, raise ethical issues because one might question whether they sufficiently respect the subject's autonomy (MacKay 2015). Proponents of opt-out consent procedures argue that they respect autonomy and promote scientific research, since opt-in procedures (i.e., procedures in which subjects are enrolled only if they sign a form stating that they want to be in a study) can result in lower enrollments and biased samples (Gulcher and Stefánsson 2000; Junghans et al. 2005; de Man et al. 2023). As mentioned in Chap. 2, Iceland implemented an opt-out consent procedure for its research database encompassing medical records and genealogical and genetic data (Weir and Olick 2004). All citizens were included in the database unless they opted out of the research. It is worth noting that the Icelandic government consulted extensively with members of the population before its parliament passed the law that enabled this research. Community engagement included town hall meetings and public education. Over 700 newspaper articles and 100 television and radio programs covered the research. Polls showed that 75% of the population approved of the research before the parliamentary vote, and that 2 years after the law was passed 90% of Icelanders who had an opinion about the research approved it (Gulcher and Stefánsson 2000). Despite widespread approval of the project, some people objected strongly to it, including members of medical organizations (Annas 2000).

One of the reasons why some people think that opt-out consent does not provide adequate respect for autonomy is that individuals (or their LAR) may not be aware of opportunities to opt out. Silence (i.e., not opting out) should not be equated with consent (MacKay 2015). For example, suppose that a hospital uses an opt-out procedure for research involving leftover biological samples similar to the proposal I defended in Sect. 5.6.1. During the admission process, patients at the hospital are presented with many forms to review and sign, such as forms dealing with billing, confidentiality protections, consent for various medical procedures, etc. Buried somewhere in the stack of paper that patients receive is a short document informing them that they may opt out of research involving their leftover biospecimens. One might argue that this consent process would not sufficiently respect autonomy because patients may not notice the opt-out form, or they may scan it quickly while reading and signing other documents. Thus, the fact that someone has not opted out of a study may not indicate that they wish to enroll in a study; it may only mean that have not paid enough attention to the study or the opt-out procedures. While opt-out studies raise the issue of whether the subject has fully consented to research participation, opt-in studies leave little room for doubt, because opting-in requires a subject take the time to read and sign a consent form.

Viewing the opt-in vs. opt-out issue from the perspective of trust, one could argue that other opt-in consent should be the default enrollment procedure, since it promotes greater transparency and awareness, and therefore trust, than opt-out. Subjects are less likely to regret being enrolled in a study if they have expended time and effort to opt into it than if they have not opted out. A person might resent the fact that they are in a study because they were too busy to opt-out, forgot to out, or did not know how to opt out. Researchers seeking approval for an opt-out procedure should provide the IRB with information related to the reasons for using the procedure (such as to opt-in is impractical or will produced biased enrollment) and the steps they will take to ensure that autonomy is respected.[11] If researchers use an opt-out consent process, they should make sure that prospective subjects are aware

[11] The researchers will also need the IRB to waive the some of the consent requirements under the provision in 45 CFR 46.116e-g.

of the study and have the opportunity to decide whether to opt out, which will help to ensure that individuals who are enrolled in the study have made a deliberate choice to participate, not just failed to opt out (MacKay 2015). With the exception of emergency research (discussed above), opt-out consent procedures should be used only for low-risk studies, such as observational research or research on patient records or biological samples, since opting out may not provide as much protection from risks as opting in (Vellinga et al. 2011). Also, out-opt procedures should not be used to enroll subjects from populations or communities that are not familiar with biomedical research, such as indigenous populations, because they may be unfamiliar with how opt-out procedures work and might therefore become very distrustful of the research (Fitzpatrick et al. 2016).

5.8 Specific, General Consent, and Tiered Consent

The sharing of data and biospecimens is important for promoting scientific progress, since science generally advances faster when researchers collaborate and pool their resources than when they work in isolation. Scientists therefore have an ethical obligation to share data and research materials. Government research sponsors, such as the NIH and NSF, also require funded researchers to share data and materials as broadly as possible (Shamoo and Resnik 2022). To promote collaboration and sharing in research, researchers frequently ask human subjects for permission to share their (usually de-identified) samples and data with other researchers working on yet to be determined future projects. This type of consent is known a general, broad, or blanket consent (Petrini 2010). Specific consent, by contrast, involves obtaining consent each time a subject's samples or data will be shared or used for a new project. Surveys indicate that about 80% of research subjects are willing to consent to general sharing and use of their data and biospecimens (Wendler 2006). General consent is common in research involving the procurement of samples or data for genetic or genomic biobanks (Weir and Olick 2004; Sanchini et al. 2016).

Some have argued that general consent is unethical because it does not adequately respect autonomy, since people may not understand what precisely they are consenting to. Subjects do not know who may use their data or biospecimens, what they may use them for, and so on (Shickle 2006). Others have argued that general consent is ethical because subjects do not need to know the specifics concerning the use of their data and biospecimens to understand that they may be used for a variety of research projects. Moreover, asking subjects for their permission each time a researcher requests samples or data (i.e., specific consent) would be inconvenient for subjects and researchers. Most people would not want to be contacted dozens of times for permission to share or use their samples or data (Wendler 2006; Petrini 2010).

The principle of trust supports broad consent, provided that researchers ensure that subjects have a good understanding of what they are consenting to. Moreover, most people are already comfortable with giving broad consent (for example, when signing a user agreement for a social media company). However, the principle of trust does not support broad consent for populations or communities that are likely to have specific concerns about the use of their data or biospecimens (Wick et al. 2024). The Havasupai tribe (see discussion in Chap. 2), for example, objected to using their blood samples for research on mental illness, incest, and the tribe's origins. They also did not want to allow their samples to be shared with other researchers. Other research subjects may object to allowing their biospecimens or data to be used for commercial purposes. Researchers who are studying communities or populations should try to determine whether they are likely to have specific concerns about the use of their data or biospecimens. Community engagement may not be necessary for every type of study, but it would be appropriate if the researchers have reasons to believe that the community or population may have concerns regarding the use of data or biospecimens. For example, the researchers studying the Havasupai tribe could have approached tribal representatives to learn about any concerns they might have had about the use of their data or blood samples.

To accommodate individuals, communities, or populations with unique concerns related to the sharing and use of their data and biospecimens, investigators may decide to give research subjects a menu of options (known as tiered consent). Options could include specific consent, general consent, consent for research only on certain types of diseases related to their disease, or consent only for non-commercial research. Subjects could also indicate that they refuse to allow their data or biospecimens be used for certain types of research (such as human cloning) (McGuire and Beskow 2010; Master and Resnik 2013). While the tiered approach has the advantage of accommodating different concerns that individuals, communities, or populations may have about the use of their data and specimens, a potential drawback of this approach is that it may be confusing for research subjects. However, this problem can be overcome by developing simplified consent forms that offer clear choices to subjects (McGuire and Beskow 2010). Another problem with the tiered approach is that it may be difficult for researchers to keep track of what subjects have consented to, which may cause them to unintentionally violate the consent agreement. However, this problem can be managed through good record-keeping practices (Master and Resnik 2013).

5.9 Deception in Research

The Milgram case (see discussion in Chap. 2) illustrates some of the ethical issues that may arise when researchers deceive subjects in research. Deception can compromise informed consent and respect for autonomy, since it may involve withholding some types of information from subjects or even lying to them (Wendler and Miller 2008). For example, Milgram did not inform the subjects who were instructed to administer punishments that they were not really giving electric shocks to their "victims". Milgram argued that if he had disclosed this information, this would have undermined validity of the study, since the experiment would not be a genuine test of punishers' willingness to obey an authority figure (Milgram 1974).[12]

One way of enhancing the consent process in experiments involving deception is to inform subjects that the research may involve deception without telling them the precise nature of the deception. Wendler and Miller (2008) refer to this procedure as authorized deception. For example, double-blinded, placebo-controlled clinical trials are designed to prevent research subjects or researchers from discovering who is receiving the experimental treatment and who is receiving a placebo. The purpose of double-blinding is to control for the placebo effect (Wendler and Miller 2008). The placebo effect is a well-documented phenomenon in which a patient's belief that they are receiving an effective treatment influences their response to therapy (Finniss et al. 2010). For double-blinding to be effective, the subjects and the researchers should not be able to tell the difference between the placebo and the experimental treatment. For example, pills administered to patients should be the same size, shape, and color. In some cases, the placebo may even mimic some of adverse effects of the experimental treatment that subjects and researchers may be aware of, such as nausea or recovery from surgery.[13] Blinding is a form of deception that subjects consent to.

One potential objection to authorized deception is that it still is a form of deception and therefore compromises the consent process and respect for autonomy. One might argue that for consent to be valid, subjects should know what they are consenting to. Any form of deception prevents subjects from making a fully informed choice (Wendler and Miller 2008). While this objection has some merit, it does not undermine the legitimacy of authorized deception. First, most of us would agree that authorized deception is not inherently immoral, since many human activities that we regard as ethical involve some form of deception that we agree to. For example, when you purchase a ticket to a magic show you understand that the magician will attempt to deceive you. Players in a poker game also understand that their opponents may be bluffing. Second, consent for research participation is often not fully informed because there is too much information one needs to know to be fully informed (Wendler and Miller 2008). For example, a patient in a drug study may not know information related to the chemical's toxicology, pharmacology, or biochemistry, or its potential adverse effects. Most of us would agree that the patient

[12] It is worth noting that about ten years ago researchers conducted a modified version of the Milgram experiment in which subjects could give each other mild electric shocks or a financial penalty. The subjects took turns being the punisher and the victim. The shocks and financial penalties were real and there was no deception. The subjects and experimenters were all female to control for gender effects. The researchers evaluated the subjects' willingness to obey authority by varying the conditions of the experiment. In one variation, the experimenter did not instruct the punisher to administer a punishment: the punisher was free to choose whether to punish. In another variation, the experimenters instructed the punishers to administer a punishment. The punishers were more willing to administer a punishment when they were instructed to do so, and they felt less moral responsibility for their actions. None of the subjects withdrew from the experiment or reported distress (Caspar et al. 2016).

[13] The ethics of sham surgery will be discussed in Chap. 7.

does not need to know all of this information concerning the drug as long as he or she has enough information to decide whether to participate in the study (Wendler and Miller 2008). One might argue that authorized deception can be ethical as long research subjects receive the information they need to know to make a sound decision concerning participation in a research study. IRBs should review research protocols and consent documents in studies involving authorized deception to ensure that prospective subjects will receive the information they need to make a sound decision (Wendler and Miller 2008).

A potential problem with authorized deception is that it may compromise the validity of the research. For example, if Milgram had told his subjects that the experiment involved some type of deception, many of them would have guessed that the electric shocks were not real, because they would suspect that researchers would not ask them to cause significant harm to other people. Authorized deception might undermine the validity of experiments that mimic phishing emails that attempt to induce recipients to disclose private information, such as passwords or bank account numbers. If potential recipients know in advance that they may receive a phishing attack, they may be less likely to respond to it than they would have been otherwise (Finn and Jakobsson 2007; Resnik and Finn 2018). Researchers who are planning to use authorized deception must determine whether informing subjects about the possibility of deception will compromise the validity of the research. If authorized deception is likely to adversely impact the scientific design of the study, investigators may ask the IRB to waive some of consent requirement under 45 CFR 46.116 to conduct an experiment without authorized deception.

The 2018 revision to the Common Rule includes an exemption for "benign behavioral interventions" which are: "are brief in duration, harmless, painless, not physically invasive, not likely to have a significant adverse lasting impact on the subjects, and the investigator has no reason to think the subjects will find the interventions offensive or embarrassing (45 CFR 46.104d3)." However, if the intervention involves deception, the exemption is not applicable unless "the subject authorizes the deception through a prospective agreement to participate in research in circumstances in which the subject is informed that he or she will be unaware of or misled regarding the nature or purposes of the research (45 CFR 46.104d3)." Given the language in 45 CFR 46.104d3, it is not likely that an experiment like the one Milgram did would qualify as a benign behavioral intervention.

The risks of deceptive research can be difficult to assess because they are likely to be psychological in nature (Wendler and Miller 2008). As noted in Chap. 2, some of the subjects in the Milgram experiments experienced significant distress, remorse, and anxiety as result of their participation. Milgram's experiments would probably not qualify as benign behavioral interventions because they had some lasting adverse impacts on the subjects. But were the risks of these experiments more than minimal? The federal research regulations define minimal risk as "the probability and magnitude of harm or discomfort anticipated in the research are not greater in and of themselves than those ordinarily encountered in daily life or during the performance of routine physical or psychological examinations or tests (45 CFR 46.102i)." Since Milgram's experiments were not routine physical or psychological examinations, an IRB would need to compare the risks of Milgram's research to the risks of daily life.

Most of us experience some degree of psychological harm in daily life. For example, we may face stress at work or in the home; we may have arguments with family members or coworkers; we may even experience the death of loved one or go through a divorce. An IRB would need to compare the risks of the Milgram study to the types of psychological risks we ordinarily encounter in daily life.

Thinking about deception from the perspective of trust, deception can be highly detrimental to human relationships. If we discover that a family member, friend, colleague, or merchant has lied to us, we may no longer trust that person. Deception can be one of the most significant forms of betrayal. Since deception in research may undermine trust in researchers and institutions, it is important for researchers and IRBs to take appropriate measures to minimize the damage caused by deception. Authorized deception may help to promote trust by informing subjects that they may be deceived. If subjects understand that the research may involve deception, they may experience less of a sense of betrayal when the study ends and they learn about the nature of the deception. However, if deception goes beyond what subjects expected or involves matters that they would not want to be deceived about then authorized deception may not have salutary effects. Debriefing after the experiment is over can also help promote trust (Wendler and Miller 2008). Debriefing should provide subjects with information about the nature of the deception and why researchers regarded it as necessary for conducting the experiment. Subjects should also have the opportunity to ask other questions about the research.

5.10 Assent

The federal research regulations require that children provide assent for research if they are capable of doing so (45 CFR 46.408). Assent is legally and ethically different from consent. A person who assents to research agrees to participate in a study; a person who provides consent authorizes participation in a study. While assent is part of consent, consent involves more than mere agreement: to consent to participate in research one must make an autonomous decision concerning participation (Berg et al. 2001). Thus, individuals who are not autonomous (e.g., children or mentally disabled or ill adults who lack adequate decision-making capacity) may assent to research participation but they cannot provide consent. The federal research regulations include requirements for both the child's assent and the parent or guardian's permission in pediatric research (45 CFR 46.408). IRBs determine whether children are capable of providing assent by considering their age, emotional maturity, cognitive abilities, and other factors. Seven years old is the generally accepted age for assent (Wendler 2008). Children who are older than 14 may be able to provide legally valid consent if a court rules that they are mature or emancipated minors (Berg et al. 2001).

The main reason for obtaining assent is to respect the person's dignity. Although the person providing assent is not fully autonomous, allowing them to participate in the decision and not forcing them to do something respects their dignity. Also, obtaining the person's assent may contribute to the development of their autonomy. Autonomy does not happen all at once but develops as a person matures emotionally and cognitively. Asking a child to provide assents helps to recognize his or her maturing decision-making abilities and may also help to teach the child how to make autonomous decisions (Wendler 2008).

Another reason for obtaining assent is to protect the research subject from harm (Wendler 2008). A child who participates in research against their will may suffer considerable stress or trauma. The child may resist the research procedures (such as blood draws) and harm themselves or others. Most commentators therefore recommend that children who refuse to assent not be enrolled in research, unless the research is likely to benefit them medically (Wendler 2008). The federal research regulations state that the child's assent is not necessary if the research intervention is likely to benefit the child medically and is not available outside of the study (45 CFR 46.408). For example, an 8-year-old child with cancer who does not want to participate in a clinical trial could be enrolled against their will if the trial offers them beneficial treatment that is not available outside of the trial.

It is also important to observe that obtaining assent can play an important role in building trust between research subjects, researchers, and institutions. Assent, like consent, should be viewed as a process, not merely as a form to be signed. Researchers should view assent as an opportunity to establish relationships with subjects and foster trust (Smajdor 2023). Researchers should be wary of enrolling subjects in research who refuse to provide their assent, because this could significantly damage trust.

5.11 Payment for Research Participation

Research subjects often receive financial compensation for their participation. The amount of money subjects receive varies considerably, depending on the nature of the research, the time involved in performing research activities, and the degree of pain, inconvenience, or risk associated with research procedures (Grady 2001, 2005). While subjects may receive $25 or less for completing a survey or providing a blood sample, healthy volunteers in Phase I drug safety studies may receive several thousand dollars (Grady et al. 2005). Some subjects regard themselves as "professional guinea pigs" because compensation for research participation is their primary form of income (Elliott and Abadie 2008; Abadie 2010; Fisher 2020).

There are several reasons why researchers pay research subjects. First, payment can enhance recruitment (Grady 2005). Studies show that money is an important motivating factor for many research subjects, especially for healthy volunteers, who do not expect to receive any medical benefits from participation (Tishler and Bartholomae 2002; Stunkel and Grady 2011; Fisher 2020). Second, payment can reimburse subjects for their time, inconvenience, and travel expenses. Third, payment can serve as a token of appreciation for participation in research (Grady 2005). And finally, payment is way of providing participants with a fair share of the benefits of research (Różyńska 2022).

Paying research subjects raises several ethical concerns, however (Grady 2005; Resnik 2015). The first is that financial incentives for participation may interfere with a prospective subject's ability to make a free and informed choice concerning participation. The lure of money might entice a person, especially someone who is socioeconomically disadvantaged, to enroll in a study against their better judgment (Macklin 1981; McNeill 1997). Although the federal research regulations do not address the issue of paying human subjects for their participation, they do require researchers to minimize the potential for coercion or undue influence (45 CFR 46.116). Food and Drug Administration (2018) guidance documents and Council for the International Organizations of Medical Sciences (2016) guidelines advise IRBs and investigators to ensure that payments are not so large that they constitute undue influence (or inducement) or coercion.

While financial compensation may constitute undue inducement, it is not coercion because it does not involve the use of force, intimidation, or threats. For most people, receiving money is a benefit, not harm. Although ethicists and IRB members sometimes talk about financial incentives as potentially coercive (Largent et al. 2012), this way of framing the issue confuses coercion with undue inducement (Grady 2005; Wertheimer and Miller 2008).[14]

Second, the fact that money affects a decision does not mean that it is undue inducement, since this would imply that all paid labor involves undue inducement, which is absurd. If my neighbor's 15-year-old son agrees to mow my lawn for $20, it is likely that the offer of money has influenced his decision, because he probably would not mow it for free. However, we would not say that offering to pay my neighbor's son $20 to mow my lawn would be an undue inducement (Resnik 2015). For an inducement to be undue it must have an inappropriate impact on decision-making. So, our question should boil down to: can financial incentives ever inappropriately influence the decision to participate in research?

Some have argued that financial incentives for participation can inappropriately influence decision-making by disrupting a person's judgments concerning risks and benefits. Individuals who are strongly influenced by financial motivations may disregard the risks of research so they can earn money (Macklin 1981; McNeill 1997), and some may lie about their medical conditions in order to qualify for studies or maintain enrollment (Abadie 2010; Resnik and McCann 2015; Fernandez Lynch et al. 2019).

Emanuel (2004, 2005) dismisses this concern on the grounds it does not matter whether financial incentives impact the subject's judgments concerning risks and benefits because an IRB has already approved the research and determined that the risks are reasonable in relation to the benefits. Moreover, federal guidance

[14] Millum and Garnett (2019) have argued that payment can be coercive when the person receiving the money has no alternative source of income and the payment is used to control their will. They refer to this type of situation as "subjection". I agree that situations involving "subjection" are morally problematic, but this is because they may involve exploitation (discussed below) not coercion. Because coercion is often associated with the potential for violence (e.g., assault, rape, kidnapping) I would prefer to limit the use of this word to describe situations that involve threats of harm (Wertheimer and Miller 2008).

discourages IRBs from treating money as a benefit that could offset the risks of research (Grady 2005; Food and Drug Administration 2018). Therefore, according to this *a priori* argument, financial incentives unethically induce human subjects to enroll in studies that place that involve unreasonable risks.

While Emanuel makes some important points, his reasoning is flawed. First, an IRB might approve a study without adequately reviewing the risks, which could expose subjects to unreasonable risks. In that case, the subject could play a key role in protecting themselves from risks by refusing to participate in the study. Second, the subject's judgment of what makes a risk reasonable might differ from the IRB's. The subject's judgment of what makes a study too risky is a function of their values and beliefs, including their risk tolerance. An offer of money might lead someone to take risks that are unreasonable for them, even if the IRB has judged that these risks are reasonable. Thus, financial incentives may undermine the subject's autonomous decision-making even if they do not lead them to take risks that an IRB has judged to be unreasonable (Grady 2005; Resnik 2015).

If we assume that undue inducement is a legitimate ethical concern that cannot be dismissed a priori, we need to ask whether financial incentives are likely to impact risk/benefit judgments. This is an empirical issue concerning the impact of money on human cognition. Empirical studies have shown that financial incentives can increase willingness to participate in research but usually have little impact on risk/benefit judgments. See Box 5.2.

Box 5.2: Empirical Studies of Financial Incentives and Judgments About Risk

Bentley and Thacker (2004)
The authors conducted a controlled experiment to determine the effect of level of payment on the subject's willingness to participate in research and evaluation of risk. They found that payment had positive effects on willingness to participate in research, regardless of the level of risk and that higher payments did not appear to prevent respondents from appreciating the risks of a study.

Halpern et al. (2004)
The authors conducted a controlled experiment to determine the effect of level of payment on the subject's willingness to participate in research and the risks of the research. They found that people were less likely to participate in high-risk research, regardless of the level of payment, suggesting that increasing payments does not alter risk perceptions. Interestingly, wealthier participants were more likely to say that payment was important to them than poorer participants. They concluded that although higher payment motivate research participation, they found no evidence that commonly used payment levels represent undue or unjust inducements.

Dunn et al. (2009)
The authors presented 46 people with schizoaffective disorder with vignettes about research studies with differing levels of payment, potential benefit, and risk. For all vignettes, willingness to participate was negatively associated with risk and positively associated with potential benefit. Respondents were more willing to participate in low-benefit studies if payment was increased. However, a substantial proportion of respondents were not willing to participate in the highest risk studies regardless of level of payment.

(continued)

Box 5.2 (continued)

Singer and Couper (2008)
The authors presented participants with vignettes addressing willingness to participate in a study, level of risk and level of payment and found that larger payments did not induce participants to take higher risks that they would be otherwise unwilling to accept.

Cryder et al. (2010)
The author conducted three experiments to determine willingness to participate in studies at different risk and payment levels. The study also measured risk perception. The authors found that higher paying studies were perceived to be riskier. Higher payments and perceived risks also increase the amount of time spent reviewing the risk information.

Mantzari et al. (2014)
The authors conducted a web-based experiment on 275 participants to assess the impact of financial incentives on willingness to participate in research, time spent viewing risk information, and risk-information processing, assessed by perceived-risk of taking the pill and knowledge of its side-effects. The found that incentives did not affect risk-information processing but was positively associated with time spent viewing risk material. The study found no evidence that incentives interfere with risk-information processing.

Devine et al. (2013)
The authors examined subjects' expectations for payment for research procedures and the relationship between subjects' honesty and payment expectations for participation. Subjects' expectation for payment was positively associated with greater risks. Subjects who admitted to using deception to obtain enrollment in studies were more likely to accept enrollment than those who denied using deception. Among subjects who used deception, refusal to participate was positively associated with level of risk, which suggests that subject have some level of risk aversion regardless of payment. Subjects who use deception were more risk-tolerant than those who do not but may still be able to protect themselves from risks by refusing procedures that exceed their risk tolerance.

Fernandez Lynch et al. 2019
The authors conducted experiments involving 2275 participants to determine whether there is an association between payment amount and deception frequency. While the researchers found significant levels of deception by participants, they found no association between level of payment and deception.

Halpern et al. (2021)
The authors conducted experiments involving 654 participants to determine whether level of payment affects perception of risk. They found that there were no significant effects of incentive size time spent reviewing the risk sections of consent forms, perceived research risks, trial understanding, perceived coercion, or therapeutic misconceptions. The concluded that incentives can increased participants willingness to participate but did not undermine their reasoning about risks.

Bickman et al. (2021)
The investigators did a survey involving 643 subjects and experimental vignette involving 534 participants to understand the relationship between willingness to participate and perceived risk burden. The found that incentive level is associated with willingness to participate but does not undermine risk perceptions.

Largent et al. (2022)
The authors analyzed recruitment conversations between study coordinator and research participants and conducted semi-structured interviews with participants. Many participants indicated the money was a factor in their decision to participate in the study but did not use reasoning suggesting that they were unduly induced to participate.

For example, Dunn et al. (2009) found that patients formed judgements of risk independent of the amount of money offered and were unwilling to participate in studies they perceived as too risky, even for more money; Cryder et al. (2010) found that financial incentives do not cause most subjects to discount risks but have the opposite effect: participants in this study judged higher-paying studies to be riskier than lower-paying ones and spent more time learning about studies they viewed as high-risk than those they perceived as low-risk; and Largent et al. (2022) found that money motivated participants to enroll in clinical trials but that did not undermine risk/benefit reasoning in a way that would constitute undue inducement.

While it is important for researchers to continue to explore the impact that financial incentives have on prospective research subjects' decision-making, the evidence obtained so far suggests that money does not adversely impact most subjects' risk/benefit judgments. However, it is possible that for some subjects, such as impoverished individuals, financial incentives do significantly impact risk/ benefit judgments. For this reason, it is important for researchers and IRBs to be mindful of the possibility that excessive financial incentives may unduly induce some individuals to participate in research (Resnik 2015; Gelinas et al. 2018).

Though undue inducement may be a concern with offering subjects too much money, exploitation may an issue with offering them too little (Shamoo and Resnik 2006). Exploitation involves taking unfair advantage of a person or group of people in a transaction or relationship (Wertheimer 1999, 2011). From a Kantian perspective, exploitation is unethical because it disrespects human dignity and autonomy by treating a person as mere means to obtain some goal (Sample 2003).

Exploitation may occur when one of three conditions are met: (a) the exploiter harms the exploitee in the transaction or relationship; (b) the exploitee does not consent to the transaction or relationship; and (c) the exploiter obtains more than their fair share of the benefits of the transaction or relationship. Some forms of exploitation, such as slavery, involve all three conditions (Wertheimer 2011). Exploitation can happen even when both parties consent and neither is harmed if the exploiter derives an unfair share of benefits from the transaction or relationship (Wertheimer 1999). For example, suppose a person inherits a painting of a cowboy on a horse from their grandmother and decides to sell it an antique dealer for $100. The antique dealer immediately recognizes that the painting is an original Frederick Remington work of art worth at least $500,000. We would say that the antique dealer exploited the seller because transaction, though consensual, distributed the benefits unfairly.

Although exploitation is *prima facie* wrong, we might still decide to permit some forms of exploitation because they offer important benefits to society or because banning them would interfere with economic or other rights (Wertheimer 1999). For example, one might concede that pawn brokers often exploit people who receive loans from them but decide that this business should not be banned because it offers important benefits to society and people have a right to put items up for collateral at much less than their market value.

Exploitation could occur in research with human subjects if researchers, institutions, or sponsors do not share the benefits of research with participants fairly (Resnik 2003). Human research studies often yield significant benefits for researchers (e.g., salary and career advancement), institutions (e.g., research funding), and sponsors (e.g., development of commercial products). In clinical research, subjects may receive a fair share of benefits by gaining access to medical treatment. In research on healthy volunteers who will not receive medical treatment, subjects can receive a fair share of the benefits of research by earning money for their participation. Although federal guidance discourages IRBs from treating money as a benefit, most people consider money to be beneficial (Wertheimer 2011). Failing to offer healthy research subjects enough money could therefore constitute exploitation (Shamoo and Resnik 2006). As noted earlier, some subjects may participate in Phase I drug trials on healthy volunteers because they need the money (Abadie 2010). Underpaying healthy volunteers in Phase I drug trials would be analogous to underpaying socioeconomically disadvantaged workers who perform risky manual labor (Shamoo and Resnik 2006; Elliott and Abadie 2008). Thus, when it comes to paying research subjects, researchers must avoid undue inducement (i.e., overpayment) and exploitation (i.e., underpayment) (Resnik 2015; Gelinas et al. 2018).

Exploitation could occur when researchers compensate children for participating in research, since parents may enroll children in studies to obtain money for themselves. Payments to children vary considerably, depending on the nature of the research, the procedures, the amount of time or inconvenience involved, and so on. One study found that children may receive from $1 to $1000 for their participation. Payments may take to form of cash, savings bonds, or gift cards (Weise et al. 2002). An additional concern with payments in pediatric research is that parents may coerce their children into participating in studies (Grady 2005). Some amount of financial compensation is necessary in pediatric research to encourage parents to enroll their children in studies, but compensation levels should not be excessive (Grady 2005). Some commentators have argued that payments to children should reimburse parents for their travel and time and provide children a token amount of compensation to recognize their contribution, but that payments should not be so large that they encourage parents to exploit their children for money (Committee on Drugs, American Academy of Pediatrics 1995).

A final issue related to financial incentives is that payments to research subjects may result in disproportionate (or biased) enrollment of socioeconomically disadvantaged individuals, which could undermine the generalizability of the results (Resnik 2015). Generalizability is important for enhancing the scientific and social value of research. Unless researchers have a sound scientific justification for focusing on important issues pertaining to a specific population (e.g., studying heart disease in women because it has been under-studied in this population), they should use research methods that promote the generalizability of their results. Disproportionate enrollment of socioeconomically disadvantaged individuals could affect generalizability if socioeconomically disadvantaged individuals have traits which are different from the general population. While most studies seek to achieve proportional representation of subjects with different demographic characteristics (e.g., gender, race, or ethnicity) most do not take socioeconomic status into account. There has been very little research on the relationship between payment and disproportional enrollment. However, one study found that payment does not increase the proportion of socially deprived or elderly people enrolled in clinical trials (Jennings et al. 2015).

Disproportional enrollment may also raise issues of justice if the composition of the study population results in an inequitable distribution of the risks and benefits of research (National Commission 1979). For example, one might argue that it would be unfair to include mostly socioeconomically disadvantaged individuals in a study that is likely to benefit only people who are well-off (Elliott and Abadie 2008).

There are no easy solutions to the problem of disproportionate enrollment of socioeconomically disadvantaged individuals in research. One way of dealing with this problem would be to offer a level of compensation proportional to the subjects' level of income so that people with higher incomes would earn more for their research participation. While this strategy might encourage more people with higher incomes to enroll in research, it would be blatantly unfair, because subjects would not receive equal pay for equal work (Resnik 2015). Another strategy would be to actively recruit subjects from higher income groups, but this strategy might not be very effective if members of those groups are not offered enough money to make their participation worthwhile.

It is worth mentioning that the problem of disproportionate enrollment of socioeconomically disadvantaged individuals is not likely to be as much of a problem in studies that offer subjects medical benefits, such as Phase II, III, and IV clinical trials, since patients with higher incomes are likely to be interested in enrolling studies that may benefit their health. Most diseases do not discriminate on the basis of income or wealth: rich and poor people develop cancer, heart disease, diabetes, Alzheimer's disease, and so on. The problem of disproportionate enrollment is likely to be most acute in Phase I trials and other types of research that do not provide subjects with direct medical benefits (Elliott and Abadie 2008; Fisher 2020).

Thinking about financial incentives from the perspective of trust, underpayment is likely to have more of a negative impact on trust than overpayment, because subjects who are not paid enough money may feel that they are being exploited, underappreciated, or otherwise mistreated. Ensuring that subjects are adequately compensated is a way of recognizing the value of their contributions to research. Subjects contribute their time and labor to research and often take significant risks or endure pain or other hardships in to help researchers achieve their goals. While it is important for researchers and IRBs to be mindful of how financial incentives may impact informed consent and recruitment, they should not allow these concerns to overshadow their obligation to ensure that subjects receive fair compensation for their efforts. To promote trust, consent documents should clearly describe compensation plans, e.g., how much money subjects will receive for completing different study activities, and whether payment will be pro-rated if they do not complete all study activities.

One might argue that policies which restrict the amount of money that research subjects can be paid are paternalistic and overprotective (see discussion at the end of Chap. 2). Ordinarily, we do not place limits on the amount of money that one may be paid for performing a job. If a coal company is having trouble hiring enough workers and it wants to double coal miners' wages to recruit new employees, we would have no objections to this. We would not be concerned that offering coal miners more money to perform a risky job would be an undue inducement. Indeed, we would probably be more concerned if the mining company does not offer miners enough money to perform this work, since this could be exploitative. Minimum wages laws in the U.S. and other countries were enacted to ensure that workers are adequately paid and not exploited. We have no laws concerning maximum wages. In capitalistic societies, upper limits on wages are determined by the free market, not by government regulations or ethical guidelines.

If we place limits on the amount of money that competent, adult human subjects may earn in research, we are treating research participation differently from other forms of labor. Research subjects would be protected from financial incentives that could unduly induce them to enroll in studies, but coal miners would require no such protections related to their work. We normally assume that competent adults can make sound decisions concerning payment for employment. If a coal company wants to offer miners more money, we would not say that the prospective employees cannot make sounds decisions concerning working for the company. We would usually assume that prospective employees can consider the risks and hardships associated with coal mining and decide whether the amount of money offered by the company is adequate compensation for those risks and hardships.

What, if anything, justifies treating human research participation differently from other forms of employment? Why are we concerned about paying competent, adult human subjects too much money but we are not concerned about paying coal miners too much money? There are several potential answers to these questions. The first is that participation in human research is so risky that it needs to be treated differently than other types of human labor. But this answer makes a false assumption because evidence shows that research participation is no more risky than other types of work (Wendler 2011). Indeed, in some ways participating in research is much safer than other forms of labor involving risks, such as coal mining, construction work, firefighting, etc.

The second is that the decision to participate in research is so difficult (i.e., fraught with so many complexities and conflicting values) that we need to protect adults from making poor choices. This answer also rests on a mistake because deciding whether to participate in research is no more difficult than other choices we allow competent adults to make without excessive protection, such as the decision to buy a new home, enlist in the military, marry, or donate a kidney.

The third is that paying research subjects treats the human body as a commercial product, which threatens human dignity (Radin 1996; Andrews and Nelkin 2001). If people are not paid for donating organs, they should also not be paid for providing a blood sample for research. This answer is also unsatisfactory because a great deal of human research does not involve treating the body as a commercial product in any form. For example, paying someone to answer a survey question does not involve commercializing the human body. Moreover, paying someone for providing a blood sample could be viewed as compensating them for their time or inconvenience, not for their blood.

The fourth is that we think that people should participate in research because they want to do some good for society, not because they want to earn money. Research participation should be an altruistic activity. Payments should be limited to direct reimbursement for expenses. Wertheimer (2011, p. 118) characterizes this position as the idea that "payments are a necessary evil, but an evil nonetheless." The trouble with this position is that there is no convincing ethical reason why we should think that research participation must be an altruistic activity. Subjects often participate in clinical trials for selfish reasons, such as to gain access to experimental treatments that may benefit them (Menikoff 2006). If we have no problem with patients participating in clinical trials to obtain medical benefits, why should we object to healthy volunteers participating in research to earn money? While it would be a good thing if people participated in research only for altruistic reasons, there is nothing inherently wrong with participating in research for selfish reasons, including the desire to earn money.

A fifth possible answer places our moral qualms about payment in historical context. In Chap. 2, I argued that one reason why the U.S. and other countries have adopted paternalistic and protective research regulations and guidelines is to promote public trust. The regulations and guidelines help to assure the public that rules are in place to protect the rights and welfare of human subjects and to avoid repeating the mistakes of the past. Restrictions on payments are necessary, one might argue, to assure the public that human subjects will be protected from taking excessive risks or compromising their values to earn money. The problem with this answer is that compensating research subjects for their participation is not likely to expose them to unreasonable risks or undermine their autonomy, since there are rules in place, such as requirements concerning IRB review, risk minimization, and informed consent, which are designed to protect the rights and welfare of human subjects (Emanuel 2005; Wertheimer 2011). If these rules did not exist, then paying research subjects for their participation would indeed raise major ethical concerns. But since we have such rules, it is difficult to understand why so many people have moral qualms about paying research subjects. Perhaps the payment guidelines we have developed are an overreaction to past abuses, and our concerns about paying research subjects are misplaced (Wertheimer 2011) (Box 5.3).

Box 5.3: Payment Approaches

Grady (2005) describes several different approaches to paying research subjects, each of which have advantages and disadvantages. According to the **market approach**, institutions should pay subjects according to the free market value of their labor. The amount of money paid to subjects would be a function of supply and demand. One advantage of this approach is that it would allow institutions and sponsors to adjust compensation levels to deal with problems related to recruitment. For example, if a researcher is having trouble recruiting subjects for a Phase I drug study, the sponsor could increase the payment level. Another advantage is that it allows institutions to adjust payments according to the level of risk, pain, or inconvenience of a study. For example, a study could offer more money for a skin biopsy than a blood draw. A final advantage is that this approach treats human research participation as similar to other forms of paid labor, which might encourage institutions and sponsors to offer the kinds of protections for human subjects which they provide to other paid employees, such as worker's compensation. However, since this approach places no upper or lower limits on payments, it has the potential for both undue inducement and exploitation (Resnik 2015).

According to the **reimbursement** approach, payments should only reimburse subjects for their expenses, such as travel and lost wages (Grady 2005). An advantage of this approach is that it probably would not lead to undue inducement, since subjects would not be paid more money than they are already earning. A disadvantage of this approach is that it would be unfair, since subjects with higher incomes would be paid more than those with lower incomes, due to differences in lost wages. Another disadvantage of this approach is that it would not allow researchers to increase compensation levels to boost recruitment (Resnik 2015).

According to the **token appreciation** approach, institutions should pay subjects a nominal amount of money as a sign gratitude for their help (Grady 2005). Because payment levels would be low, this approach would not lead to undue inducement, but it could lead to exploitation. Low payments levels could also hurt recruitment, especially if the research involves significant pain, discomfort, or risk.

According to the **wage-labor model**, subjects should be paid a standard wage for their time roughly equivalent to what someone would earn for unskilled labor (Grady 2005). This approach would probably not lead to undue inducement or exploitation, since compensation levels would not be very high or unfairly low (i.e., they would be a little above minimum wage). However, this approach would not allow researchers to increase compensation levels to address problems with recruitment, nor would it allow researchers to adjust payments according to the amount of pain, discomfort or risk involved in research.

It may the case the best way compensating research subjects is to use some **combination** of these different approaches, depending on the situation. For example, for pediatric research the child could receive a gift card as a token of appreciation for his or her contribution and the parent could receive reimbursement for expenses. For survey research, subjects could receive a payment based on the hourly rate for unskilled labor, which could be increased if recruitment lags. For a laboratory assay development study involving the collection of different types of biological samples, subjects could be paid per type of sample provided and reimbursed for their time and travel.

5.12 The Right to Withdraw

The right to withdraw from research participation has been an ethical requirement since it was incorporated into the Nuremberg Code in 1949. Federal research regulations require that researchers inform subjects that they "may discontinue participation at any time without penalty or loss of benefits to which the subject is otherwise entitled (45 CFR 46.116a8)." The right to withdraw is also part of the Helsinki Declaration (World Medical Association 2013) and the Council for the International Organizations of Medical Sciences guidelines (2016). Informed consent documents typically inform subjects that they can withdraw at any time for any reason without penalty and that they may refuse to participate in any research activity (Edwards 2005; Schaefer and Wertheimer 2010). Consent documents also usually inform subjects about requesting to have their biological samples destroyed and data removed from the study if they withdraw. However, it may be difficult or impossible to fulfill this request if samples or data are anonymous or have already been shared with other researchers (Melham et al. 2014). Payments may be pro-rated to ensure that subjects who withdraw receive the money that they have earned (Grady 2005). The right to withdraw is usually interpreted as inalienable, which means that one cannot voluntarily waive or restrict this right (Edwards 2005; Chwang 2008).

Although the right to withdraw is widely regarded as an important legal protection for human research subjects, one might still question its ethical justification (Schaefer and Wertheimer 2010). The first argument against this right is that it is incompatible with our understanding of the relationship between autonomy and contractual responsibility (Edwards 2005; Chwang 2008). Part of what it means to be an autonomous decision-maker is to accept responsibility for one's choices and actions. We normally hold competent adults responsible for their choices related to informal agreements or formal contracts. Part of what it means to make an agreement is to limit some of your rights so you can obtain the benefits of the agreement. The other party has the right deny you certain benefits or penalize you if you do not live up to your end of the bargain. If the contract is legally valid, the courts will enforce it. For example, if you sign a contract to rent an apartment and do not pay your rent on time, the landlord may take steps to evict you. If you break the lease before it expires, you may face a financial penalty. Informed consent can be viewed as an agreement between the subject and researcher (Berg et al. 2001; Manson and O'Neill 2007). If people are allowed to limit their rights in other types of contracts, they should also be able to limit their rights in research agreements, including theirs right to withdraw. Thus, the right to withdraw should alienable (Chwang 2008).

The first argument is not a compelling reason to reject an inalienable right to withdraw, because other considerations, such as protecting human life or welfare, may sometimes take precedence over allowing people to alienate some of their rights (McConnell 2010). One might argue that states can place some limits on the ability to alienate one's right to life to preserve life and protect people from harm (Feinberg 1986). For example, dueling, Russian roulette, euthanasia, and assisted suicide are illegal in many states. As John Stuart Mill famously argued, one should not be allowed to alienate one's right to liberty by selling oneself into slavery (Mill 1978). Though Mill claimed that selling oneself into slavery should not be permitted because it would undercut the very foundation of liberty, one could also argue that it should not be permitted because it would threaten one's life or health. Likewise, one might argue that we should not allow people to alienate their right to withdraw to protect them from harm (McConnell 2010).

The second argument against an inalienable right to withdraw is that sometimes withdrawing may place the research subject at risk (Resnik and Ness 2012). For example, if a subject has been taking an experimental medication in a clinical trial, and stopping the medication suddenly is likely to result in adverse withdrawal effects, then they may place their own health at risk if they stop participating in the trial. In a case like this, the right to withdraw could be restricted to protect subjects from harm. Subjects would be allowed to stop participating in a study only if they follow a medication withdrawal protocol.

The second argument is not compelling reason for rejecting an inalienable right to withdraw, since subjects could taper off their medications without participating in the study. Patients who want to withdraw would be allowed to take the experimental medication under a doctor's supervision even though they would no longer be research subjects.

A third argument is that withdrawing from research may also place other people at risk in some rare situations (Resnik and Ness 2012). For example, consider the xenotransplantation research discussed in Chap. 1. A subject who receives a pig organ and then withdraws from the study would increase their risk of transmitting zoonoses to the public because they would not longer be monitored. At the very least, it seems reasonable to agree to continued monitoring for zoonoses even if they withdraw from the study.

A fourth argument is that withdrawing from research may compromise the study itself if loss of a research subject will undermine the statistical power of the study (Chwang 2008). This is a weak argument against the right to withdraw since (a) the loss of one subject is not likely to significantly affect the statistical significance of the results; and (b) it may be possible to enroll additional subjects to meet the target sample size.

Given that there are some arguments against an inalienable right to withdraw from research, what are the arguments in favor of it? One might be tempted to say that granting subjects a right to withdraw is necessary to protect them from coercion. A person should not face a threat of harm or loss of benefit if they decide to stop participating in a study. However, this position confuses coercion with the enforcement of prior agreements (Chwang 2008). For example, if I threaten to damage your car if you do not give me ride to my home, this would be coercive. However, if I agree to pay you for gas if you give me a ride home and you demand gas money once we arrive, this would not be coercive because you are enforcing our prior agreement. One could argue that penalizing a subject for withdrawing is a way of enforcing this agreement: it is not coercion.

A better justification is that an inalienable right to withdraw is necessary to protect subjects from harm. Subjects may not be able to anticipate how they will react to different research interventions, medications, or procedures and they may change their minds about participation when they experience some undesirable effects, such as pain, discomfort, nausea, dizziness, or claustrophobia (Edwards 2005; Schaefer and Wertheimer 2010). If a subject develops intolerable symptoms or becomes anxious about possible adverse effects, he or she should be able to withdraw without penalty. If a subject faces a penalty for withdrawing, he or she may place his or her own health at risk by remaining in the study or not telling the researcher about the problems he or she is experiencing.

Approaching these issues the perspective of trust, we can construct another rationale for the right to withdraw: the right is justified because it helps to promote trust between research subjects and researchers and institutions, as well as the public's trust in the research enterprise (Schaefer and Wertheimer 2010). Subjects may want to be assured that they can withdraw from a study without penalty, especially if they are experiencing adverse effects. This assurance can enhance their trust in the investigator and the institution. The right to withdraw can promote the public's trust in the research enterprise by helping researchers, institutions, and sponsors to avoid potentially disastrous situations involving harm to human subjects. For example, the people could become outraged if they learn that a subject died in an experiment because they remained in the study to avoid a financial penalty for withdrawing. To avoid this adverse negative publicity associated with type of potential tragedy, subjects should be able to withdraw at any time without penalty.

Before concluding this section, it is worth noting that by granting human subjects an inalienable right to withdraw we are providing them with more protection from harm than people normally receive outside of the research context. Outside of the research context, we allow competent adults to face financial penalties for breaking contracts. In some cases, people may place themselves at some risk of harm to avoid these penalties. For example, a person who has developed an allergy to the carpeting in his apartment might remain in the dwelling until their lease expires to avoid a financial penalty. Likewise, a professional athlete might place their health at risk in order to penalties built into his contract. Though some might regard these added safeguards for human subjects as paternalistic and overprotective (Chwang 2008), I have argued that they can justified as a means of protecting subjects from harm and promoting trust.

5.13 Conclusion

In this chapter, I have applied my trust-based approach to a variety of issues related to informed consent in research, including disclosure, documentation, consent by parties other than the subject, research without consent, opt-out consent, general concept, deception, assent, payment for research participation, and the right to withdraw. Although the federal regulations provide considerable guidance pertaining to informed consent, ethical issues still arise due to controversies concerning the interpretation of the regulations, conflicts between ethical principles (e.g., respect for human dignity vs. beneficence) and the need to acknowledge local customs and traditions concerning consent. I have argued that trust plays a key role throughout the consent process and that reflecting on the importance of promoting trust can help us address ethical issues pertaining to consent. In the next chapter, I will examine the relationship between trust and privacy and confidentiality.

References

Abadie, R. 2010. *The professional Guinea pig: Big pharma and the risky world of human subjects*. Durham: Duke University Press.

Afifi, R.Y. 2007. Biomedical research ethics: An Islamic view part II. *International Journal of Surgery* 5 (6): 381–383.

Albala, I., M. Doyle, and P.S. Appelbaum. 2010. The evolution of consent forms for research: A quarter century of changes. *IRB* 32 (3): 7–11.

Andrews, L., and D. Nelkin. 2001. *Body bazaar: The market for human tissue in the biotechnology age*. New York: Crown.

Annas, G. 2000. Rules for research on human genetic variation—Lessons from Iceland. *New England Journal of Medicine* 342 (24): 1830–1833.

Appelbaum, P.S., and C.W. Lidz. 2008. The therapeutic misconception. In *The Oxford handbook of clinical research ethics*, ed. E.J. Emanuel, C. Grady, R.A. Crouch, R.K. Lie, F.G. Miller, and D. Wendler, 633–644. New York: Oxford University Press.

Appelbaum, P.S., L.H. Roth, and C. Lidz. 1982. The therapeutic misconception: Informed consent in psychiatric research. *International Journal of Law and Psychiatry* 5 (3–4): 319–329.

Appelbaum, P.S., L.H. Roth, C.W. Lidz, P. Benson, and W. Winslade. 1987. False hopes and best data: Consent to research and the therapeutic misconception. *Hastings Center Report* 17 (2): 20–24.

Bathe, O.F., and A.L. McGuire. 2009. The ethical use of existing samples for genome research. *Genetics in Medicine* 11 (10): 712–715.

Beauchamp, T.L., and J.F. Childress. 2012. *Principles of biomedical ethics*. 7th ed. New York: Oxford University Press.

Bentley, J.P., and P.G. Thacker. 2004. The influence of risk and monetary payment on the research participation decision making process. *Journal of Medical Ethics* 30 (3): 293–298.

Berg, J.W., P.S. Appelbaum, L.S. Parker, and C.W. Lidz. 2001. *Informed consent: Legal theory and clinical practice*. 2nd ed. New York: Oxford University Press.

Beskow, L.M., L. Lin, C.B. Dombeck, E. Gao, and K.P. Weinfurt. 2017. Improving biobank consent comprehension: A national randomized survey to assess the effect of a simplified form and review/retest intervention. *Genetics in Medicine* 19 (5): 505–512.

Bickman, L., H.J. Domenico, D.W. Byrne, R.N. Jerome, T.L. Edwards, M. Stroud, L. Lebo, K. Mcguffin, C.H. Wilkins, and P.A. Harris. 2021. Effects of financial incentives on volunteering for clinical trials: A randomized vignette experiment. *Contemporary Clinical Trials* 110: 106584.

Bothun, L.S., S.E. Feeder, and G.A. Poland. 2021. Readability of participant informed consent forms and informational documents: From phase 3 COVID-19 vaccine clinical trials in the United States. *Mayo Clinic Proceedings* 96 (8): 2095–2101.

Brock, D.W. 2008. Philosophical justifications of informed consent in research. In *The Oxford handbook of clinical research ethics*, ed. E.J. Emanuel, C. Grady, R.A. Crouch, R.K. Lie, F.G. Miller, and D. Wendler, 606–612. New York: Oxford University Press.

Buchanan, A.E., and D.W. Brock. 1990. *Deciding for others: The ethics of surrogate decision-making*. Cambridge: Cambridge University Press.

Capron, A. 2008. Legal and regulatory standards of informed consent in research. In *The Oxford handbook of clinical research ethics*, ed. E.J. Emanuel, C. Grady, R.A. Crouch, R.K. Lie, F.G. Miller, and D. Wendler, 613–632. New York: Oxford University Press.
Caspar, E.A., J.F. Christensen, A. Cleeremens, and P. Haggard. 2016. Coercion changes the sense of agency in the human brain. *Current Biology* 26: 1–8.
Chwang, E. 2008. Against the inalienable right to withdraw from research. *Bioethics* 22 (7): 370–378.
Committee on Drugs, American Academy of Pediatrics. 1995. Guidelines for the ethical conduct of studies to evaluate drugs in pediatric populations. *Pediatrics* 95 (2): 286–294.
Corbie-Smith, G., S.B. Thomas, M.V. Williams, and S. Moody-Ayers. 1999. Attitudes and beliefs of African Americans toward participation in medical research. *Journal of General Internal Medicine.* 14 (9): 537–546.
Corbie-Smith, G., S.B. Thomas, and D.M. St George. 2002. Distrust, race, and research. *Archives of Internal Medicine* 162 (21): 2458–2463.
Council for the International Organizations of Medical Sciences. 2002. *International ethical guidelines for biomedical research involving human subjects*. Geneva: Council for the International Organizations of Medical Sciences.
———. 2016. *International ethical guidelines for health-related research involving humans*. Geneva: Council for the International Organizations of Medical Sciences.
Courtland, S.D., G. Gauss, and D. Schmidtz. 2022. Liberalism. In *Stanford encyclopedia of philosophy*. Available at: https://plato.stanford.edu/entries/liberalism/. Accessed 21 June 2024.
Cryder, C.E., J.A. London, K.G. Volpp, and G. Loewenstein. 2010. Informative inducement: Study payment as a signal of risk. *Social Science and Medicine* 70 (3): 455–464.
de Man, Y., Y. Wieland-Jorna, B. Torensma, K. de Wit, A.L. Francke, M.G. Oosterveld-Vlug, and R.A. Verheij. 2023. Opt-in and opt-out consent procedures for the reuse of routinely recorded health data in scientific research and their consequences for consent rate and consent bias: Systematic review. *Journal of Medical Internet Research* 25: e42131.
de Melo-Martín, I., and A. Ho. 2008. Beyond informed consent: The therapeutic misconception and trust. *Journal of Medical Ethics* 34 (3): 202–205.
Devine, E.G., M.E. Waters, M. Putnam, C. Surprise, K. O'Malley, C. Richambault, R.L. Fishman, C.M. Knapp, E.H. Patterson, O. Sarid-Segal, C. Streeter, L. Colanari, and D.A. Ciraulo. 2013. Concealment and fabrication by experienced research subjects. *Clinical Trials* 10 (6): 935–948.
Dickert, N.W., and J. Sugarman. 2005. Ethical goals of community consultation in research. *American Journal of Public Health* 95 (7): 1123–1127.
———. 2007. Getting the ethics right regarding research in the emergency setting: Lessons from the PolyHeme study. *Kennedy Institute of Ethics Journal* 17 (2): 153–169.
Dickert, N.W., N. Eyal, S.F. Goldkind, C. Grady, S. Joffe, B. Lo, F.G. Miller, R.D. Pentz, R. Silbergleit, K.P. Weinfurt, D. Wendler, and S.Y.H. Kim. 2017. Reframing consent for clinical research: A function-based approach. *American Journal of Bioethics* 17 (12): 3–11.
Dunn, L.B., D.S. Kim, I.E. Fellows, and B.W. Palmer. 2009. Worth the risk? Relationship of incentives to risk and benefit perceptions and willingness to participate in schizophrenia research. *Schizophrenia Bulletin* 35 (4): 730–737.
Edwards, S.J. 2005. Research participation and the right to withdraw. *Bioethics* 19 (2): 112–130.
Elliott, C., and R. Abadie. 2008. Exploiting a research underclass in phase 1 clinical trials. *New England Journal of Medicine* 358 (22): 2316–2317.
Emanuel, E.J. 2004. Ending concerns about undue inducement. *Journal of Law, Medicine & Ethics* 32 (1): 100–105.
———. 2005. Undue inducement: Nonsense on stilts? *American Journal of Bioethics* 5 (5): 9–13.
Faden, R.R., T.L. Beauchamp, and N.M.P. King. 1986. *A history and theory of informed consent*. New York: Oxford University Press.
Feinberg, J. 1986. *Harm to self*. New York: Oxford University Press.

Fernandez Lynch, H., S. Joffe, H. Thirumurthy, D. Xie, and E.A. Largent. 2019. Association between financial incentives and participant deception about study eligibility. *JAMA Network Open* 2 (1): e187355.

Finn, P., and M. Jakobsson. 2007. Designing ethical phishing experiments. *IEEE Technology and Society Magazine* 26: 46–58.

Finniss, D.G., T.J. Kaptchuk, F. Miller, and F. Benedetti. 2010. Biological, clinical, and ethical advances of placebo effects. *Lancet* 375 (9715): 686–695.

Fisher, J.A. 2020. *Adverse events: Race, inequality, and the testing of new pharmaceuticals*. New York: NYU Press.

Fitzpatrick, E.F., A.L. Martiniuk, H. D'Antoine, J. Oscar, M. Carter, and E.J. Elliott. 2016. Seeking consent for research with indigenous communities: a systematic review. *BMC Medical Ethics* 17 (1): 65.

Flory, J., and E. Emanuel. 2004. Interventions to improve research participants' understanding in informed consent for research: A systematic review. *Journal of the American Medical Association* 292 (13): 1593–1601.

Food and Drug Administration. 2013. Exception from Informed Consent Requirements for Emergency Research Guidance for Institutional Review Boards, Clinical Investigators, and Sponsors. Available at: https://www.fda.gov/media/80554/download. Accessed 29 June 2024.

———. 2018. Payment and reimbursement to research subjects: Guidance for institutional review boards and clinical investigators. Available at: https://www.fda.gov/regulatory-information/search-fda-guidance-documents/payment-and-reimbursement-research-subjects. Accessed 30 June 2024.

Gelinas, L., A. Wertheimer, and F.G. Miller. 2016. When and why is research without consent permissible? *Hastings Center Report* 46 (1): 1–9.

Gelinas, L., E.A. Largent, G. Cohen, S. Kornetsky, B.E. Bierer, and H. Fernandez Lynch. 2018. A framework for ethical payment to research participants. *New England Journal of Medicine* 378 (8): 766–771.

Gelsinger, P., and A.E. Shamoo. 2008. Eight years after Jesse's death: Are human research subjects any safer? *Hastings Center Report* 38 (2): 25–27.

Glannon, W. 2006. Phase I oncology trials: Why the therapeutic misconception will not go away. *Journal of Medical Ethics* 32 (5): 252–255.

Grady, C. 2001. Money for research participation: Does in jeopardize informed consent? *American Journal of Bioethics* 1 (2): 40–44.

———. 2005. Payment of clinical research subjects. *Journal of Clinical Investigation* 115 (7): 1681–1687.

Grady, C., N. Dickert, T. Jawetz, G. Gensler, and E. Emanuel. 2005. An analysis of U.S. practices of paying research participants. *Contemporary Clinical Trials* 26 (3): 365–375.

Grady, C., G. Touloumi, A.S. Walker, M. Smolskis, S. Sharma, A.G. Babiker, N. Pantazis, J. Tavel, E. Florence, A. Sanchez, F. Hudson, A. Papadopoulos, E. Emanuel, M. Clewett, D. Munroe, E. Denning, and INSIGHT START Informed Consent Substudy Group. 2017. A randomized trial comparing concise and standard consent forms in the START trial. *PLoS One* 12 (4): e0172607.

Guillemin, M., L. Gillam, E. Barnard, P. Stewart, H. Walker, and D. Rosenthal. 2016. "We're checking them out": Indigenous and non-Indigenous research participants' accounts of deciding to be involved in research. *International Journal of Equity Health* 15: 8.

Gulcher, J.R., and K. Stefánsson. 2000. The Icelandic Healthcare Database and informed consent. *New England Journal of Medicine* 342 (24): 1827–1830.

Hadden, K.B., L.Y. Prince, T.D. Moore, L.P. James, J.R. Holland, and C.R. Trudeau. 2017. Improving readability of informed consents for research at an academic medical institution. *Journal of Clinical Translational Science* 1 (6): 361–365.

Halpern, S.D., J.H. Karlawish, D. Casarett, J.A. Berlin, and D.A. Asch. 2004. Empirical assessment of whether moderate payments are undue or unjust inducements for participation in clinical trials. *Archives of Internal Medicine* 164 (7): 801–803.

Halpern, S.D., M. Chowdhury, B. Bayes, E. Cooney, B.L. Hitsman, R.A. Schnoll, S.F. Lubitz, C. Reyes, M.S. Patel, S.R. Greysen, A. Mercede, C. Reale, F.K. Barg, K.G. Volpp, J. Karlawish, and A.J. Stephens-Shields. 2021. Effectiveness and ethics of incentives for research participation: 2 randomized clinical trials. *JAMA Internal Medicine* 181 (11): 1479–1488.

Henderson, G.E., L.R. Churchill, A.M. Davis, M.M. Easter, C. Grady, S. Joffe, N. Kass, N.M. King, C.W. Lidz, F.G. Miller, D.K. Nelson, J. Peppercorn, B.B. Rothschild, P. Sankar, B.S. Wilfond, and C.R. Zimmer. 2007. Clinical trials and medical care: Defining the therapeutic misconception. *PLoS Medicine* 4 (11): e324.

Hill, E.M., E.L. Turner, R.M. Martin, and J.L. Donovan. 2013. "Let's get the best quality research we can": Public awareness and acceptance of consent to use existing data in health research: A systematic review and qualitative study. *BMC Medical Research Methodology* 13:72.

Jennings, C.G., T.M. MacDonald, L. Wei, M.J. Brown, L. McConnachie, and I.S. Mackenzie. 2015. Does offering an incentive payment improve recruitment to clinical trials and increase the proportion of socially deprived and elderly participants? *Trials* 16: 80.

Juengst, E.T. 1998. Groups as gatekeepers to genomic research: Conceptually confusing, morally hazardous, and practically useless. *Kennedy Institute of Ethics Journal* 8 (2): 183–200.

Junghans, C., G. Feder, H. Hemingway, A. Timmis, and M. Jones. 2005. Recruiting patients to medical research: Double blind randomised trial of "opt-in" versus "opt-out" strategies. *British Medical Journal* 331 (7522): 940.

Karlawish, J.H. 2008. Emergency research. In The Oxford handbook of clinical research ethics, ed. E.J. Emanuel, C. Grady, R.A. Crouch, R.K. Lie, F.G. Miller, and D. Wendler, 280–289. New York: Oxford University Press.

King, J.S., and B. Moulton. 2006. Rethinking informed consent: The case for shared medical decision-making. *American Journal of Law and Medicine* 32 (4): 429–501.

Krogstad, D.J., S. Diop, A. Diallo, F. Mzayek, J. Keating, O.A. Koita, and Y.T. Touré. 2010. Informed consent in international research: The rationale for different approaches. *American Journal of Tropical Medicine and Hygiene* 83 (4): 743–747.

Kudaibergenova, T. 2024. The ethics of research informed consent from the Kyrgyz perspective: A qualitative study. *Developing World Bioethics*. May 14. https://doi.org/10.1111/dewb.12451. Epub ahead of print.

Largent, E.A., C. Grady, F.G. Miller, and A. Wertheimer. 2012. Money, coercion, and undue inducement: Attitudes about payments to research participants. *IRB* 34 (1): 1–8.

Largent, E.A., W. Eriksen, F.K. Barg, S.R. Greysen, and S.D. Halpern. 2022. Participants' perspectives on payment for research participation: A qualitative study. *Ethics and Human Research* 44 (6): 14–22.

Larson, E., G. Foe, and R. Lally. 2015. Reading level and length of written research consent forms. *Clinical and Translational Science* 8 (4): 355–356.

Lewis, C., M. Clotworthy, S. Hilton, C. Magee, M.J. Robertson, L.J. Stubbins, and J. Corfield. 2013. Consent for the use of human biological samples for biomedical research: A mixed methods study exploring the UK public's preferences. *BMJ Open* 3 (8): e003022.

Liamputtong, P., ed. 2008. *Doing cross-cultural research: Ethical and methodological perspectives*. Dordrecht: Springer.

Lidz, C.W., K. Albert, P. Appelbaum, L.B. Dunn, E. Overton, and E. Pivovarova. 2015. Why is therapeutic misconception so prevalent? *Cambridge Quarterly of Healthcare Ethics* 24 (2): 231–241.

Macklin, R. 1981. On paying money to research subjects: 'Due' and 'undue' inducements. *IRB* 3 (5): 1–6.

MacKay, D. 2015. Opt-out and consent. *Journal of Medical Ethics* 41 (10): 832–835.

Mandava, A., C. Pace, B. Campbell, E. Emanuel, and C. Grady. 2021. The quality of informed consent: Mapping the landscape. A review of empirical data from developing and developed countries. *Journal of Medical Ethics* 38 (6): 356–365.

Manson, N.C., and O. O'Neill. 2007. *Rethinking informed consent in bioethics*. Cambridge: Cambridge University Press.

Mantzari, E., F. Vogt, and T.M. Marteau. 2014. Does incentivising pill-taking 'crowd out' risk-information processing? Evidence from a web-based experiment. *Social Science and Medicine* 106: 75–82.

Master, Z., and D.B. Resnik. 2013. Incorporating exclusion clauses into informed consent for biobanking. *Cambridge Quarterly of Healthcare Ethics* 22 (2): 203–212.

McConnell, T. 2010. The inalienable right to withdraw from research. *Journal of Law, Medicine & Ethics* 38 (4): 840–846.

McGuire, A.L., and L.M. Beskow. 2010. Informed consent in genomics and genetic research. *Annual Review of Genomics and Human Genetics* 11: 361–381.

McNeill, P. 1997. Paying people to participate in research: Why not? A response to Wilkinson and Moore. *Bioethics* 11 (5): 390–396.

Melham, K., L.B. Moraia, C. Mitchell, M. Morrison, H. Teare, and J. Kaye. 2014. The evolution of withdrawal: Negotiating research relationships in biobanking. *Life Science and Social Policy* 10: 16.

Menikoff, J. 2006. *What the Doctor didn't say: The hidden truth about medical research*. New York: Oxford University Press.

Milgram, S. 1974. *Obedience to authority*. New York: Harper and Rowe.

Mill, J.S. 1978 [1859]. *On liberty*, ed E. Rapaport. Indianapolis: Hackett.

Miller, F.G. 2008. Research on medical records without informed consent. *Journal of Law, Medicine, and Ethics* 36 (3): 560–566.

Miller, F.G., and H. Brody. 2002. What makes placebo-controlled trials unethical? *American Journal of Bioethics* 2 (2): 3–9.

Miller, A.D., and R. Perry. 2012. The reasonable person. *New York University Law Review* 97 (2): 323–392.

Millum, J., and D. Bromwich. 2021. Informed consent: What must be disclosed and what must be understood? *American Journal of Bioethics* 21 (5): 46–58.

Millum, J., and M. Garnett. 2019. How payment for research participation can be coercive. *American Journal of Bioethics* 19 (9): 21–31.

Montalvo, W., and E. Larson. 2014. Participant comprehension of research for which they volunteer: A systematic review. *Journal of Nursing Scholarship* 46 (6): 423–431.

Morse, B., K.K. Kim, Z. Xu, C.G. Matsumoto, L.M. Schilling, L. Ohno-Machado, S.S. Mak, and M.S. Keller. 2023. Patient and researcher stakeholder preferences for use of electronic health record data: A qualitative study to guide the design and development of a platform to honor patient preferences. *Journal of the American Medical Informatics Association* 30 (6): 1137–1149.

National Commission for the Protection of Human Subjects of Biomedical or Behavioral Research. 1979. *The Belmont report: Ethical principles and guidelines for the protection of human subjects of research*. Washington, DC: Department of Health, Education, and Welfare.

National Institutes of Health. 2004. Clinical research and the HIPAA privacy rule. Available at: https://privacyruleandresearch.nih.gov/clin_research.asp. Accessed 26 June 2024.

Nishimura, A., J. Carey, P.J. Erwin, J.C. Tilburt, M.H. Murad, and J.B. McCormick. 2013. Improving understanding in the research informed consent process: A systematic review of 54 interventions tested in randomized control trials. *BMC Medical Ethics* 14: 28.

Odwazny, L.M., and B.E. Berkman. 2017. The "reasonable person" standard for research informed consent. *American Journal of Bioethics* 17 (7): 49–51.

Onvomaha, T.P., N. Kass, and P. Akweongo. 2006. The informed consent process in a rural African setting: A case study of the Kassena-Nankana district of Northern Ghana. *IRB* 28 (3): 1–6.

Paasche-Orlow, M.K., H.A. Taylor, and F.L. Brancati. 2003. Readability standards for informed-consent forms as compared with actual readability. *New England Journal of Medicine* 348 (8): 721–726.

Page, S.A., K.P. Manhas, and D.A. Muruve. 2016. A survey of patient perspectives on the research use of health information and biospecimens. *BMC Medical Ethics* 17 (1): 48.

Pandiya, A. 2010. Readability and comprehensibility of informed consent forms for clinical trials. *Perspectives in Clinical Research* 1 (3): 98–100.

Petrini, C. 2010. "Broad" consent, exceptions to consent and the question of using biological samples for research purposes different from the initial collection purpose. *Social Science and Medicine* 70 (2): 217–220.

Pietrzykowski, T., and K. Smilowska. 2021. The reality of informed consent: Empirical studies on patient comprehension—Systematic review. *Trials* 22 (57). https://doi.org/10.1186/s13063-020-04969-w.

Plöckinger, U., and U. Auga. 2022. The "four principles" of Western medical bioethics and the bioethics of Shīʿī Islam in Iran—is the claim of universality by both justified? *Religions* 13 (11): 1118.

Ploug T. 2020. In Defence of informed consent for health record research - why arguments from 'easy rescue', 'no harm' and 'consent bias' fail. *BMC Medical Ethics* 21 (1): 75.

Princewill, C.W., A.S. Jegede, K. Nordström, B. Lanre-Abass, and B.S. Elger. 2017. Factors affecting women's autonomous decision making in research participation amongst Yoruba women of western Nigeria. *Developing World Bioethics* 17 (1): 40–49.

Radin, M. 1996. *Contested commodities*. Cambridge, MA: Harvard University Press.

Rashad, A.M., F. MacVane Phipps, and M. Haith-Cooper. 2004. Obtaining informed consent in an Egyptian research study. *Nursing Ethics* 11 (4): 394–399.

Resnik, D.B. 2003. Exploitation in biomedical research. *Theoretical Medicine and Bioethics* 24 (3): 233–259.

———. 2009a. The clinical researcher-subject relationship: A contextual approach. *Philosophy, Ethics, and Humanities in Medicine* 4: 16.

———. 2009b. Do informed consent documents matter? *Contemporary Clinical Trials* 30 (2): 114–115.

———. 2015. Bioethical issues in providing financial incentives to research participants. *Medicolegal and Bioethics* 5: 35–41.

———. 2018. Ethics of community engagement in field trials of genetically modified mosquitoes. *Developing World Bioethics* 18 (2): 135–143.

———. 2019. Two unresolved issues in community engagement for field trials of genetically modified mosquitoes. *Pathogens and Global Health* 113 (5): 238–245.

———. 2021. Informed consent, understanding, and trust. *American Journal of Bioethics* 21 (5): 61–63.

Resnik, D.B., and P.R. Finn. 2018. Ethics and phishing experiments. *Science and Engineering Ethics* 24 (4): 1241–1252.

Resnik, D.B., and J. Hecking. 2020. Cultural variation in informed consent for clinical research participation. In *Handbook of ethnography in healthcare research*, ed. P. Hackett and C. Hayre, 33–46. New York: Routledge.

———. 2022. Cultural variation in informed consent for clinical research participation. In *Handbook of ethnography in healthcare research*, ed. P. Hackett and C. Hayre, 33–46. New York, NY: Routledge.

Resnik, D.B., and D.J. McCann. 2015. Deception by research participants. *New England Journal of Medicine* 373 (13): 1192–1933.

Resnik, D.B., and E. Ness. 2012. Participants' responsibilities in clinical research. *Journal of Medical Ethics* 38 (12): 746–750.

Resnik, D.B., S. Peddada, J. Altilio, N. Wang, and J. Menikoff. 2008. Oncology consent forms: Failure to disclose off-site treatment availability. *IRB* 30 (6): 7–11.

Różyńska, J. 2022. The ethical anatomy of payment for research participants. *Medicine, Health Care, and Philosophy* 25 (3): 449–464.

Sample, R.J. 2003. *Exploitation: What it is and why it's wrong*. Lanham: Rowman and Littlefield.

Sanchini, V., G. Bonizzi, D. Disalvatore, M. Monturano, S. Pece, G. Viale, P.P. Di Fiore, and G. Boniolo. 2016. A trust-based pact in research biobanks. From theory to practice. *Bioethics* 30 (4): 260–271.

Schaefer, G.O., and A. Wertheimer. 2010. The right to withdraw from research. *Kennedy Institute of Ethics Journal* 20 (4): 329–352.
Seibert, P.S., P. Stridh-Igo, and C.G. Zimmerman. 2002. A checklist to facilitate cultural awareness and sensitivity. *Journal of Medical Ethics* 28 (3): 143–146.
Shamoo, A.E., and D.B. Resnik. 2006. Strategies to minimize risks and exploitation in phase one trials on healthy subjects. *American Journal of Bioethics* 6 (3): W1–W13.
———. 2022. *Responsible conduct of research*. 4th ed. New York: Oxford University Press.
Shickle, D. 2006. The consent problem within DNA biobanks. *Studies in History and Philosophy of Biological and Biomedical Research* 37: 503–519.
Singer, E., and M.P. Couper. 2008. Do incentives exert undue influence on survey participation? Experimental evidence. *Journal of Empirical Research on Human Research Ethics* 3 (3): 49–56.
Smajdor, A. 2023. Reification and assent in research involving those who lack capacity. *Journal of Medical Ethics* 49 (7): 474–480.
Smischney, N.J., J. Pannu, R.F. Hinds, and J.B. McCormick. 2018. Community consultation for planned emergent use research: Experiences from an academic medical center. *JMIR Research Protocols* 7 (5): e10062.
Stunkel, L., and C. Grady. 2011. More than the money: A review of the literature examining healthy volunteer motivations. *Contemporary Clinical Trials* 32 (3): 342–352.
Tam, N.T., N.T. Huy, T.B. Thoale, N.P. Long, N.T. Trang, K. Hirayama, and J. Karbwang. 2015. Participants' understanding of informed consent in clinical trials over three decades: Systematic review and meta-analysis. *Bulletin of the World Health Organization* 93 (3): 186–98H.
Thong, I.S., M.Y. Foo, M.Y. Sum, B. Capps, T.S. Lee, C. Ho, and K. Sim. 2016. Therapeutic misconception in psychiatry research: A systematic review. *Clinical Psychopharmacological Neuroscience* 14 (1): 17–25.
Tishler, C.L., and S. Bartholomae. 2002. The recruitment of normal healthy volunteers: A review of the literature on the use of financial incentives. *Journal of Clinical Pharmacology* 42 (4): 365–375.
United Nations. 1948. Universal Declaration of Human Rights. Available at: https://www.un.org/en/about-us/universal-declaration-of-human-rights. Accessed 21 June 2024.
US Constitution. 1789. Available at: https://uscode.house.gov/static/constitution.pdf. Accessed 21 June 2024.
Vellinga, A., M. Cormican, B. Hanahoe, K. Bennett, and A.W. Murphy. 2011. Opt-out as an acceptable method of obtaining consent in medical research: A short report. *BMC Medical Research Methodology* 11: 40.
Warner, T.D., C.J. Weil, C. Andry, H.B. Degenholtz, L. Parker, L.J. Carithers, M. Feige, D. Wendler, and R.D. Pentz. 2018. Broad consent for research on biospecimens: The views of actual donors at four U.S. medical centers. *Journal of Empirical Research in Human Research Ethics* 13 (2): 115–124.
Weijer, C., and E.J. Emanuel. 2000. Ethics. Protecting communities in biomedical research. *Science* 289 (5482): 1142–1144.
Weir, R.F., and R.S. Olick. 2004. *The stored tissue issue: Biomedical research, ethics, and law in the era of genomic medicine*. New York: Oxford University Press.
Weise, K.L., M.L. Smith, K.J. Maschke, and H.L. Copeland. 2002. National practices regarding payment to research subjects for participating in pediatric research. *Pediatrics* 110 (3): 577–582.
Wendler, D. 2006. One-time general consent for research on biological samples. *BMJ* 332 (7540): 544–547.
———. 2008. The assent requirement in pediatric research. In *The Oxford textbook of clinical research ethics*, ed. E.J. Emanuel, C. Grady, R.A. Crouch, R.K. Lie, F.G. Miller, and D. Wendler, 661–669. New York: Oxford University Press.
———. 2011. What we worry about when we worry about the ethics of clinical research. *Theoretical Medicine and Bioethics* 32 (3): 161–180.

———. 2021. The ethics of mandatory retention of clinical biospecimens for research. *Journal of General Internal Medicine* 36: 2818–2819.

Wendler, D., and F.G. Miller. 2008. Deception in research. In *The Oxford textbook of clinical research ethics*, ed. E.J. Emanuel, C. Grady, R.A. Crouch, R.K. Lie, F.G. Miller, and D. Wendler, 315–324. New York: Oxford University Press.

Wertheimer, A. 1999. *Exploitation*. revised ed. Princeton: Princeton University Press.

———. 2011. *Rethinking the ethics of clinical research: Widening the lens*. New York: Oxford University Press.

Wertheimer, A., and F.G. Miller. 2008. Payment for research participation: A coercive offer? *Journal of Medical Ethics* 34 (5): 389–392.

Whitney, S.N., A.L. McGuire, and L.B. McCullough. 2004. A typology of shared decision making, informed consent, and simple consent. *Annals of Internal Medicine* 140 (1): 54–59.

Wick, M., D. Erickson, J. Hoffman, L. Johnson, and T. Angradi. 2024. Navigating university openness in research policy inconsistent with indigenous data sovereignty: A case analysis. *Ethics & Human Research* 46 (2): 2–15.

World Health Organization. 2020. Community engagement. Available at: https://www.who.int/publications/i/item/9789240010529. Accessed 19 June 2024.

World Medical Association. 2013. Declaration of Helsinki: Ethical principles for medical research involving human subjects (2013 revision). Available at: https://www.wma.net/policies-post/wma-declaration-of-helsinki-ethical-principles-for-medical-research-involving-human-subjects/. Accessed 3 July 2024.

Wu, C., N. Wang, Q. Wang, C. Wang, Z. Wei, Z. Wu, S. Yu, and X. Jiang. 2024. Participants' understanding of informed consent in clinical trials: A systematic review and updated meta-analysis. *PLoS One* 19 (1): e0295784.

Yamanaka, M., M. Suzuki, and K. Sato. 2021. Patient perspectives on research use of residual biospecimens and health information: On the necessity of obtaining societal consent by creating a governance structure based on value-sharing. *Research Ethics* 17 (1): 103–119.

Chapter 6
Privacy and Confidentiality

Abstract This chapters explores a variety of issues related to protecting privacy and confidentiality in research including the moral and legal basis for privacy and confidentiality; the relationship between trust and privacy and confidentiality; the difference between privacy and confidentiality; the Privacy Rule of the Health Information Portability and Accountability Act (HIPPA); the European General Data Protection Regulation; methods and strategies for protecting privacy and confidentiality; the conflict between sharing data and biospecimens and protecting privacy and confidentiality; data security, data use agreements; the difference between identified and de-identified data and biospecimens; the Tarasoff case; privacy in genetic and genomic research involving family members; reporting of suspected abuse and neglect of children or vulnerable adults; and reasons for breaching privacy and confidentiality in some situations. The chapter also argues that recognizing the importance of trust in research can provide investigators with useful guidance and help solve some ethical dilemmas involving privacy and confidentiality.

Keywords Privacy · Confidentiality · Trust · Ethics · Data sharing · Biospecimen sharing · Law · Regulation

In the previous chapter, I examined some key ethical dilemmas related to informed consent in research and argued that thinking about the importance of trust can help us address these issues. In this chapter I will continue some of the themes developed in the previous chapter by reflecting on the relationship between trust and protection of privacy and confidentiality. I will also consider how the trust-based approach deals with some ethical dilemmas related to protecting privacy and confidentiality in research.

D. B. Resnik, *The Ethics of Research with Human Subjects*, The International Library of Bioethics 111, https://doi.org/10.1007/978-3-031-82757-0_6

6.1 Privacy, Confidentiality, and Trust

Privacy and confidentiality are distinct concepts in ethics and the law. Privacy refers to freedom from unwanted intrusion into one's private affairs. Privacy includes physical privacy (i.e., freedom from unwanted invasions of one's body or private space), decisional privacy (i.e., freedom from unwanted interference in one's private decisions), and informational privacy (i.e., freedom from unwanted disclosures of one's private information). Private information is information linked directly to the individual for which there is a reasonable expectation of privacy. Private information includes medical records, research records, social security numbers, bank and credit card statements, and personnel records. A social scientist who observes customers and merchants interacting at a grocery store would not be invading privacy because the store is a public area and there is no reasonable expectation of privacy in this setting. Likewise, information which is in the public domain, such as birth and death records, is not considered private. Confidentiality refers to protection of private information. Thus, confidentiality addresses one aspect of the broader notion of privacy, i.e., informational privacy (Forster 2006; Hodge and Gostin 2008).

As argued in Chap. 4, protecting an individual's privacy and confidentiality is important not only for respecting their human dignity and preventing harm to them but also for promoting their trust. We usually share our secrets only with individuals whom we trust, such as close friends, family members, doctors, lawyers, priests, or therapists (Bok 1989). Someone who discloses a secret without permission violates the trust that has been placed in them and damages the relationship. Most professional codes of ethics include a duty of confidentiality. The Declaration of Helsinki states that "Every precaution must be taken to protect the privacy of research subjects and the confidentiality of their personal information (World Medical Association 2024)." The CIOMS (2016) guidelines include extensive guidance on protecting confidentiality in research involving the storage, analysis, and sharing of health information and biospecimens.

Breaches of confidentiality or privacy can seriously damage trust in investigators, institutions, and the research enterprise. For example, the Havasupai case (discussed in Chap. 2) involved not only a violation of the community's consent but also of its privacy, since blood samples were shared without permission. Some of other troubling cases of privacy breaches involving medical research include:

- In February 2008, someone stole a laptop computer containing medical information on 2500 patients enrolled in a clinical trial conducted at the National Institutes of Health (NIH). The laptop, which was stolen from an employee's locked automobile trunk, contained information pertaining to diagnoses, clinical trial data, and laboratory test results. The person who stole the computer potentially had access to private information because the data on the laptop had not been encrypted, which was a violation of the NIH's computer security policy. NIH officials did not inform the patients of the security breach until nearly a month after it occurred because they did not feel that the patients were at immediate risk, and they did not want to unduly alarm them (Nakishima and Weiss 2008).

- In 2010, mammography patients and community members were outraged when they learned that someone had hacked into a University of North Carolina computer server containing the personal information of 180,000 mammography patients from around the state. The university determined that the investigator in charge of database security, radiology professor Bonnie Yankaskas, neglected her duties and should be terminated or demoted. Faculty members came to her defense, however, and argued that the university was trying to make her into a scapegoat. Yankaskas claimed that she did not have the technical expertise or training needed to keep the research database secure (Mangan 2010). The University and Yankaskas reached a legal settlement in 2011. Under the terms of settlement, the university agreed to restore Yankaskas status as a full professor and pay her legal fees, and Yankaskas agreed to retire by the end of 2011 (University of North Carolina at Chapel Hill 2011).
- Cybersecurity is an ongoing concern in healthcare and biomedical research. Almost every day there is a cybersecurity attack at a hospital or medical center somewhere in the world. Most of the security breaches involve hacking or ransomware attacks (She et al. 2020). According to the U.S. Office of Civil Rights, from 2009 to 2023, there were 5887 healthcare data breaches of 500 or more records affecting a total of 519,935,970 healthcare records (Alder 2024). The largest attack was against Anthem, Inc. in 2015 and impacted nearly 79 million patients (Alder 2024).

6.2 Privacy Laws and Regulations

As mentioned in Chap. 2, federal regulations require that research protocols include appropriate provisions for protecting privacy and confidentiality and informing participants about these provisions. Although the regulations do not spell out these provisions, the Common Rule does mention that "The Secretary of HHS will, after consultation with the Office of Management and Budget's privacy office and other Federal departments and agencies that have adopted this policy, issue guidance to assist IRBs in assessing what provisions are adequate to protect the privacy of subjects and to maintain the confidentiality of data (45 CFR 46.111a7)." However, the Secretary of HHS has not (as of the writing of this book) issued guidance to assist IRBs with making decisions concerning the adequacy of confidentiality protections. Some best practices for protecting privacy and confidentiality include (see Forster 2006; Hodge and Gostin 2008):

1. Limiting access to data or biospecimens to members of the research team or other authorized users;

2. Implementing robust and validated strategies for protecting the security of data and specimens, including cybersecurity measures, such as password protection, encryption, and storage of electronic data behind firewalls;
3. Training research staff on confidentiality and privacy rules and procedures and cybersecurity practices;
4. Removing personal identifiers from data or biospecimens and marking them with a code;
5. Obtaining a Certificate of Confidentiality (CoC) from the Department of Health and Human Services, which allows investigators to resist court orders for access to data.[1]
6. Informing subjects about the procedures that will be used to protect privacy and confidentiality as well as their limitations; for example, use of a phrase like "we will protect your privacy and confidentiality to the extent allowable by law" can be helpful.

Prior to the advent of the privacy rule of the Health Insurance Portability and Accountability Act (HIPAA) in 2003, medical confidentiality in the U.S. was covered only by state medical malpractice or confidentiality laws (Hodge and Gostin 2008). However, many patient advocates and attorneys argued that these laws did not provide enough protection for privacy and confidentiality and a strong, federal law was needed.

The HIPAA privacy rule includes numerous rules and procedures for protecting the confidentiality of private health information (PHI) (Department of Health and Human Services 2002, 2017). PHI is information that is associated with at least one of 17 types of personal identifies, such as name, address, social security number, telephone number, and medical record number. Hospitals, clinics, and other entities covered by HIPAA are required not to disclose PHI without the patient's permission. Researchers who are working for or with covered entities may obtain PHI if patients sign a consent document granting them access it. As noted in Chap. 5, a privacy board or an IRB can allow investigators to have access to PHI without the patient's consent for certain types of research, such as reviewing or analyzing medical records or determining the feasibility of a clinical trial. Additionally, covered entities may disclose PHI without consent for public health and criminal law purposes, including reporting of abuse or neglect, infectious diseases, and stabbings or gunshot wounds (Department of Health and Human Services 2017). Criminal (e.g., deliberate) violations of HIPAA are punishable by fines of up to $250,000 and a maximum of ten years in prison.

[1] The extent of confidentiality protections provided by CoCs is not well-established, since very few cases have tested the limits of CoCs. In one case, a court allowed a criminal defendant to gain access to confidential research records protected by a CoC to obtain evidence that could be used to discredit the testimony of a witness for the prosecution. The defendant's attorney wanted to know whether the witness was in a research study protected by a CoC (Beskow et al. 2008).

Countries that belong to the European Union (EU)[2] have adopted laws that conform to the General Data Protection Regulation (GDPR), which became effective in 2018 (Marelli and Testa 2018). The scope of GDPR is very broad and encompasses the processing of any information that can directly or indirectly identify an individual, such name, address, phone number, but also gender, ethnicity, biometric data, web cookies, and even political opinions. Processing includes any action performed on data, including collecting, recording, analyzing, storing, using, or sharing data. GDPR applies to data collected for a variety of purposes, including health care delivery, health care insurance, research, law, government, and business. Fines for deliberately violating GDPR are very steep, up to 10 million dollars (Wolford 2024). Data processing is allowed when: (1) data processing is necessary to prepare or enter into a contract with the individual (for example, doing a background check on a person to determine whether you should lease them an apartment); (2) data processing is required by law (for example, to report gunshots wounds or infectious diseases); (3) data processing is in the public interest or necessary to fulfill an official function (for example, processing data related to water use); (4) the individual has consented to data processing (for example, data processing as part of medical treatment or research); or (5) the processor has some other legitimate interest in processing the data. Consent-based data processing poses a significant challenge for research with human subjects because consent must be freely given, unambiguous, documented, and **specific** (Wolford 2024). As discussed in Chap. 5, specific consent poses some significantly practical challenges for conducting scientific research. The GDPR does include an exception for scientific research:

> It is often not possible to fully identify the purpose of personal data processing for scientific research purposes at the time of data collection. Therefore, data subjects should be allowed to give their consent to certain areas of scientific research when in keeping with recognised ethical standards for scientific research. Data subjects should have the opportunity to give their consent only to certain areas of research or parts of research projects to the extent allowed by the intended purpose (General Data Protection Regulation 2018).

While this exception is helpful, it is not clear how broad or general consent can be because it is limited to "certain areas of scientific research" for an "intended purpose." Because this section of the GDPR leaves a great deal open to interpretation, it remains to be seen whether the regulation will help or hinder scientific research (Slokenberga et al. 2021).

Researchers should inform the IRB and Information Technology Officers as soon as possible about breaches of data privacy, confidentiality, or security so they can take effective action to deal with these problems. Although breaches of privacy, confidentiality, or security can be embarrassing for researchers and institutions, it is far better to inform subjects in a timely fashion about these problems when they arise, instead of allowing them to fester. Subjects may be understandably upset if

[2] Current EU states are Austria, Belgium, Bulgaria, Croatia, Republic of Cyprus, Czech Republic, Denmark, Estonia, Finland, France, Germany, Greece, Hungary, Ireland, Italy, Latvia, Lithuania, Luxembourg, Malta, Netherlands, Poland, Portugal, Romania, Slovakia, Slovenia, Spain and Sweden.

investigators do not inform them of breaches in a timely fashion, or if they learn about breaches from a third party (such as the media).

Communications with subjects should describe the breach, what is being done to prevent it from happening again, and whether subjects are at increased risk of harm from the breach. Such communications, if conveyed respectfully and contritely, can promote trust. Although the duty to protect privacy and confidentiality has a strong basis in ethics and the law, it may sometimes conflict with other legal or ethical duties, such as the obligation to share data or research materials or the obligation to protect individuals or the public from harm. This chapter will examine some of the dilemmas that can arise when the duty to protect privacy/confidentiality conflicts with other obligations.

6.3 Sharing of Data and Biospecimens

Openness—the sharing (and publishing) of data, materials, methods, and results—is essential for scientific collaboration and progress (Shamoo and Resnik 2022). U.S. government agencies, including the NIH and NSF, require funded investigators to share data and materials with other researchers and publish their data and results (National Institutes of Health 2023; National Science Foundation 2024). Most scientific journals require investigators to make supporting data, methods, and materials available to other researchers as a condition of publication (Resnik et al. 2019). Professional codes of ethics also endorse open sharing of data and materials and publication of data and results. The FAIR (Findable, Accessible, Interoperable, Reusable) principles are some useful and widely accepted guidelines for storing, managing, and sharing scientific data (Wilkinson et al. 2016).

Although openness is a key part of science's ethos, it may conflict with protecting the privacy and confidentiality of human research participants when people outside the research team have access to personal and sensitive information about research participants. For many years, researchers resolved this dilemma by removing personal identifiers from data and biospecimens that they shared with other researchers or published (McGuire and Gibbs 2006; Rothstein 2010). From 2007 to 2009, the NIH required funding investigators to deposit de-identified human genomic data on public websites, such as the database of Genotypes and Phenotypes (dbGaP) (Resnik 2010). All of this changed, however, when statisticians developed methods for re-identifying individuals in de-identified databases (Lin et al. 2004; Homer et al. 2008; Benitez and Malin 2010). Today, artificial intelligence (AI) can also be used to re-identify individuals (Rocher et al. 2019; Platzer 2022).

There are several different types of re-identification that raises privacy issues. The first type involves using information that is known to identify an individual, such as genomic information, and identifying that individual in a de-identified dataset, which may contain other information, such as medical information (Lin et al. 2004; McGuire and Gibbs 2006; Lowrance and Collins 2007; Homer et al. 2008). This method of re-identification has valuable applications in the military, disaster recovery, and in criminal law (Bukyya et al. 2021). For example, this method has been used to identify individuals who were killed during the September 11, 2001, terrorist attacks on the World Trade Center in New York City (CBS News 2023).

The second type of re-identification goes in the opposite direction, from a de-identified piece of information (such as a DNA sample) to identified information. For example, police used DNA left at crime scenes to link Joseph DeAngelo to 13 murders and 51 rapes committed in California between 1974 and 1986. The criminal investigators asked a lab to sequence the DNA and attempted it match it to DNA in criminal databases. When they failed to turn up a match, they asked forensic genealogist Barbara Rae Venter to help with the case. Venter was able to match it to the DNA of six of DeAngelo's family members contained in genealogy databases (Ancestry.com and 23andMe).[3] She then was able to identify DeAngelo as the most likely suspect because he had blue eyes and the person who left DNA at the crime scenes had genes for blue eyes. DeAngelo confessed to the crimes (St. John 2020).

Although much of the bioethics discussion of re-identification has focused on using genomic information to re-identify individuals, individuals can also be re-identified based on other types of information, such as demographic, geospatial, purchasing, or medical information. (Lowrance and Collins 2007; Lubarsky 2017; Rocher et al. 2019). For example, if one knows that a research subject is a 53-year, female physics professor at Harvard University, it may be possible to identify that person in a faculty listing based on this information. Advances in machine learning and other forms of artificial intelligence are likely to make it easier to re-identify people.

The upshot of recent work on re-identifying individuals is that de-identification no longer guarantees confidentiality of human research data (Resnik 2010). Although de-identification is still an important strategy for protecting privacy and confidentiality (National Institutes of Health 2024a), scientists have begun pursuing other strategies for protecting confidentiality and privacy of genomic data. One of these is to use contracts for sharing data or biospecimens with investigators and institutions. A data use agreement (DUA) is contract for sharing data which spells out conditions for using and storing the data. DUAs usually prohibit recipients from sharing data without permission or attempting to re-identify individuals in de-identified datasets. A material transfer agreement (MTA) is a contract for sharing research materials, such as cells, tissues, animals, or chemical reagents, which spells out conditions for using the materials. MTAs may also prohibit recipients from sharing samples without permission or attempting to re-identify individuals in de-identified samples (Resnik 2010). For example, the NIH requires institutions to sign DUAs to obtain access to dbGaP data (National Institutes of Health 2024a, b). Although DUAs and MTAs can help protect privacy and confidentiality they increase administrative burdens and paperwork for investigators and institutions. DUAs and MTAs do not provide unqualified protection for confidentiality and

[3] This case raises interesting consent issues because people who provided DNA for these genealogy databases may not have understood that their genetic information could be used for law enforcement purposes (Berkman et al. 2018).

privacy, since investigators or institutions might violate them. Violators would still be liable under contract law, but it is not clear that these legal implications would always deter violators, since institutions might not sue for breach of contract.

Another way of protecting privacy and confidentiality is to alter or "mask" the data to make it more difficult to re-identify individuals (Lowrance and Collins 2007; Richardson et al. 2015; National Academies of Sciences, Engineering, and Medicine 2023). For example, researchers could share genomic data after removing sequences that uniquely identify individuals. While altering the data has some value as a means of protecting privacy, it may also seriously compromise the quality and scientific utility of the data. Additionally, researchers continue to develop new ways of identifying individuals in degraded data (Resnik 2010; Rocher et al. 2019; Platzer 2022).

Some researchers have responded to the conflict between openness and confidentiality by asking genomic research subjects to forego traditional confidentiality and privacy protections. To participate in the Personal Genome Project (PGP), subjects must agree to make their personally identified genomic, demographic, and medical data available on a public website (Church 2005; Lunshof et al. 2008). The project was initially piloted on a group of subjects with a background in genetics or genetic counseling, who were motivated to participate by their desire to advance scientific research and probably understood the implications of waiving traditional confidentiality and privacy protections (Church 2005). The project has expanded its enrollment since then to include subjects without a background in genetics or genetic counseling. Over three thousand participants are enrolled in the study (Ball et al. 2014). One of the benefits of the project to participants is that it provides them with access to their genetic test results (Lunshof et al. 2008). Asking participants to forego traditional confidentiality protections does not violate the federal research regulations, since the regulations only require that confidentiality and privacy protections where they are appropriate (45 CFR 46.111a7). One could argue that confidentiality and privacy protections are not needed if one of the aims of the study is to make data widely available and the subjects agree to waive these protections (Lunshof et al. 2008).

Risks related to the public disclosure of private information are an important ethical concern with the PGP. Research subjects could experience discrimination, stigma, or embarrassment as result of such disclosure. Presumably, the subjects are aware of these risks when they consent and decide to accept them to help advance the goals of the study (Ball et al. 2014). However, the family members of research subjects also face these risks, since it may be possible to predict that a family member is likely to have a genetic trait or medical condition, based on their genetic relationship to the participant (Resnik 2010). To address this issue, the PGP requires that if someone with a monozygotic (i.e., genetically identical) twin wants to enroll in the study their sibling must also agree to enroll (Church 2005). However, other closely-related family members could also face risks. For example, if one knows that someone is the mother of a participant in the PGP, and the participant has a genetic mutation associated with an 80% lifetime risk of breast cancer, then one could predict that the mother has a 50% chance of having this mutation. One might argue that PGP is unethical because it imposes significant risks on family members of the participants without their consent (Resnik 2010). Relatives of Henrietta Lacks (see case discussion in Chap. 1) objected to the publication of the HeLa cell genome sequence without their consent, since the genome contains genetic information that can be linked to family members (Yandell 2013). (I will discuss risks to third parties in greater depth in Chap. 7.)

As we saw in the Havasupai case discussed in Chap. 2, protecting privacy and confidentiality can also be an important concern when investigators plan to publish research on communities or distinct populations or share biospecimens (Resnik and Kennedy 2010; Harding et al. 2012). Even before the Havasupai incident occurred, members of Native American and indigenous communities had been concerned about the use of their data and biospecimens and in some cases had accused scientists of stealing them (McCartney et al. 2022; Sabatello et al. 2022). Today, Native American, indigenous, and other communities are asserting their rights to control the use of their data and biospecimens to protect the privacy and their sovereignty (Hudson et al. 2023; Ravindran 2024). Many researchers who work with indigenous communities and populations follow CARE (collective benefit, authority to control, responsibility, and ethics) principles for data governance (Research Data Alliance 2024).

It is also worth noting that publishing photographs or images of human subjects raises privacy concerns. To protect privacy and confidentiality when publishing photographs or images of research subjects, researchers must obtain consent from the subjects, and they should use pseudonyms to refer to subjects and obscure identifying features, such as faces or tattoos. Sometimes research subjects may permit or even encourage investigators to use their real names or to publish photos or images with identifying features. If subjects freely consent to the publication of identifying information, there is nothing wrong with this practice.

The principle of trust can help resolve the issues discussed in this section. To promote the trust in research that involves the sharing of human data and biospecimens, it is important not only to inform participants about the practices that will be implemented to safeguard their privacy and confidentiality, but also to solicit community and public opinion about these practices (Chang 2019). The Havasupai and Henriette Lacks case demonstrate that there is often a chasm between researchers' expectations concerning these topics and the expectations of communities, populations, and the public. Researchers who are studying communities or populations should discuss their publication plans with representatives from those communities or populations, understand their concerns, and respect their wishes concerning the use of data and biospecimens. Although openness is essential for scientific progress, publication and sharing plans should protect the community's interests (Resnik and Kennedy 2010). Partnering with leaders of populations and communities is especially important when conducting research on Native American and indigenous communities (Oster and Lightning 2022; McCartney et al. 2022; Sabatello et al. 2022; Hudson et al. 2023; Ravindran 2024).

6.4 Justifying Breaches of Privacy and Confidentiality

Although researchers have ethical and legal duties to protect privacy and confidentiality, situations may arise in which they may decide to breach privacy or confidentiality to protect human subjects, communities, third parties, or the public from harm. Because breaching confidentiality is a serious violation of the subject's autonomy that can negatively impact their trust in researchers and the institution, the default position should be to maintain confidentiality unless there is a compelling moral reason to breach it. Discussed below are some situations in which researchers may have reasons to breach confidentiality.

6.5 Harm to Self

Sometimes investigators must decide what to do when they discover that research subjects intend to harm themselves (Vannoy et al. 2010; Resnik et al. 2015). Suppose that during an interview a psychiatric research subject expresses a strong desire to kill himself. He also talks about how, when, and why he would commit suicide. In this case, the investigator would face a conflict between beneficence (i.e., the duty to protect the subject's life/health) and respect for autonomy (i.e., the duty to honor the subject's right to control private information about himself) and trust (i.e., the duty to promote the subject's trust). One could argue that in this case, beneficence should take precedence over autonomy and trust, because a person who expresses a strong desire to kill himself may be suffering from a mental or emotional affliction, such as severe depression or psychosis, which compromises decision-making. Since the subject may not be fully autonomous, some form of paternalistic intervention may be warranted. The intervention could include offering to take the person to a local crisis center or emergency department, or contacting the person's family, depending on the situation. If the person conducting the interview is a staff member, she should inform her supervisor about the situation and take appropriate action while waiting for further instruction. Investigators and staff members must exercise discretion and good judgment when dealing with subjects who express suicidal thoughts, since people sometimes talk about suicide but do not intend to go through with it. Investigators and staff members should follow standard operating procedures for responding to suicidal ideation by research subjects (Sarkhel et al. 2023; Vannoy et al. 2010). If an investigator determines that a person who expresses suicidal thoughts is not a high risk for suicide, then it may not be appropriate to breach that person's confidentiality and betray their trust. A wiser course of action might be to ask the person if they would be interested in a referral for counseling or the phone number of a suicide hotline.

6.6 Harm to Others

Investigators also face moral dilemmas when they discover that research subjects intend to harm other people (Resnik et al. 2015). In 2012, researchers from the National Institute of Environmental Health Sciences (NIEHS) conducted a study on the health risks associated with participation in cleanup activities related to an oil spill in the Gulf of Mexico, which occurred in 2010 because of an explosion at British Petroleum's Deepwater Horizon oil rig that killed 11 workers. During interviews with NIEHS researchers, some of the subjects who lived in the area affected by the oil spill said they were contemplating suicide (see discussion above) and some threatened to bomb British Petroleum offices (Resnik et al. 2015). Fortunately, the researchers had anticipated that this type of problem might occur, based on the dire socioeconomic conditions of the population and the large sample size (approximately 33,000 adults) and they had developed standard operating procedures for the interviewers to follow, which included reporting threats to their supervisors, who would make reports to the IRB and (if appropriate) law enforcement. The researchers conducting the interviews and their supervisors faced a conflict between duties related to preventing innocent people from being harmed on the one hand and respecting the subject's rights to privacy and confidentiality on the other.

A famous court case addressed the issue of breaching confidentiality to prevent harm to others. In the summer of 1969, Prosenjit Poddar, a graduate student at the University of California at Berkeley, confided to his psychotherapist, Dr. Lawrence Moore, that he wanted to kill Tatiana Tarasoff when she returned from a trip, because he was angry that she did not want to have a romantic relationship with him. Dr. Moore called the campus police and recommended that Poddar be civilly committed because he was suffering from a mental illness (schizophrenia) and was dangerous to himself or others. The police detained Poddar but determined he was not dangerous and released him shortly thereafter. Moore's supervisor, Dr. Harvey Powelson, told Moore not to have Poddar detained again. Neither Moore, nor Powelson, nor the campus police told Tarasoff about Poddar's threat to kill her. On October 27, 1969, Poddar stabbed Tarasoff to death when she returned from her trip. Tarasoff's parents sued Moore and the university for negligence, arguing that they had failed to exhibit reasonable care by failing to warn Tarasoff about the threat. After lower courts ruled in favor of the plaintiffs, the defendants appealed the case to the California Supreme Court, which upheld decisions made by the lower courts. The court ruled that Dr. Moore had a duty to warn Tarasoff about Poddar's threat and held that the professional duty of confidentiality does not allow one to ignore threats to public health and safety (*Tarasoff v. Regents of the University of California* 1976).

The Tarasoff case has had a significant impact on the law concerning professional confidentiality and the duty to protect the public from harm. Numerous state courts have followed or cited the Tarasoff ruling, and some state legislatures have adopted laws that require psychotherapists or other health professionals to breach patient confidentiality to protect the public from harm (Walcott et al. 2001).

However, the case remains controversial. Some courts have rejected the ruling in *Tarasoff*, while others have limited professional obligations to protect the public from harm. For example, some courts have ruled that for there to be a duty to warn others the person must communicate a specific threat to harm a specific person (Walcott et al. 2001). Organizations representing professional psychologists and psychiatrists have objected to the Tarasoff decision on the grounds that requiring therapists to breach confidentiality to protect the public from harm would interfere with clinical practice and undermine patients' trust in their doctors (Felthous 1999; Walcott et al. 2001).

A key issue relevant to breaching confidentiality to protect other people from harm is evaluating the nature of the risk. People sometimes threaten to harm others when they are angry or distraught, and it can be difficult to decide whether these threats are genuine, especially if one does not have a longstanding relationship with the person. If an investigator overreacts to an idle threat made by a research subject and warns the authorities, he or she may damage the subject's trust irreparably and cause the research subject unnecessary harm (such as the stress of being questioned or arrested by the police). Investigators should therefore carefully consider the nature of the risk that the subject poses to others before taking action. To breach confidentiality when a research subject threatens someone else, an investigator should have credible evidence[4] that the subject intends to seriously harm a specific victim (or victims) and that the danger is imminent (Felthous 1999; Beauchamp and Childress 2012). Investigators and staff members should develop and follow standard operating procedures for responding to threats made by research subjects (Resnik et al. 2015).

6.7 Suspected Abuse/Neglect

Investigators may sometimes discover evidence of child abuse/neglect when they examine or interview pediatric research subjects or during data and sample collection activities in private residences. All U.S. states have laws that mandate reporting of suspected child/abuse neglect to the local authorities (Resnik and Randall 2018). Some laws only mandate reporting for certain professions, such as physicians, nurses, or educators. North Carolina's child abuse statute requires any person or institution who suspects child abuse or neglect to report it to the department of social services. A person (or institution) who knowingly or wantonly fails to report suspected child abuse or neglect, or prevents someone else from reporting it, may be liable for a misdemeanor (North Carolina General Statutes 2013).

Because children are vulnerable to harm or exploitation, there is little ethical controversy concerning the duty to report child abuse/neglect. When the department

[4] In the law, credible evidence is evidence that is "worthy of belief; trustworthy (Black's Law Dictionary 2001:250)."

of social services receives a report of child abuse or neglect, it will conduct an investigation and may contact other authorities (such as the police) and take temporary custody of the child (or children). Since these actions can be very disruptive and stressful for the family, investigators and research staff should report child abuse/neglect only when they have a reasonable suspicion[5] that it has occurred, so that they do not cause unnecessary harm to families. Researchers who work with children should have appropriate training and experience in recognizing the signs of child abuse or neglect, so that they can report it appropriately. They should also follow standard operating procedures for dealing with suspected child abuse or neglect and include information in consent and assent documents about mandated reporting when they are conducting research in which there is a reasonable chance they will obtain evidence of suspected abuse or neglect (Resnik and Randall 2018).

Investigators face similar issues when they discover evidence of abuse of elderly or disabled adults during examinations, interviews, or in-home visits. Most states also have laws mandating the reporting of abuse, neglect, or exploitation of elderly or disabled adults. Because elderly or disabled adults may be more vulnerable to harm or exploitation than other adults, there is little ethical controversy concerning the duty to report abuse or neglect of these individuals. The main issue pertains to knowing how to identify signs of abuse or neglect. False reports of abuse or neglect can be stressful for caregivers and research subjects. Investigators and research staff should report abuse/neglect of older or disabled adults only when they have a reasonable suspicion that this has occurred. Researchers who work with older or disabled adults should have appropriate training and experience in recognizing the signs of abuse or neglect, so that they can report it appropriately. They should also follow standard operating procedures for dealing with suspected abuse or neglect of elderly or disabled research subjects.

6.8 Communicable Disease Reporting

Sometimes investigators perform tests on research subjects which may reveal that they have a dangerous communicable disease. Many states and countries have laws requiring health care professionals to report certain types of communicable diseases to public health authorities. For example, North Carolina requires health care professionals to report 73 different diseases to the Division of Public Health, including acquired immune deficiency syndrome (AIDS), human immunodeficiency virus (HIV), botulism, gonorrhea, hepatitis, influenza leading to death, Lyme disease, malaria, plague, Rocky Mountain spotted fever, syphilis, and yellow fever (North Carolina Administrative Code 2015a, b). Disclosures must include the person's name, address, and clinical and epidemiologic information. The Division of Public

[5] In the law, reasonable suspicion is defined as a suspicion which is "supported by specific and articulable facts (Black's Law Dictionary 2001:584)."

Health uses this information to conduct disease surveillance, investigate epidemics, and ask the infected person whom he or she has had contact with. For example, if a person is diagnosed with HIV/AIDS in North Carolina, public health authorities will ask them for the names of people they have had sexual relations with or used intravenous drugs with. They will then contact these sexual and drug partners and inform them that they have had contact with someone with HIV/AIDS (this procedure is known as contact tracing). They will also advise these partners to be tested for HIV.

Communicable disease reporting laws present investigators with a conflict between protecting confidentiality, respecting autonomy, and promoting trust on the one hand and obeying the law and protecting public health on the other. Since research subjects with communicable disease may infect other people and threaten public health, there is little controversy concerning the obligation to breach confidentiality to comply with communicable disease reporting laws. To avoid undermining subjects' trust, investigators should inform research participants about testing for communicable diseases (such as HIV) and their obligations to report these diseases to the public health authorities. They should also inform them that the public health authorities will protect their confidentiality to the extent allowable under reporting laws. If subjects test positive for communicable diseases, investigators should inform them that they pose a risk to the public health and provide them with information about how to avoid transmitting those diseases. They should also provide them with a referral for treatment. The Icahn School of Medicine IRB, for example, requires researchers who are planning to test for HIV to include the following language in consent forms:

> To take part in this research study, your blood will be tested for evidence of HIV, the virus that causes AIDS. People can get HIV through unprotected sexual contact with someone who has HIV, and through contact with blood (as in sharing needles including for piercing, tattooing, and injecting drugs). People who are pregnant with HIV infections can transmit HIV to their infants during pregnancy, delivery or while breastfeeding. There are treatments for HIV/AIDS that can help people stay healthy. People with HIV/AIDS can adopt safe practices to protect uninfected and infected people in their lives from getting HIV or getting infected with a different strain of HIV. By law, positive test results for HIV/AIDS (as well as other communicable diseases such as hepatitis B, hepatitis C, and syphilis) are reported to the NYS Department of Health so they can study how people get and transmit the disease and notify sexual or needle-sharing partners they may have been exposed. If you wish to be tested anonymously for HIV/AIDS, the research team can refer you to a public testing center, but you will not be able to be in this study. New York State law protects the confidentiality of HIV test results and other related information. It is illegal to discriminate against a person based on their HIV status and services are available to help if this happens. You are free to refuse to get an HIV test, but if you refuse you cannot be part of this research study (Icahn School of Medicine 2020).

In some cases, researchers may learn that subjects with communicable diseases do not intend to comply with procedures designed to protect the public health. For example, suppose a research subject with HIV admits that he does not intend to tell his wife and he plans to continue having unprotected sex with her. In many states and nations, it is a crime for an HIV-infected person to knowingly have unprotected sex (Ahmed and Hull 2011). When a researcher learns that a participant with HIV does not intend to take precautions to protect their sexual partners, the researcher must decide whether to inform those partners under the duty to warn rationale articulated in the *Tarasoff* ruling. Investigators should use good judgment and balance conflicting moral values or principles in situations like these (Stanard and Hazler 1995). Normally, the researcher will be able to protect the subject's partners by making a report to the public health authorities, who will initiate contact tracing. However, an investigator might be concerned that further action is needed if they have reasons to believe that the public health authorities are not likely to take effective and timely action to deal with situation. Since these other steps may badly damage the subject's trust, researchers should avoid them, if at all possible. Instead, they should inform the subject about the risks that they pose to their partners and encourage them to take reasonable precautions to avoid infecting them (Stanard and Hazler 1995).

6.9 Sharing Genetic Information with Family Members

Researchers sometimes acquire genetic or genomic information that may be useful for promoting the health or well-being of subjects' family members. Researchers may acquire this information deliberately, i.e., when they are testing for genetic variants known or thought to be associated with particular diseases, or inadvertently, i.e., when they happen to discover such variants while testing for something else (see discussion of intended vs. incidental findings in Chap. 7). In either case, the researchers might feel that they have a duty to breach confidentiality to protect or help family members, who may also have a potentially harmful genetic variant or other condition (Dugan et al. 2003; Falk et al. 2003; Liao 2009). Consider the following scenarios:

- **Huntington's disease.** During a study of the genetics of dementia, investigators discover that Tom, a 45-year-old male research subject, has tested positive for that gene that causes Huntington's disease, a progressive and incurable genetic condition involving degeneration of nerve cells in the brain. Onset of the illness typically occurs when a person is in their 40s. Victims develop dementia and lose control of their muscles. People who have only a single copy of the Huntington gene have nearly a 100% chance of developing disease (Mayo Clinic 2024). Tom has a wife and 22-year-old daughter. Investigators share the test result with Tom, and he indicates that he does not want to tell his family members that he has the Huntington gene. Investigators are considering whether to share Tom's test result with his daughter, who may use this information to seek effective treatment (if it becomes available), prepare for the impact of the disease, or make reproductive choices (e.g., decide not to have children).
- **Breast cancer genetics.** During a study of the relationship between genetics, environment, and cancer, investigators discover that Marsha, a 35-year-old

female, has tested positive for BRCA1 mutations. These mutations increase the risk of developing breast cancer by age 70 from 12% to 55–65% and ovarian cancer from 1.3% to 11–17% (National Cancer Institute 2020). Marsha's mother died from breast cancer at age 60. When Marsha learns her results, she says she does not want to share them with her 39-year-old sister or 55-year-old aunt, neither of whom have breast cancer. Investigators are considering whether to share this information the sister and aunt so they can be tested for BRCA1 mutations and take steps to reduce their risks of developing breast cancer or detecting it early (if they test positive).
- **Heart disease.** Investigators are conducting research on the genetics of heart disease among men and women without any evidence of heart disease by age 50. During their investigation, they discover a genetic variant that increases the risk of developing the disease to 57% for men and 44% for women. Men who are free from heart disease at age 50 normally have a 52% chance of developing this condition by age 95 and women have a 39% chance (Lloyd-Jones et al. 2006). They are planning to inform all the research subjects who test positive for this gene about their increased risk of developing heart disease, and they are considering whether to inform the subjects' family members, so they can be tested.
- **Misattributed paternity.** Researchers who are conducting a study on the genetic and environmental factors related to colorectal cancer discover that a 21-year-old woman and her purported father are not genetically related. They are trying to decide whether to inform the subjects about this finding (Lowe et al. 2017).

The above scenarios involve conflicts between protecting the subject's confidentiality and preventing harm to other people. Researchers might believe, following the *Tarasoff* reasoning discussed above, that they have a duty to inform affected[6] family members about the risks they or their potential offspring face from having a genetic predisposition for a disease and that this duty justifies breaching confidentiality (Keeling 2004).

However, there are some important differences between these situations and the Tarasoff case that should make investigators reluctant to breach confidentiality.

First, none of these research subjects are intending to harm their family members. The potential harms are related to genetic variants the subjects may share with the family members, not the subjects' intentions or motives. Although the subjects' intent does not necessarily affect the risk of harm, it does impact their culpability, and one might argue that subjects who do not have malicious motives deserve greater confidentiality protections than those who do.

Second, none of these situations involve the imminent risk of severe harm. In the first three cases, there are risks of severe harm, but these are not imminent. In the misattributed paternity case, it is not at all clear whether not knowing one's correct

[6] By "affected" I mean family members who have a significant chance of having the genetic variant(s). For example, if a man has the Huntington gene, his daughter but not his wife would be affected. If a woman has BRCA1 mutations associated with increased cancer risk, her daughter(s), sister(s), grandmother(s), niece(s), aunt(s) and mother could all be affected.

paternity is harmful or whether knowing one's correct paternity would be beneficial. Indeed, disclosing misattributed paternity could disrupt family relationships and do more harm than good (Lowe et al. 2017).

Third, in the Huntington's case the harm is not currently preventable. Since there is currently no treatment for this disease, informing family members will not prevent them from developing the disease or halting its progression and may cause them needless stress. It may, however, help them to avoid passing the gene on to the next generation if they decide not to reproduce.

Fourth, in the last case three, the probability of developing the disease because of the genetic predisposition is low. The heart disease genetic variant is associated only with a 5% increased chance of developing the heart disease by age 95 for men or women. Also, numerous risk factors other than this genetic variant, such as smoking, obesity, diabetes, poor diet, sedentary lifestyle, family history, and stress, can increase the risk of developing heart disease.

Given these considerations, one could argue that the case for breaching confidentiality to share a subject's genetic test result with affected family members is strongest when disclosure is essential for preventing harms that are (a) severe, (b) highly probable, (c) imminent, and (d) preventable. None of these cases meet all four of these conditions, although the BRCA1 cases meet three of them (severity, probability, and preventability). Researchers who are considering breaching confidentiality should first inform research subjects about the risks affected family members face and encourage them to share their results with them before violating confidentiality and undermining trust (*Pate v. Threlkel* 1995).

6.10 Conclusion

It is important for investigators to protect subjects' privacy and confidentiality to respect their autonomy, protect them from harm, and promote their trust. Investigators should inform subjects of the measures they will take to protect confidentiality and privacy. Potential threats to privacy and confidentiality are likely to continue to evolve in response to advance in science, technology, and statistics. For example, it may become much easier to identify individuals in databases and to sequence human DNA from bodily fluids or skin left in public places. Restricting access to genomic data and samples by means of DUAs and MTAs represents a workable compromise that protects privacy and confidentiality, promotes trust, and allows science to advance. Only DNA sequence information that has been completely stripped of personal identifiers should be published or placed on publicly available websites.[7] Situations may sometimes arise when investigators have a duty to breach confidentiality to prevent harm or promote public health. In some cases, investigators may be legally required to breach confidentiality so they can report suspected

[7] This may be easier said than done.

abuse/neglect or communicable diseases to government officials. Where appropriate, investigators should inform research subjects about their legal obligations under these reporting laws. They should use good judgment when deciding whether to go beyond what it legally required to protect individuals or the public from harm.

References

Ahmed, A., and B. Hull. 2011. Sex and HIV disclosure. *American Bar Association: Human Rights* 38: 2. Available at: http://www.americanbar.org/publications/human_rights_magazine_home/human_rights_vol38_2011/human_rights_spring2011/sex_and_hiv_disclosure.html. Accessed 9 July 2024.

Alder, S. 2024. Healthcare data breach statistics. The HIPAA Journal, 20 June. Available at: https://www.hipaajournal.com/healthcare-data-breach-statistics/#:~:text=Largest%20Healthcare%20Data%20Breaches%20(2009,records%20of%2078.8%20million%20individuals. Accessed 4 July 2024.

Ball, M.P., J.R. Bobe, M.F. Chou, T. Clegg, P.W. Estep, J.E. Lunshof, W. Vandewege, A. Zaranek, and G.M. Church. 2014. Harvard personal genome project: Lessons from participatory public research. *Genome Medicine* 6 (2): 10.

Beauchamp, T.L., and J.F. Childress. 2012. *Principles of biomedical ethics*. 7th ed. New York: Oxford University Press.

Benitez, K., and B. Malin. 2010. Evaluating re-identification risks with respect to the HIPAA privacy rule. *Journal of the American Medical Informatics Association* 17 (2): 169–177.

Berkman, B.E., W.K. Miller, and C. Grady. 2018. Is it ethical to use genealogy data to solve crimes? *Annals of Internal Medicine* 169 (5): 333–334.

Beskow, L.M., L. Dame, and E.J. Costello. 2008. Research ethics. Certificates of confidentiality and compelled disclosure of data. *Science* 322 (5904): 1054–1055.

Black's Law Dictionary, 2nd Pocket Edition. 2001. St. Paul: West Publishing.

Bok, S. 1989. *Secrets: On the ethics of concealment and revelation*. New York: Vintage Books.

Bukyya, J.L., M.L.A. Tejasvi, A. Avinash, P. Chanchala, P. Talwade, M.M. Afroz, A. Pokala, P.K. Neela, T.K. Shyamilee, and V. Srisha. 2021. DNA profiling in forensic science: A review. *Global Medical Genetics* 8 (4): 135–143.

CBS News. 2023. Remains of 2 people killed in 9/11 attack on World Trade Center identified with DNA testing. CBS News, 11 September. Available at: https://www.cbsnews.com/news/911-remains-2-people-killed-world-trade-center-identified-dna/. Accessed 7 July 2024.

Chang, C.T. 2019. Bank on we the people: Why and how public engagement is relevant to biobanking. *Michigan Technology Law Review* 25: 239–300.

Church, G.M. 2005. The personal genome project. *Molecular Systems Biology* 1 (2005): 0030.

Council for the International Organizations of Medical Sciences. 2016. *International ethical guidelines for health-related research involving humans*. Geneva: Council for the International Organizations of Medical Sciences.

Department of Health and Human Services. 2002. 45 CFR parts 160 and 164: Standards for privacy of individually identifiable health information; final rule. *Federal Register* 67 (157): 53182–53273.

———. 2017. Health information privacy. Available at: https://www.hhs.gov/hipaa/for-professionals/special-topics/research/index.html. Accessed 21 Oct 2024.

Dugan, R.B., G.L. Wiesner, E.T. Juengst, M. O'Riordan, A.L. Matthews, and N.H. Robin. 2003. Duty to warn at-risk relatives for genetic disease: Genetic counselors' clinical experience. *American Journal of Medical Genetics C: Seminars in Medical Genetics* 119C (1): 27–34.

Falk, M.J., R.B. Dugan, M.A. O'Riordan, A.L. Matthews, and N.H. Robin. 2003. Medical geneticists' duty to warn at-risk relatives for genetic disease. *American Journal of Medical Genetics A* 120A (3): 374–380.
Felthous, A.R. 1999. The clinician's duty to protect third parties. *Psychiatric Clinics of North America* 22 (1): 49–60.
Forster, D.G. 2006. Privacy and confidentiality. In *Institutional review board management and function*, ed. E.A. Bankert and R.J. Amdur, 140–146. Boston: Jones and Bartlett.
General Data Protection Regulation. 2018. Consent to certain area of scientific research. Available at: https://gdpr.eu/Recital-33-Consent-to-certain-areas-of-scientific-research/. Accessed 5 July 2024.
Harding, A.B., D. Harper, C. Stone, P. O'Neill, S. Harris Berger, and J. Donatuto. 2012. Conducting research with tribal communities: Sovereignty, ethics, and data-sharing issues. *Environmental Health Perspectives* 120 (1): 6–10.
Hodge, J.G., and L.O. Gostin. 2008. Confidentiality. In *The Oxford handbook of clinical research ethics*, ed. E.J. Emanuel, C. Grady, R.A. Crouch, R.K. Lie, F.G. Miller, and D. Wendler, 673–681. New York: Oxford University Press.
Homer, N., S. Szelinger, M. Redman, D. Duggan, W. Tembe, J. Muehling, J.V. Pearson, D.A. Stephan, S.F. Nelson, and D.W. Craig. 2008. Resolving individuals contributing trace amounts of DNA to highly complex mixtures using high-density SNP genotyping microarrays. *PLoS Genetics* 4 (8): e1000167.
Hudson, M., S.R. Carroll, J. Anderson, D. Blackwater, F.M. Cordova-Marks, J. Cummins, D. David-Chavez, A. Fernandez, I. Garba, D. Hiraldo, M.B. Jäger, L.L. Jennings, A. Martinez, R. Sterling, J.D. Walker, and R.K. Rowe. 2023. Indigenous peoples' rights in data: A contribution toward indigenous research sovereignty. *Frontiers in Research Metrics and Analytics* 8: 1173805.
Icahn School of Medicine. 2020. Guidance: HIV testing in research. Available at: https://icahn.mssm.edu/files/ISMMS/Assets/Research/PPHS/HRP-905-Guidance-HIV%20Testing%20in%20Research%20(12.08.2022).pdf. Accessed 9 July 2024.
Keeling, S.L. 2004. Duty to warn of genetic harm in breach of patient confidentiality. *Journal of Law and Medicine* 12 (2): 235–253.
Liao, S.M. 2009. Is there a duty to share genetic information? *Journal of Medical Ethics* 35 (5): 306–309.
Lin, Z., A.B. Owen, and R.B. Altman. 2004. Genetics. Genomic research and human subject privacy. *Science* 305 (5681): 183.
Lloyd-Jones, D.M., E.P. Leip, M.G. Larson, R.B. D'Agostino, A. Beiser, P.W. Wilson, P.A. Wolf, and D. Levy. 2006. Prediction of lifetime risk for cardiovascular disease by risk factor burdenat 50 years of age. *Circulation* 113 (6): 791–798.
Lowe, G., J. Pugh, G. Kahane, L. Corben, S. Lewis, M. Delatycki, and J. Savulescu. 2017. How should we deal with misattributed paternity? A survey of lay public attitudes. *AJOB Empirical Bioethics* 8 (4): 234–242.
Lowrance, W.W., and F.S. Collins. 2007. Ethics. Identifiability in genomic research. *Science* 317 (5838): 600–602.
Lubarsky, B. 2017. Re-identification of "anonymized" data. *Georgetown Law and Technology Review* 202 (1) Available at: https://georgetownlawtechreview.org/re-identification-of-anonymized-data/GLTR-04-2017/. Accessed 5 July 2024.
Lunshof, J.E., R. Chadwick, D.B. Vorhaus, and G.M. Church. 2008. From genetic privacy to open consent. *Nature Reviews Genetics* 9 (5): 406–411.
Mangan, 2010. Chapel Hill researcher fights demotion after security breach. Chronicle of Higher Education, 5 October. Available at: https://www.chronicle.com/article/chapel-hill-researcher-fights-demotion-after-security-breach/. Accessed 4 July 2024.
Marelli, L., and G. Testa. 2018. Scrutinizing the EU general data protection regulation: How will new decentralized governance impact research? *Science* 360 (6388): 496–498.

Mayo Clinic. 2024. Huntington's disease. Available at: http://www.mayoclinic.org/diseases-conditions/huntingtons-disease/basics/definition/con-20030685. Accessed 9 July 2024.
McCartney, A.M., J. Anderson, L. Liggins, M.L. Hudson, M.Z. Anderson, B. TeAika, J. Geary, R. Cook-Deegan, H.R. Patel, and A.M. Phillippy. 2022. Balancing openness with indigenous data sovereignty: An opportunity to leave no one behind in the journey to sequence all of life. *Proceedings of the National Academy of Sciences* 119 (4): e2115860119.
McGuire, A.L., and R.A. Gibbs. 2006. Genetics. No longer de-identified. *Science* 312 (5772): 370–371.
Nakishima, E., and R. Weiss. 2008. Patients' data on laptop stolen. *Washington Post*, 24 March:A1.
National Academies of Sciences, Engineering, and Medicine. 2023. *Opportunities and challenges for digital twins in biomedical research: Proceedings of a workshop*. Washington, DC: National Academies Press.
National Cancer Institute. 2020. BRCA1 and BRCA2: Cancer risk and genetic testing. Available at: http://www.cancer.gov/about-cancer/causes-prevention/genetics/brca-fact-sheet. Accessed 9 July 2024.
National Institutes of Health. 2023. Final NIH policy for data management and sharing (NOT-OD-21-013). Available at: https://grants.nih.gov/grants/guide/notice-files/NOT-OD-21-013.html. Accessed 5 July 2024.
———. 2024a. Protecting participant privacy when sharing scientific data. Available at: https://sharing.nih.gov/data-management-and-sharing-policy/protecting-participant-privacy-when-sharing-scientific-data/principles-and-best-practices-for-protecting-participant-privacy. Accessed 8 July 2024.
———. 2024b. Data use certification agreement. Available at: https://dbgap.ncbi.nlm.nih.gov/aa/wga.cgi?view_pdf&stacc=phs000674.v1.p1. Accessed 7 July 2024.
National Science Foundation. 2024. NSF's data sharing policy. Available at: https://new.nsf.gov/funding/data-management-plan#nsfs-data-sharing-policy-1c8. Accessed 5 July 2024.
North Carolina Administrative Code. 2015a. Chapter 41—Health: Epidemiology; Subchapter 41A.0202. Control Measures—HIV, amended November 1, 2007.
———. 2015b. Chapter 41—Health: Epidemiology; Subchapter 41A.0101.
North Carolina General Statutes. 2013. 7B-301. Duty to report abuse, neglect, dependency, or death due to maltreatment.
Oster, R.T., and P. Lightning. 2022. Commentary: Developing relationships through trust in indigenous health research. *Healthcare Policy* 17 (4): 56–62.
Pate v Threlkel. 1995. 661 So 2d 278 (Fla 1995).
Platzer, M. 2022. AI-based re-identification attacks—And how to protect against them. Mostly AI, 22 April. Available at: https://mostly.ai/blog/synthetic-data-protects-from-ai-based-re-identification-attacks. Accessed 5 July 2024.
Ravindran, S. 2024. Open with care. *Science* 386 (6720): 372–375.
Research Data Alliance. 2024. CARE principles for indigenous data governance. Available at: https://www.rd-alliance.org/app/uploads/2024/03/CARE20Principles20for20Indigenous20Data20Governance_OnePagers_FINAL20Sept2006202019.pdf. Accessed 7 July 2024.
Resnik, D.B. 2010. Genomic research data: Open vs. restricted access. *IRB* 32 (1): 1–6.
Resnik, D.B., and C.E. Kennedy. 2010. Balancing scientific and community interests in community- based participatory research. *Accountability in Research* 17 (4): 198–210.
Resnik, D.B., and D.C. Randall. 2018. Reporting suspected abuse or neglect in research involving children. *Journal of Medical Ethics* 44 (8): 555–559.
Resnik, D.B., A.K. Miller, R.K. Kwok, L.S. Engle, and D.P. Sandler. 2015. Ethical issues in environmental health research related to public health emergencies: Reflections in the GuLF study. *Environmental Health Perspectives* 123 (9): A227–A231.
Resnik, D.B., M. Morales, R. Landrum, M. Shi, J. Minnier, N.A. Vasilevsky, and R.E. Champieux. 2019. Effect of impact factor and discipline on journal data sharing policies. *Accountability in Research* 26 (3): 139–156.

Richardson, D.B., M.P. Kwan, G. Alter, and J.E. McKendry. 2015. Replication of scientific research: Addressing geoprivacy, confidentiality, and data sharing challenges in geospatial research. *Annals of GIS* 21 (2): 101–110.

Rocher, L., J.M. Hendrickx, and Y.A. de Montjoye. 2019. Estimating the success of re-identifications in incomplete datasets using generative models. *Nature Communications* 10: 3069.

Rothstein, M.A. 2010. Is deidentification sufficient to protect health privacy in research? *American Journal of Bioethics* 10: 3–11.

Sabatello, M., D.O. Martschenko, M.K. Cho, and K.B. Brothers. 2022. Data sharing and community-engaged research. *Science* 378 (6616): 141–143.

Sarkhel, S., V. Vijayakumar, and L. Vijayakumar. 2023. Clinical practice guidelines for management of suicidal behaviour. *Indian Journal of Psychiatry* 65 (2): 124–130.

Shamoo, A.E., and D.B. Resnik. 2022. *Responsible conduct of research*. 4th ed. New York: Oxford University Press.

She, A.H., M. Zarour, M. Alenezi, A.K. Sarkar, A. Agrawal, R. Kumar, and R.A. Khan. 2020. Healthcare data breaches: Insights and implications. *Healthcare (Basel)* 8 (2): 133.

Slokenberga, S., O. Tzortzatou, and J. Reichel, eds. 2021. *GDPR and biobanking: Individual rights, public interest and research regulation across Europe*. Cham: Springer.

St. John, P. 2020. The untold story of how the Golden State Killer was found: A covert operation and private DNA. *Los Angeles Times*, 8 December. Available at: https://www.latimes.com/california/story/2020-12-08/man-in-the-window. Accessed 5 July 2024.

Stanard, R., and R. Hazler. 1995. Legal and ethical implications of HIV and the duty to warn for counselors: Does Tarasoff apply? *Journal of Counseling & Development* 73 (4): 397–400.

Tarasoff v. Regents of the University of California. 1976. 17 Cal. 3d 425, 551 P.2d 334, 131 *Cal. Rptr*. 14 (Cal. 1976).

University of North Carolina at Chapel Hill. 2011. UNC-Chapel Hill, Professor Yankaskas reach settlement. Press release, April 11, 2011. Available at: http://uncnewsarchive.unc.edu/2011/04/15/unc-chapel-hill-professor-yankaskas-reach-settlement-2/. Accessed 11 Aug 2017.

Vannoy, S., U. Whiteside, and J. Unützer. 2010. Current practices of suicide risk management protocols in research. *Crisis* 31 (1): 7–11.

Walcott, D.M., P. Cerundolo, and J.C. Beck. 2001. Current analysis of the Tarasoff duty: An evolution towards the limitation of the duty to protect. *Behavioral Science and Law* 19: 323–343.

Wilkinson, M.D., M. Dumontier, I.J. Aalbersberg, G. Appleton, M. Axton, A. Baak, N. Blomberg, J.W. Boiten, L.B. da Silva Santos, P.E. Bourne, J. Bouwman, A.J. Brookes, T. Clark, M. Crosas, I. Dillo, O. Dumon, S. Edmunds, C.T. Evelo, R. Finkers, et al. 2016. The FAIR guiding principles for scientific data management and stewardship. *Scientific Data* 3: 160018. Erratum in: *Scientific Data* 6(1):6.

Wolford, B. 2024. What is GDPR, the EU's new data protection law? Available at: https://gdpr.eu/what-is-gdpr/. Accessed 21 Oct 2024.

World Medical Association. 2024. Declaration of Helsinki: Ethical principles for medical research involving human subjects (2024 revision). Available at: https://www.wma.net/policies-post/wma-declaration-of-helsinki-ethical-principles-for-medical-research-involving-human-subjects/. Accessed 3 July 2024.

Yandell, K. 2013. Privacy and the HeLa genome. The Scientist, March 26, 2013. Available at: http://www.the-scientist.com/?articles.view/articleNo/34840/title/Privacy-and-the-HeLa-Genome/. Accessed 7 July 2024.

Chapter 7
Risks and Benefits

Abstract This chapter explores a variety of issues related to managing risks and benefits in research with human subjects, including defining risks and benefits; minimizing risks; the relationship between scientific study design issues and risks and benefits; assessing risks and benefits; balancing risks and benefits; the role of empirical evidence and intuition in risk/benefit decision-making; social benefits and risks; harms to communities, family members, and other third parties; the ethics of randomly assigning patients to different treatment groups; the use of placebos in research; the relevance of equipoise to clinical trial design and problems with the equipoise concept; challenge studies involving healthy human subjects; phase I clinical studies on healthy volunteers and cancer patients; limits for risks in studies involving healthy volunteers; controversies concerning the definition of minimal risk; alternative study designs, such as uncontrolled trials; sharing individualized results with research participants; providing ancillary medical care to research participants; post-trial access to medications; and fair sharing of the benefits of research.

Keywords Ethics · Human subjects · Benefits · Risks · Randomization · Controlled trials · Risk assessment · Risk management · Healthy volunteers · Risk minimization · Trust

In the previous two chapters, I applied the trust-based approach to ethical issues related to protected rights to autonomy and privacy in research. In this chapter, I will shift the focus away from protection of rights toward the protection of welfare by examining issues related to managing risks and benefits in research with human subjects. In the first edition of the book, I divided risks and benefits into separate chapters, but I have decided to combine them in this (revised) edition to enhance discussion of both topics because risks are almost always understood and assessed in relation to benefits (Resnik 2018b).

As discussed in Chap. 2, two of the criteria for IRB approval including in the Common Rule address risks and benefits (45 CFR 46.111a):

D. B. Resnik, *The Ethics of Research with Human Subjects*, The International Library of Bioethics 111, https://doi.org/10.1007/978-3-031-82757-0_7

(1) **Risks to subjects are minimized**:

(i) By using procedures that are consistent with sound research design and that do not unnecessarily expose subjects to risk, and
(ii) Whenever appropriate, by using procedures already being performed on the subjects for diagnostic or treatment purposes.

(2) **Risks to subjects are reasonable in relation to anticipated benefits**, if any, to subjects, and **the importance of the knowledge** that may reasonably be expected to result. In evaluating risks and benefits, the IRB should consider only those risks and benefits that may result from the research (as distinguished from risks and benefits of therapies subjects would receive even if not participating in the research). **The IRB should not consider possible long-range effects of applying knowledge gained in the research** (e.g., the possible effects of the research on public policy) as among those research risks that fall within the purview of its responsibility (45 CFR 46.111, emphasis added)

The Common Rule also requires researchers to inform subjects about the reasonably foreseeable risks and benefits of the research (45 CFR 46.116b2 and b3).

The Belmont Report discusses addresses risks and benefits under the principle of beneficence:

> Persons are treated in an ethical manner not only by respecting their decisions and protecting them from harm, but also by making efforts to secure their well-being. Such treatment falls under the **principle of beneficence**...Two general rules have been formulated as complementary expressions of beneficent actions in this sense: **(1) do not harm and (2) maximize possible benefits and minimize possible harms...**However, even avoiding harm requires learning what is harmful; and, in the process of obtaining this information, persons may be exposed to risk of harm...The problem posed by these imperatives is to decide when it is justifiable to seek certain benefits despite the risks involved, and when the benefits should be foregone because of the risks. **The obligations of beneficence affect both individual investigators and society at large, because they extend both to particular research projects and to the entire enterprise of research.** In the case of particular projects, investigators and members of their institutions are obliged to give forethought to the maximization of benefits and the reduction of risk that might occur from the research investigation. In the case of scientific research in general, members of the larger society are obliged to recognize the longer term benefits and risks that may result from the improvement of knowledge and from the development of novel medical, psychotherapeutic, and social procedure (National Commission 1979: 5; emphasis added).

The Helsinki Declaration says the following about risks and benefits:

> Medical **research involving human subjects may only be conducted if the importance of the objective outweighs the risks and burdens to the research subjects**...All medical research involving human subjects must be preceded by careful assessment of predictable risks and burdens to the individuals and groups involved in the research in comparison with foreseeable benefits to them and to other individuals or groups affected by the condition under investigation. **Measures to minimise the risks must be implemented.** The risks must be continuously monitored, assessed and documented by the researcher...Physicians may not be involved in a research study involving human subjects unless they are confident that the risks have been adequately assessed and can be satisfactorily managed (World Medical Association 2013).

While these three important sources of guidance[1] agree that risks should be minimized and that risks should be reasonable in relation to benefits, they disagree about the scope of risk/benefit evaluation. The Common Rule focuses on "risks to subjects" and admonishes IRB members to not consider "the possible long-range effects of applying knowledge gained in research." Likewise, the Helsinki Declaration also focuses on "risks and burdens to research subjects." However, the Belmont Report asserts that "the obligations of beneficence…extend both to particular research projects and to the entire enterprise of research." Below we will discus issues related to the scope of risk/benefit evaluation.

7.1 What Are Risks and Benefits?

A risk is a chance that an outcome regarded as harmful (or bad) will occur. A benefit is a chance that an outcome regarded as beneficial (or good) will occur. The degree of risk or benefit is a function of the **probability** and the **magnitude** the outcome. For example, a medical procedure with a 1% probability of death is much riskier than one with a 70% probability of a headache and no chance of death, because death is much worse than a headache (Levine 1988). Likewise, a medical procedure with a 20% of curing blindness is more beneficial than a procedure with 70% of relieving a headache. Rid et al. (2010) have developed a seven-point scale for categorizing different types of harm associated with clinical research based on the duration and magnitude of the harm, which is discussed in Sect. 7.10 of this chapter. To the best of my knowledge, no one has developed a scale for ranking benefits. Because judgments about risks and benefits incorporate judgments about probabilities and judgments about the value of outcomes, **judgments about risks and benefits are inherently normative** (Shrader-Frechette 1991; Resnik 2017a, 2021a).

Since risks and benefits are partly a function of probability, it will be useful to distinguish between different types of probability. A statistical probability is one based on the observed frequency of an event. For example, if you flip a coin 1000 times and it comes up heads 700 times, you can infer that the probability of flipping the coin and getting heads is 70%, + or − the standard error for this experiment (Weiss 2011). A mathematical probability is an estimate calculated by dividing the number of times a specified outcome can occur by total number of possible outcomes. For example, a mathematical estimate of the probability of flipping a coin and getting heads is 50%, since there is one specified outcome (heads) and two possible outcomes (heads, tails). A subjective probability is an educated guess that an outcome will occur. For example, one could place a bet on a horse race based on one's best guess that a particular horse will win. A best guess may be a highly

[1]Other ethics guidelines include statements about risks and benefits similar to the found in the Common Rule, Belmont Report, and Helsinki Declaration (see Nuremberg Code 1949; Council for the International Organizations of Medical Sciences 2016).

educated guess, but we would still call it a subjective probability because it is not based objective, empirical data or mathematical relationships (Resnik 2021b).

Some commentators have argued that we should avoid using subjective probability estimates in science and medicine, because subjective estimates can be biased by one's background beliefs, assumptions, values, and opinions (Earman 1992). If we are unable to objectively assess the probability that a negative outcome related to research will occur, and we do not want to use a subjective estimate, we could say that the negative outcome is a possible risk of research without assigning it a numerical estimate (Resnik 2021b). If we think that a study may involve risks that we are not currently aware of, we could say that there may be some unknown risks related to the research. Our ability to reliably estimate risks based on scientific evidence is an important issue in deciding whether risks are reasonable in relation to benefits (discussed below) because it may be unreasonable to take a significant risk if we are uncertain about the probability that it will occur (Resnik 2021b).

The risks of research can be divided into two basic categories: risks to the subjects and risks to others. Risks to the subject may include (Levine 1988):

- Medical risks, such as death, disability, injury, toxicity, nausea, shortness of breath, or adverse drug reactions;
- Psychological risks: such as pain, discomfort, distress, anxiety, remorse, or shame;
- Social risks: such as discrimination, stigma, bias, identity or theft;
- Financial risks: such as having to cover the costs of medical bills or loss of income due to injury or disability;
- Legal risks: such as liability for child support as a result of a paternity discovered during genetic testing.

Risks to others may include:

- Risks to identifiable third parties, such as fetuses of pregnant participants and family members
- Risks to researchers or research staff
- Risks to communities or populations
- Risks to society including the environment

The benefits of research can also be divided into two basic categories: benefits to the subjects and benefits to others. Benefits to the subjects may include (Levine 1988):

- Medical benefits, such as treatment or prevention of disease, restoration of function, and prolongation of life;
- Information about one's health, such as test results
- Psychological benefits, such as relief of pain or suffering and satisfaction derived from making a valuable contribution to society by participating in research;
- Financial benefits, such as earning money from participating in research

Benefits to others may include:

- Advancement of human knowledge
- Development of medical products

- Promotion public health
- Development of the economy
- Development of research infrastructure

As will be discussed below, some of these risks and benefits are controversial. For now, I am more interested in listing things that people might view as risks and benefits instead of considering whether these things ought to be considered as risks or benefits or how we ought to think about them.

7.2 Risks, Benefits, and Trust

Risks and benefits have important implications for trust. Poorly managed risks can negatively impact trust in researchers, institutions, IRBs, and the research enterprise since people expect researchers and institutions to take appropriate steps to protect them from harm. The Tuskegee Study (discussed in Chap. 2) is paradigmatic case of how poor risk management destroys trust. In the study, the researchers mismanaged risks to the 399 African America men in the control group by preventing them from getting effective treatment for their disease when it became available. This harmful conduct, combined with the human rights abuses arising from consent violations, had a negative impact on African Americans' trust in biomedical research (Corbie-Smith et al. 2002).

Serious harms to healthy volunteers can be especially detrimental to trust because we do not expect healthy people to be harmed in research (Resnik 2012). When a terminally ill cancer patient dies from taking an experimental medication that could save their life, their death is regrettable, but not unexpected. Jesse Gelsinger's death in a Phase I trial (discussed in Chap. 2) significantly damaged the public's trust in gene therapy research because he was relatively healthy (and young) (Yarborough and Sharp 2009). If Gelsinger had not died in the experiment, the other problems with study, such as undisclosed risks and conflicts of interest and lax adverse event reporting, may have appeared to be less serious. Because he died, however, these problems became magnified in the public's mind (Yarborough and Sharp 2009). Other healthy volunteer studies that have negatively impacted public trust in research include Ellen Roche's death (discussed in Chap. 2), a Phase I trial of TGN1412 (an antibody) conducted by Paraxel in 2006 in London's Northwick Park Hospital in which six healthy volunteers developed severe immune reactions and organ failure (Nature 2006; Goodyear 2006), and a Phase I study of a mood disorder medication conducted by Biotrial in Rennes, France in 2016, in which six subjects were hospitalized, three suffered permanent brain damage, and one was pronounced brain-dead (i.e., permanent loss of all brain functions) (Chan 2016).

Psychosocial, cultural, and other non-medical harms can also negatively impact trust. In the Havasupai case (discussed in Chap. 2) the subjects were not physically injured, but many suffered psychological and cultural harm from the misuse of their blood samples and said the researchers and the university had broken their trust. Data security breaches (discussed in Chap. 6) can cause subjects psychological distress and lead to identity theft undermine trust in researchers and institutions.

While poorly managed risks can undermine trust, benefits can strengthen trust. One way of strengthening trust is to conduct community-based participatory research, also known as community-engaged research (CER). In CER, researchers form partnerships with community members and community organizations to design, plan, and conduct research that is relevant to the needs of the community. Community members can help researchers develop aims and objectives, provide critical feedback on study documents (such as surveys and consent forms), and help recruit participants (O'Fallon and Dearry 2002; Resnik and Kennedy 2010). CER can promote communication, mutual understanding and trust between communities, researchers, and research institutions (Khodyakov et al. 2017). Although many research projects do not easily fit within the CER paradigm, most do offer tangible, trust-building benefits to participants, communities, and public. Below, I will discuss some different ways of providing benefits to participants and communities, including sharing the results of research, providing ancillary care to and post-trial access to medications to participants, developing drugs, and helping to build biomedical and research infrastructures.

7.3 Scientific Design

Before an IRB can determine whether the risks of a study are acceptable, it must be assured that the study is well-designed. Sound research design is widely recognized as a key principle of research ethics, since poorly designed studies are not likely to generate data or results that benefit society (Emanuel et al. 2000). Although the federal research regulations do not say much about research design other than that procedures to minimize risk should be consistent with sound research design (45 CFR 46.111a), they clearly imply that studies should be well-designed, since the risks of research cannot be reasonable if a study is not likely to yield important benefits (Emanuel et al. 2000). The Nuremberg Code (1949), Helsinki Declaration (World Medical Association 2024), and Council for the International Organizations of Medical Sciences (2016) guidelines also address the importance of good research design. While an in-depth discussion of the principles of sound research design is beyond the scope of this book, some important guidelines include (see Gallin and Ognibene 2012; Binik and Hey 2019; Shamoo and Resnik 2022):

1. The aims and objectives are scientifically and/or pragmatically justified; for example, the research is expected to address important scientific questions, facilitate the development of medical treatments or public health interventions; or provide information useful for public policy.
2. The aims, objectives, hypotheses, methods, tests, experiments, survey instruments, questionnaires, procedures, endpoints, variables, study population, data

analysis plans, theoretical frameworks, recruitment and consent methods, risks and benefits are clearly described and appropriate for the subject matter.
3. The research has a solid foundation in prior work, including, animal or human experiments, epidemiology, or basic scientific research.
4. The research includes strategies, methods, and procedures to minimize and control bias.
5. If applicable,[2] the research is statistically sound; for example, the sample size is adequate to achieve statistically significant results.
6. If applicable,[3] the research is expected to be reproducible.
7. Data, results, and biospecimens will be published and shared to the extent that this is consistent with protection of human research subjects (see Chap. 6).
8. The data and biospecimens will be securely stored.
9. The research team is well-qualified to conduct the study.
10. The institution has adequate facilities, personnel, and funding to support the research.

Although IRBs are supposed to focus on ethical/legal issues, they can also address scientific design issues to the extent that these impact risks and benefits (Emanuel et al. 2000; Binik and Hey 2019). Thus, scientific and ethical issues are not completely separable and often intersect in research review (Levine 1988). The use of placebo control groups (discussed below) is a clear example of this interaction, but there are many others. IRBs often include members with expertise in the relevant scientific areas who are qualified to comment on study design issues. IRBs may also consult with outside experts on scientific or ethical issues. Ideally, studies should undergo some form of scientific review prior to IRB review to ensure that research design issues are adequately addressed and allow the IRB to focus on ethical/legal issues (see, for example, National Institutes of Health 2023).

7.4 Managing Risk

Investigators, institutions, sponsors, and IRBs are charged with managing the risks of research (Wendler and Miller 2007, 2008; Rid and Wendler 2011). Risk management consists of five distinct but related processes:

1. **Risk identification**, i.e., listing and describing risks, including known risks and possible risks (or hazards). While there is widespread agreement that risks to subjects must be identified, there is less agreement about how far to extend risk identification (for example, to communities or society). See discussion below.

[2] "If applicable" is used here because some types of research, such as pilot studies, ethnography, and focus groups, are not intended to produce statistically significant results.

[3] "If applicable" is used here because some types of research, including clinical case reports, ethnography, and focus groups, are not intended produce reproducible results.

2. **Risk minimization, mitigation, and monitoring**, i.e. various steps taken to reduce the probability that harms will occur or reduce their impact when they do occur. See discussion below.
3. **Risk evaluation,** i.e., determining the probability and magnitude of possible harms, which involves a) assigning probabilities to harms, to the extent that these are known, and b), making judgments concerning the severity of harm pertaining to risks; that is, how bad the risk is (Shrader-Frechette 1991). A key problem in risk evaluation is that investigators and IRBs may lack enough evidence to make objective and reliable probability estimates (Resnik 2017a, 2021a).
4. **Risk/benefit balancing**, i.e., deciding whether the risks are worth taking (or reasonable), given the benefits. Note: risk balancing and evaluation both involve moral judgment (Rossi and Nelson 2012). See discussion below.
5. **Risk communication**, i.e., disclosing risks to human subjects and other impacted parties, such as communities (Resnik 2017b). This aspect of risk management occurs mostly during the consent process, which was discussed in Chap. 5.

The research proposal submitted by the investigator to the IRB should contain information related to these different aspects of risk management. If the proposal does not include this necessary information, the IRB may table it and send it back to the investigator with a request for more risk information. The IRB may also consult with outside experts concerning the risks of the study or conduct its own literature review. As noted above, it may not always be possible to obtain objective probability estimates of risk, due to lack of evidence, and there may be some risks which are not known (Resnik 2021a).

7.5 Minimizing, Mitigating, and Monitoring Risk

Determining how best to minimize, mitigate, and monitor risks is a fact-intensive investigation that depends on various details related to the study, such as the aims and objectives, target population, procedures, methods, tests, and interventions (Levine 1988). Nevertheless, there are some general considerations that apply to many different studies, such as:

- Conducting a review of the published literature (including animal and human studies) related to the research to better understand the risks (the investigator and IRB were faulted were not doing an adequate job of this in the Ellen Roche case, discussed in Chap. 2).
- For Phase I studies involving novel drugs or biologics, beginning the study by dosing only a few subjects at first and dosing additional subjects after an evaluation of the safety data shows that it is appropriate to do so (see TGN1412 mentioned above, see Nature 2006; Kenter and Cohen 2006).
- For Phase I dosing studies, beginning with a low dose, and gradually increasing the dose to a tolerable level, while monitoring data for safety (Kimmelman 2004; Resnik and Tinkle 2007).

- Excluding research subjects from the study who may be at increased risk due to their medical condition, concurrent or recent enrollment in another study, age, etc. (Levine 1988).[4]
- Carefully monitoring subjects when they are in the study and following up with them after they have completed the study (Morse et al. 2001; Kimmelman 2008).
- For clinical trials, using a data and safety monitoring board (DSMB) to evaluate preliminary data from the study and determine whether the study should be modified or stopped to protect subjects because the data indicate that an experimental treatment is ineffective or dangerous (Morse et al. 2001).[5]
- For studies involving medications, making sure the subjects have no known drug allergies.
- For studies involving experimental medications, making sure that these are prepared according to standards of quality and purity.
- Using a study design, such as active control groups or a crossover trial, to minimize risks and maximize benefits.[6]
- Reporting unanticipated problems, non-compliance, adverse events to the IRB, institution, and sponsor in a timely fashion (Levine 1988).
- For studies involving radiation, consulting with a radiation safety committee to protect subjects from unnecessary radiation exposures.
- For studies involving medications, following appropriate guidelines for dosing and administration (Morse et al. 2001).
- For studies involving the collection of blood, taking no more blood than is necessary to achieve the study's aims and not taking too much blood during a particular period, e.g., no more than 10.5 mL/kg subject body weight or 550 mL, whichever is less, per 8 weeks in healthy adults (University of Michigan 2022).
- For studies involving biopsies, taking no more tissue than is required to achieve the study's aims and providing appropriate wound care (Kimmelman et al. 2017).
- For studies involving magnetic resonance imaging, ensuring that subjects have no metal on or in their body which could be attracted to the magnet.
- For studies involving medical procedures that require sedation, ensuring that subjects follow proper safeguards, such as not eating or taking certain types of medications prior to sedation and waiting an appropriate time in the recovery area.
- For studies involving drug washout periods, close-monitoring of subjects to protect their welfare (Shamoo and Keay 1996; DuVal 2004).[7]

---

[4]This strategy can raise issues of justice and fairness, which will be discussed in Chap. 8.

[5]A DSMB could stop a study because a treatment has been proven so be so effective that it should be made immediately available to all patients who need it (Morse et al. 2001; Friedman and Schron 2008). To minimize ethical dilemmas related to stopping a study, DSMBs usually develop statistically-based stopping rules.

[6]The choice of a control group for a clinical trial often raises complex questions related to scientific rigor and ethics, which will be discussed below.

[7]During a washout period, subjects are asked to stop using medications for a period time, usually about 30 days, prior to the study to avoid dangerous drug interactions and confounding effects.

- For studies involving significant medical risk, using appropriate facilities and having access to appropriate resources (such as medical supplies, a crash cart, oxygen) and personnel (doctors, nurses) to treat injuries or other problems.
- For studies which ask subjects questions about stressful events they have experienced, such as trauma or abuse, making counseling services available for subjects who experience distress due to recalling these experiences.
- For studies involving deception, debriefing subjects after the study is over, i.e., explaining how and why deception took place.
- For all types of research, ensuring that researchers and staff have the appropriate experience, knowledge, and training to perform study procedures safely and effectively.
- For research involving medical risks to participants, ensuring subjects who are injured during the study receive appropriate medical care (Walker et al. 2022).
- For all types of research, developing and following standard operating procedures (SOPs) for performing interventions, procedures, and tests.
- For all types of research, implementing procedures to protect privacy and confidentiality (see Chap. 6).
- Taking steps, such as entering data into registries of subjects in Phase I studies, to ensure that subjects do not enroll in two different studies at the same (or within a safe window of time) without the investigator's knowledge or consent. Registries are important for minimizing risks to subjects and promoting the integrity of the study data (Resnik and Koski 2011).

As one can see from this list, there are many ways of minimizing and mitigating the risks of research. The list is but a small sample of the types of measures that investigators might use to protect human subjects from harm.

7.6 Risks to People Other than the Human Subject

Human subjects are not the only people who may be placed at risk by research. Research risks often impact many people other than the human subject, for example:

- Clinical trials that enroll pregnant or lactating women may expose fetuses or nursing infants to dangerous drugs (Grady and Denny 2008).
- Vaccination experiments that involve transmissible biological agents may create risks of infection for researchers, family members or others who come into contact with the research subjects (Resnik and Sharp 2006).
- Asthma studies involving pesticide treatments in the home to control insect allergens may create health hazards for occupants who come into contact with the chemicals (Resnik and Zeldin 2008).
- Thyroid cancer studies that require patients to ingest radioactive iodine may expose family members or others to hazardous radiation emanating from the patient (Wald 2010).

- Community-based research may generate information that could lead to stigma, bias, or discrimination against community members (Resnik and Kennedy 2010).
- Xenotransplantation experiments could introduce zoonoses into research subjects which could infect researchers, family members, and the human population (see discussion in Chap. 1; Melo et al. 2001).
- Research involving the administration of chemotherapy could pose risks to the environment from waste disposal or excretion of drugs into the water supply (Resnik and Tinkle 2007; Castellano-Hinojosa et al. 2023).
- Research on genetic factors related to intelligence could exacerbate discrimination or prejudice against certain racial or ethnic groups (Tabery 2015).

As one can see from these examples, research often involves risks to many people who are not research subjects, including members of the research team, the subject's family members, communities, populations, and even public health and the environment (Kimmelman 2005, 2020; Shah et al. 2018; Hanser 2020; Rothstein et al. 2024; Pratt 2024). In this book, I have already addressed some issues involving risks to others, including risks to communities (see discussion of community issues in Chap. 5) and risks to family members (see discussion of confidentiality issues in Chap. 6). As mentioned previously, with the exception of Subpart B of the Common Rule, which deals with risks to fetuses, federal research regulations instruct IRBs and investigators to focus on risks to the human subject and not on risks associated with the long-range effects of applying research.[8] Although the regulations do not prohibit IRBs from considering risks to other parties, such as family members or communities, they also do not require them to do so. This situation creates a serious regulatory gap with respect to certain kinds of risks (Kimmelman 2005; Resnik and Sharp 2006; Secretary's Advisory Committee on Human Research Protections 2022). In this section, I will consider managing risks to people other than the research subject in more depth.[9]

The federal research regulations notwithstanding, there are several compelling moral and legal arguments for investigators, IRBs, and institutions to manage risks to people other than the research subject, including:

- The principle of beneficence implies obligations to minimize risks, irrespective of whom they fall upon (See Chap. 3). According to the *Belmont Report* "Risks and benefits of research may affect the individual subjects, the families of the individual subjects, and society at large…In balancing these different elements, the risks and benefits affecting the immediate research subject will normally carry special weight. On the other hand, interests other than those of the subject may on some occasions be sufficient by themselves to justify the risks involved

[8]The U.S. federal regulations are an anomaly with respect to the topic. Rothstein et al. (2024) studied human subjects regulations from 22 countries and found that 8 require IRBs or RECs to consider the long-term social implications of research, 3 permit it, 10 do not explicitly permit, require it, or prohibit it, and only the U.S. prohibits it.

[9]These other people are sometimes referred as bystanders (Kimmelman 2005) or third parties (Resnik and Sharp 2006).

in the research, so long as the subjects' rights have been protected. Beneficence thus requires that we protect against risk of harm to subjects and also that we be concerned about the loss of the substantial benefits that might be gained from research (National Commission 1979; 8)."

- Prominent moral theories, including virtue ethics, utilitarianism, Kantianism, and natural law theory, hold that we should not intentionally harm other people or place them at unreasonable risk of harm (see Chap. 3).
- Since federal regulations require IRBs to consider the social benefits of research in their risk/benefit balancing (i.e., "the importance of the knowledge" see discussion below), they should also consider social risks, because benefits must be understood in relation to risks.
- According to negligence law, we have a legal duty not to impose unreasonable risks on others; if we violate this duty and someone is harmed as a result, we can be held liable for negligence (Kionka 2020).
- The CIOMS guidelines hold that investigators and IRBs should manage risks to communities, populations, and members of the research team (Council for the International Organizations of Medical Sciences 2016).
- The Secretary's Advisory Committee on Human Research Protections recommended that investigators, IRBs and institutions manage risks to non-subjects, such as family members, research staff, and other identifiable individuals placed at risk by research (Secretary's Advisory Committee on Human Research Protections 2022).
- The principle of trust implies that investigators and IRBs should address some types of risks beyond those that that impact human subjects, since harms to family members, communities, or the public may erode trust.

There are also solid arguments against expanding the scope of risk management beyond the human subject, including:

- Expanding the scope of risk management is a form of regulatory overreach that may trigger hostility, resentment, and non-compliance among investigators (Gunsalus et al. 2005; Schneider 2015).
- Expanding the scope of risk management diverts attention from the management of the most important risks, i.e., risks to human subjects (Gunsalus et al. 2005).
- IRBs, investigators, and institutions lack the expertise and experience to adequately deal with risks to communities, populations, or the public (Klitzman 2013).
- The IRB is a local review committee and expanding the scope of risk management requires it to deal with controversial social or political issues that are best handled by state or national governments (Gunsalus et al. 2005; Jordan and Gray 2017).
- Some risks, such as risks to research staff, are better addressed by organizations other than the IRB, such as the institution's occupational health and safety office.

Although I do not think these objections undermine the arguments in favor of managing risks to people other than research subjects, they do give us reasons to think carefully about how IRBs, investigators, and institutions should do this. IRBs, investigators, and institutions should address risks that are clearly within the scope of the expertise and are not adequately addressed by other organizations. Managing risks to human subjects should be prioritized in practice and policies. The following are some key points, offered as commonsense advice:

1. Risks to research staff are best addressed by principal investigators and organizations other than the IRB, such as the institution's occupational health and safety office. Where appropriate (for example, experiments involving infecting human subjects with biological agents that can be transferred to other people), the research protocol should address management of risks to research staff, but in most situations, this should not be the IRB's concern.
2. Risks to identifiable individuals (such as members of the subject's household), organizations (such as a business if a study involve collection of data from its employees or on its property), and communities (for example, when conducting community-engaged research) should be addressed in the protocol and appropriately managed.
3. In many cases, risks to identifiable individuals can be managed by informing the subject about these risk and advising them to take appropriate steps to minimize them. For example, if a subject will be receiving radioactive iodine therapy, they should be advised of the risk this creates for people they encounter and how to protect them from this risk.
4. There may be cases where it is appropriate to take steps to prevent harm to individuals who cannot be identified. For example, if a research subject receives anesthesia as part of a procedure and drives before they are fully recovered, they could have an accident and seriously harm themselves or other people, who cannot be identified in advance. This risk can be managed by required informing the subject about the risk and requiring them to use public transportation or receive a ride home from a family member or friend after the procedure.
5. In some cases, risk management may require an authorization from organizational or community leaders. For example, to study occupational health risks at factory, it will be necessary to obtain approval from owners or managers of the factory. In some types of community-based studies, it may be necessary to consult with and obtain approval and cooperation from community leaders. It may also be necessary to consult with leaders concerning decisions related to publication and data sharing (see discussions research with communities in Chaps. 4 and 5).
6. In some cases, a person may need to be excluded from a study or be withdrawn from it to manage risks. For example, if an experimental medication received as part of study poses a risk to a lactating infant, women who are nursing their babies should be told that they must agree to stop nursing to be included in the study (and find another means of feeding their baby) and if they do not agree to stop nursing they will be excluded from study (Food and Drug Administration 2019). (Issues related to including pregnant women in research will be discussed in more depth in Chap. 8.)

7. For xenotransplantation experiments, subjects may need to agree to long-term monitoring to minimize the risk of transmitting zoonoses to the human population (see discussion in Chap. 1).
8. IRBs may consider the social, political, public health, and environment risks of research but should be wary of giving them undue influence in decision-making because these risks are often highly politicized, uncertain, and far-removed from the study being considered for approval (Jordan and Gray 2017). Consider the following:

    - An investigator is proposing to conduct a study of an experimental medication for uterine cancer. Animal studies suggest that it may an effective abortifacient. Should the IRB not approve this study because if the drug is approved by the FDA, it could be used (off-label) to abort fetuses?
    - An investigator is proposing to conduct a study of a medication to prevent congenital hearing loss. However, members of deaf community oppose the study because they are concerned that widespread use of the medication would lead to discrimination against them. Should the IRB not approve the study because of the concerns raised by the deaf community?
    - A private company wants to dose human volunteers with a pesticide to determine whether existing regulations for use of the chemical are overly restrictive. Should the IRB not approve the study because the company might use the data it acquires from the study to convince the regulatory agency to weaken restrictions on the pesticide, which could place public health and the environment at risk?

While these social, political, public health, and environmental risks are important, they are usually best handled by organizations equipped to deal with them in a fair, reasonable, and democratic way, such as national committees sponsored by government agencies, or legislative bodies, or professional organizations (Moreno et al. 1998). IRBs and research institutions are ill-equipped to deal with these issues and may therefore do more harm than good if they disapprove of studies because they are concerned about social, political, public health, or environmental risks.

7.7 Social Benefits

Few people would dispute the idea that research with human subjects ought to yield social benefits (Emanuel et al. 2000; Wendler and Rid 2017; Resnik 2018a; London 2022). Support for this idea, known as the social value principle, occurs in the federal regulations, which require that risks are reasonable in terms of the benefits to the subjects and the "importance" of the knowledge expected to be gained, where importance can be equated with social value or benefit (45 CFR 46.111a). The social value principle also appears prominently in the Nuremberg Code (1949), which states that "The experiment should be such as to yield fruitful results for the good of society;" the Helsinki Declaration, which states that "Medical research involving human subjects may only be conducted if the importance of the objective outweighs the risks and burdens to the research subjects (World Medical Association 2024);" and the CIMOS guidelines, which state that "The ethical justification for undertaking health-related research involving humans is its scientific and social value (Council for the International Organizations of Medical Sciences 2016:1)." Additionally, one could argue that social value is important in fostering the public's trust in research, especially for studies that are funded with the public's money (Resnik 2018a).

Although almost all commentators agree that research ought to have social value, there is much less agreement about what this means in particular cases, because people disagree about what makes research socially valuable and whether social value is a necessary condition for research to be ethical or only an important desiderata (Wertheimer 2015; Resnik and Elliott 2016; Resnik 2018a). Which of these projects would or would not produce socially valuable results(?):

- Research on a new abortifacient
- Research on using fetal tissue to treat Parkinson's disease
- Research on using human embryonic stem cells to repair spinal chord fissures
- Research on genetic factors related to intelligence
- Research on genetically enhancing human athletic performance
- Research on gun control
- Research to reverse male pattern baldness
- Research whose primary goal is to help a company market a new drug

Claims about social benefits, like claims about social risks (discussed above), are sometimes disputed because they are based on controversial assumptions about fundamental moral, social, economic, or political values. As noted in Chap. 1, London (2022) argues that research with human subjects should promote egalitarian political ideals, but libertarians or utilitarians would disagree with this view. Supporters of a woman's right to abortion argue that research on abortifacients, abortion procedures, and abortion access is socially valuable, but abortion opponents would deny this claim (Schott et al. 2023).

Fortunately, the vast majority of studies that are submitted to a typical IRB have undisputed social value. Few people would question that social value of a clinical trial comparing two different treatment regimens for breast cancer or a study of the genetic and environmental factors related to the development of schizophrenia. While it will rarely be the case that IRB members will dispute the social value of a study submitted to the board, a more common scenario is that members might ask whether a study has enough social value to justify the risks that will be imposed on human participants. For example, one of the key ethical questions related to the COVID-19 challenge studies (see discussion in Chap. 1) was whether they were valuable <u>enough</u> to justify exposing human subjects to a dangerous disease (Rosenheck 2022).

7.8 Is Payment for Participation a Benefit?

If you ask the average person on the street whether money is beneficial, they may look at you like you're asking a stupid question and respond, "of course it is." Although our commonsense intuitions support the idea that money is a benefit, there is a general consensus among bioethicists that the IRB should not treat it as a benefit when weighing the risks and benefits of a study (Gordon et al. 2006). The FDA (2018) advises IRBs not to treat payment for participation as a benefit when it reviews research. The main argument for this position is that treating payment for participation as a benefit could allow an IRB to approve a high-risk study with few social benefits (Wertheimer 2013). However, this argument is not very convincing because, IRBs are not likely to do this if they take steps to ensure that risks will be minimized and benefits to society will be realized. Only a very irresponsible IRB would approve a high-risk study in which the chief benefit is that the subjects will be paid. Although I will not dispute the collective wisdom that the IRB should not treat money as a benefit, I do believe that the IRB should nevertheless consider the positive impact that payment may have on participants when it reviews research proposals. As discussed in Chap. 5, payments can make subjects feel like their contributions are appreciated and that they are being treated fairly, and therefore can help foster trust. From the subject's perspective, payment is a benefit and sometimes it's the most important benefit.

7.9 Risk Evaluation

As noted above, risk evaluation involves scientific and moral judgment, since it requires the IRB to assign probabilities to different outcomes and values (from good to bad) to these outcomes (Levine 1988; Rid et al. 2010; Rid and Wendler 2011; Rossi and Nelson 2012; Resnik 2017a). For example, suppose that the IRB is reviewing a study on healthy volunteers and the main risks of the study are related to a bronchoscopy that will be performed on the subjects. The risks of the bronchoscopy are coughing, discomfort, chest pain, bleeding, and in rare cases, pneumothorax, i.e., air between the lungs and chest wall, and in very rare case, death (Mayo Clinic 2024). To decide whether the benefits of the study (e.g., the knowledge gained) justify the risks of the bronchoscopy to the subjects, the IRB needs to know the probabilities associated with these harms and it must be able to determine the seriousness of these harms. For example, if the risk of death from a bronchoscopy is only 1/10,000, then this risk might be reasonable but if it is 1/100, then it might not be.

In making risk evaluations, IRBs should distinguish between the risks of the research and the risks of the clinical care that the patients/subjects would receive even if they were not participating in research (Wendler and Miller 2007, 2008). For example, consider a hypothetical study whose goal is to determine whether genetic factors impact tolerance for a kidney transplant. The study enrolls patients who are receiving a kidney transplant. The investigators collect an additional 50 ml of blood when blood is being drawn for clinical testing. The investigators perform genetic analyses on DNA extracted from the blood and correlate their findings with patient outcomes. The main risk of this research would be the risk of collecting an additional 50 ml of blood, not the risk of the kidney transplant, since the patients are receiving the transplant as part of their clinical care.

Risk evaluation should consider the net risks of all the study activities (see Table 7.1). For example, suppose a pulmonary function study includes: a medical history and exam; 30 ml blood draw once a week for 6 weeks, urine sample collection once a week for 6 weeks; sputum sample collection once week for 6 weeks; pulmonary function testing once every 6 weeks; two chest x-rays (radiation dose equivalent to 20 days of normal background radiation); and two bronchoscopies with conscious sedation at week 2 and week 6. The risks of this study would be the risks of all study activities combined (Wendler and Miller 2007, 2008). For studies involving exposure to ionizing radiation, it may be necessary to consider not only the radiation risk from the study but also how this risk will increase the person's lifetime radiation risk from other studies and activities (Resnik 2017b). See Table 7.1. Rid et al. (2010) have developed a relative to scale for rating the magnitude of harms of clinical research. The scale is based on input from ethicists, clinicians, and IRB chairs. See Table 7.2.

7.10 Risk/Benefit Balancing, Moral Values, and Trust

Risk benefit/balancing involves synthesizing the information from the first three steps of risk management and deciding whether the risks are reasonable or acceptable, given the benefits. This step can be the most complex and controversial aspect of the ethics and oversight of research with human subjects for several reasons:

1. **People may disagree about what should be counted as a risk or a benefit.** Should we consider payment for participation to be a benefit? Should risks to society or the environment be counted as risks? See discussion below.
2. **There may be insufficient evidence to reliably and accurately estimate the probabilities pertaining to risks and benefits** (Kimmelman 2004; Resnik 2017a, b, 2021a).
3. **People may still disagree about the evaluation of risks and benefits, i.e., the magnitude or risks or benefits.** Is anxiety worse than nausea? Is permanent toe numbness worse than losing a finger? I access to medical information as beneficial (sometimes) as access to medical treatment? (Kimmelman 2004).
4. **Even when people agree on all 1, 2, and 3 then may disagree about how to balance risks and benefits.** Should we infect healthy volunteers with COVID-19 to accelerate vaccine development? (See discussion in Chap. 1). Should we allow children to participate in risky research that does not benefit them as individuals

Table 7.1 Estimated risks associated with some research procedures or tests in healthy adults

Procedure or test	Risk of a serious adverse event[a]	Reference
Collecting less than 550 ml of blood in an 8- week period	Negligible	OHRP (1998)
Collecting urine, sputum, or check scrapings	Negligible	OHRP (1998)
Medical imaging routinely used in clinical examinations, such as x-rays, MRI (without contrast agent), EKG, EEG, DEXA scan, and sonography	Negligible[b]	OHRP (1998)
Interviews, surveys, focus groups	Negligible	OSHR (1998)
Skin, muscle, or fat tissue biopsy	Negligible	Wang et al. (2011)
Allergy skin testing	1/100,000	Bernstein et al. (2004)
Procedure involving general anesthesia	3/100,000	Schiff et al. (2014)
MRI with contrast agent	7.5/100,000	Schmidt et al. (2011)
Exercise stress testing	8/100,000	Gibbons et al. (1989)
Collecting 550 ml of blood at one time	2/1000	Crocco and D'Elia (2007)
Bronchoscopy	5/1000	Pue and Pacht (1995)
Pharmacokinetic testing and drug dosing	6.4/1000	Emanuel et al. (2015)
Lumbar puncture	1.5/100	Thomas et al. (2000)

[a]Serious adverse events (SAEs) include worsening of one's illness, acute toxicity, hospitalization, disability or permanent loss of function, life-threatening condition, congenital or birth defect, or death (FDA 2023). Non-serious adverse events (AEs) include many different things, such bruising or bleeding at site of a blood draw, headache, nausea, dizziness, mild fever, and shortness of breath.
[b]Although the risk per imaging test is negligible, risks for ionizing radiation exposure are cumulative (Kritsaneepaiboon et al. 2018)

Table 7.2 Scale for rating the magnitude of harms in clinical research (from Rid et al. 2010)

Magnitude	Example	Duration
Negligible	Mild nausea, skin bruise or irritation, minor bleeding	Minutes to hours
Small	Headache, congestion, fever, anxiety	Hours to days
Moderate	Uncomplicated bone fracture, insomnia for several weeks	Weeks
Significant	Torn ligament with joint instability, major depressive episode	Weeks to months
Major	Psychotic episode, loss of finger, loss of sexual function	Months to permanent
Severe	Paraplegia, loss of limb, loss of hearing or vision	Permanent
Catastrophic	Severe dementia, permanent coma, death	Permanent

but may benefit other children? (See discussion in Chap. 8). Should we compromise consent requirements to do important public health research? (See discussion in Chap. 5) (Kimmelman 2004).

The upshot of this is that moral judgment is woven throughout risk/benefit balancing and risk management generally (Shrader-Frechette 1991; Kimmelman 2004). As emphasized in Chap. 1–4, research with human subjects involves fundamental moral conflicts between the rights and welfare individuals and the greater social good. Nowhere do these conflicts appear more clearly than in risk/benefit balancing, since this step often require investigators, IRBs, and institutions to make difficult choices among competing values based on philosophical principles, religious traditions, or cultural norms.

Because risk/benefit balancing involves moral judgment and deliberation, IRB decision-making often exhibits inconsistency.[10] As we saw in Chap. 3, inconsistency is also a feature of our moral experience. That is, different people may draw different moral conclusions from the same set of facts and principles. IRB members, as well as IRBs, also behave this way because they interpret and apply the rules differently (Resnik 2015a). They behave more like juries than electronic calculators.

Numerous studies have documented significant differences in risk/benefit evaluations, consent requirements, and other determinations when different IRBs review the same research protocol used in a multisite study (Lux et al. 2000; Silverman et al. 2001; Hirshon et al. 2002; Green et al. 2006; Silberman and Kahn 2011). Green et al. (2006), for example, found significant variation in IRB risk/benefit evaluations of the same study reviewed at 43 different research sites. Ten IRBs gave the study expedited review because they judged it as minimal risk, 31 gave it full board review because they viewed it as more than minimal risk, one declared that the study was exempt from review, and one refused to approve the study on the grounds that it was too risky (Green et al. 2006). Shah et al. (2004), found significant variation in IRB members' judgments about risks. They found that 48% of 188 responding IRB chairs judged a magnetic resonance imaging scan with no sedation to be minimal risk, 35% said it was a minor increase over minimal risk, 9% said it was more than a minor increase over minimal risk, and 8% said that they didn't know its risk. 23% classified allergy skin testing as minimal risk, 43% classified it as a minor increase over minimal risk, 27% classified it as more than a minor increase over minimal risk, and 17% said they were unable to determine the risk (Shah et al. 2004). Reducing inconsistency of IRB review is one of the reasons why the 2018 revision of the Common Rules include requirement for a single IRB in multisite research (Wolinetz and Collins 2017).[11]

Researchers have complained about IRB inconsistency for decades (Reynolds 2002; Edwards et al. 2004; Silberman and Kahn 2011; Friesen et al. 2019) and these

[10] This is not the only explanation of inconsistency, but it is clearly an important one. Other plausible explanations include difference in expertise, experience, and resources (Silberman and Kahn 2011; Resnik 2017a, b; Serpico 2024).

[11] The other main reason is to reduce regulatory burden (Wolinetz and Collins 2017).

criticisms have a considerable degree of merit (Resnik 2017a, b; 2021a; Serpico 2024). To address this issue, as well as other epistemological problems related to IRB judgment and decision-making, some researchers and ethicists have argued that IRB determinations should be based on empirical evidence from the peer-reviewed literature and the historical record of IRB decisions, as well as quantitative reasoning methods, such as statistical inference, cost-benefit analysis, and decision theory (Martin et al. 1995; Anderson et al. 2012; Pritchard 2011; Bernabe et al. 2012; Kimmelman and Henderson 2016; Resnik 2017a, b, 2021a; Seykora et al. 2021; Serpico 2024). Elsewhere (Resnik 2017a, b, 2021a) I have defended evidence-based and quantitative IRB decision-making, within in limits. While I think that researchers, IRBs, institutions, and sponsors should make use of the best available evidence pertaining the ethical oversight of human research protocols, and that researchers should conduct empirical studies that can be useful in making these judgments and decisions,[12] I also believe that there are limits to this approach because of the inherent moral character of the decisions being made. Scientific research can provide us with information about the likely risks and benefits of a proposed study, such as a COVID-19 vaccine challenge trial (see Chap. 1), but it cannot tell us whether we <u>should</u> take these risks, i.e., whether the risks are reasonable or acceptable in relation to the benefits (Resnik 2017a, b).[13]

As argued in Chap. 3, unless one accepts a crude form of utilitarianism, moral decisions are not quantifiable. If this is the case, it follows that risk/benefit determinations in research with human subjects (and other IRB decisions) are also not quantifiable, because they have an inherent moral dimension. The authors of the Belmont Report understood this issue and put the matter this way:

> It is commonly said that benefits and risks must be 'balanced' and shown to be 'in a favorable ratio.' The metaphorical character of these terms draws attention to the difficulty of making precise judgments. Only on rare occasions will quantitative techniques be available for the scrutiny of research protocols (National Commission 1979, pp. 8–9).

The value-laden nature of IRB decision-making has important implications for notions of trust. Because IRB decision-making is influenced by moral values, it is very important for IRBs to take into account local interests, concerns, beliefs, language, and cultural and religious traditions (Klitzman et al. 2019; Taylor et al. 2019; Eder 2023). If an IRB does not do this (or not do it well), it may lose the trust of potential research participants and the local community by undermining their values. For example, an IRB that reviews research being conducted on a Native American tribe, such as the Havasupai (see discussion in Chap. 2), should take into account the tribe's beliefs about the sacredness of blood when it reviews the protocol and consent document and manages the risks and benefits of the research. If an IRB reviews an emergency research study (with no consent) that will take place in

[12] A good example of this approach is the research on the effects of financial incentives on subjects' judgment and decision-making discussed in Chap. 5.

[13] This point is related to the idea, discussed by Hume and Moore, that it is a fallacy to infer normative statements from descriptive ones. See Chap. 3.

a low-income, urban minority community, it should take into account the community's concerns about exploitation and historical injustice when it reviews the study. Because local context is so important for IRB review, IRBs should support and encourage meaningful participation of community members at meetings and single-IRBs for multisite studies should have mechanisms for incorporating meaningful local input into their decisions (Klitzman 2012; Kane et al. 2019; Eder 2023).

7.11 Research on Healthy Volunteers

Now that we have discussed some foundational issues for thinking about risk management in human subjects research, we will consider some specific topics. The first one we will address is research on healthy volunteers, because this can raise significant issues concerning public trust (Miller and Grady 2001; Resnik 2007, 2012; Bambery et al. 2016; Bull et al. 2020; Jamrozik et al. 2021). Healthy volunteers are involved in research in a variety of ways, such as:

- Participating in social/behavioral studies, such as surveys, interviews, focus groups, or behavioral experiments;
- Serving as healthy controls for epidemiological and clinical research
- Serving as experimental subjects for Phase I testing of drugs or biologics for safety, pharmacokinetics, or toxicity.
- Undergoing physiological, environmental, or pathologic challenges, such as engaging in strenuous exercise, breathing polluted air, or being exposed to industrial chemical or pathogens (a.k.a. "challenge studies).

As one can see, these studies range in degree of risk from very low to potentially very serious. In Walter Reed's malaria studies (see Chap. 2), two healthy volunteers died and most became very ill with malaria. In the COVID-19 challenge studies (see Chap. 1) healthy volunteers were exposed a pathogen, SARS-Cov-2, that has killed 20 million people since the fall of 2019.

As mentioned earlier, the occurrence of serious adverse events (such as death or hospitalization) in healthy volunteer studies can significantly undermine public trust, which is one of the reasons why these studies have received considerable ethical scrutiny in recent years (and deservedly so).[14] The Ellen Roche, Jesse Gelsinger,[15] TGN1412, and Biotrial cases illustrate how the public can react negatively when healthy people are seriously harmed in research. Thus, safeguarding public trust should be a key consideration in ethical guidance and policy for research involving healthy volunteers (Resnik 2012, 2015b).

[14] The notorious experiments conducted by German and Japanese researchers in World War Two involved healthy, coerced subjects.

[15] Although Gelsinger was not a healthy volunteer in the technical sense of the term, he was a fairly healthy 18-year youth who probably would have lived a longer time if he had not succumbed to the gene therapy vector infused into his liver.

While cases like those mentioned above attract a great deal of publicity, there is evidence that they are probably outliers and that risks in Phase I studies are very low. For example, Emanuel et al. (2015) conducted a meta-analysis of adverse events in 431 non-oncology Phase I drug studies involving 11,028 participants conducted between 2004 and 2011 at Pfizer test sites in Belgium, Singapore, and the U.S. They found that while 7028 subjects (63.7%) experienced adverse events, 84.6% of these events were mild or 14.4% were moderate and there were not deaths. However, 1% were serious and 0.31% were severe. Tranter et al. (2013) analyzed safety data from Phase I clinical trials involving monoclonal antibodies (similar to those used in the TGN1412 case). They estimated that the risk of death or a life-threatening event was between 1/425 and 1/1700 per volunteer trials; however, all of these death/life-threatening event data were from the TGN1412 study. When these data were removed, the risk went down 1/100000 to 1/1,000,000 per volunteer-trials.

The prospect of serious harm to healthy volunteers raises the issue of whether there should be any limits on the risks they are permitted to face in research (Miller and Joffe 2009; Resnik 2012; Różyńska 2015; Eyal et al. 2020). Although the federal regulations place some limits on the risks that children, pregnant women, fetuses, neonates, and prisoners, can face in research (see Chap. 8), they place no limits on the risks that healthy, adult volunteers may be exposed to in research; they only require that risks be reasonable in relation to expected benefits (45 CFR 46.111). The only research ethics guideline that mentions limitations on risks for healthy volunteers is the Nuremberg Code (1949), which states that: "No experiment should be conducted, where there is an a priori reason to believe that death or disabling injury will occur; except, perhaps, in those experiments where the experimental physicians also serve as subjects." This guidance is not very useful, however, because it does not define "a priori reason" and it would probably prohibit most oncology clinical trials in which research participants must face the risk of a serious adverse event but also have the opportunity to receive effective treatment for their disease (Resnik 2012).

Healthy volunteer studies benefit society in many ways (Miller and Joffe 2009). Phase I studies play a crucial role in the development of new drugs and biologics, and human challenge studies provide scientists and regulators with important data pertaining to risks related to exposure to chemicals and pathogens (Resnik 2012; Bambery et al. 2016). Though controversial, COVID-19 challenge studies have also provided useful information about how to control the spread of this disease (Rosenheck 2022). Although healthy volunteers do not receive direct medical benefits from participation, they may receive other benefits, such information about their health, money, and a sense of self-worth from knowing that they have made a contribution to the advancement of science. Although the main reason why healthy people volunteer for risky studies is to earn money (Stunkel and Grady 2011; Fisher 2020), federal agency guidance discourages IRBs from treating financial compensation as benefit in its risk/benefit calculus (Wertheimer 2013; see discussion above).

Since healthy volunteer studies offer important benefits to society and to the participants, the main ethical issue is whether risks should be limited to prevent other types of harms, such as harm to the research enterprise if a volunteer is seriously injured or killed (Resnik 2012; Różyńska 2015). Thus, the issue of whether to limit the risks that healthy, adult volunteers can face in research involves a conflict between respect for dignity/autonomy and beneficence on the one hand and non-maleficence on the other. Respect for dignity/autonomy and beneficence justify the position that there should be no limitations on the risks that healthy adults can face in research. Healthy adults, one might argue, should be allowed to make decisions concerning their risk exposure without outside interference (Różyńska 2015). If we respect healthy adults' decisions to engage in risky occupations (such as logging or coal mining) and risky recreational activities (such as sky diving or mountain climbing), we should also respect their informed decisions to participate in risky research (Miller and Wertheimer 2007).

The principle of non-maleficence supports the position that there should be some limits on the risks that healthy adults can take in research (Resnik 2012; Różyńska 2015). We enact many different policies to that protect healthy adults from exposures to risks, such as laws pertaining to seatbelt and motorcycle helmet use, food and drug safety, workplace safety, and so on. The reasoning that justifies prohibitions on selling or possessing cocaine, methamphetamine, and other dangerous drugs, also justifies preventing healthy adults from participating in experiments that society deems to be too risky (Miller and Wertheimer 2007; Resnik 2012). While these policies are paternalistic (see discussion in Chap. 2), one might argue that they are needed to protect people from harm.

If we think of this issue from the perspective of trust, catastrophic harms to human subjects (such as death) can significantly undermine the public's trust in the research enterprise (Resnik 2012). As discussed previously, the deaths of Gelsinger and Roche elicited a negative public reaction, as did the disastrous Phase I trials conducted by Paraxel and Biotrial. One could argue that some limits on the risks that healthy volunteers can face in research are necessary not only to protect individuals from harm but also to protect the scientific enterprise (Miller and Wertheimer 2007; Resnik 2012; Różyńska 2015).

This argument also applies to the case of self-experimentation, in which scientists/physicians take serious risk to advance human knowledge. One of the most famous cases of self-experimentation occurred in 1983, when Barry Marshall, a professor of clinical microbiology at the University of Western Australia, ingested a solution containing *Heliobacter pylori* bacteria to test his hypothesis that these pathogens can cause ulcers. Marshall developed an ulcer, which he was able to treat successfully with antibiotics. He shared the 2005 Nobel Prize in Physiology and Medicine with Robin Warren for this discovery (Azad 2014). Not all self-experimenters are as careful or responsible as Marshall, however. "Do-it-yourself," citizen scientists have injected themselves with unregulated chemicals, biologic compounds, and gene therapy vectors with the goal of discovering new vaccines and therapies or even augmenting themselves (Keulartz and van den Belt 2016; Kolodziejczyk 2017). Although scientists who experiment on themselves usually have highly honorable motivations, these sorts of studies should be regulated to protect research subjects and the scientific enterprise (Davis 2003).

Assuming that some limits on risks to healthy volunteers are morally justified, the next issue to address is establishing an acceptable risk level or "maximum" risk (minimal risk is discussed below). Since moral theory does not tell us much specifically about what makes a risk acceptable (see Chap. 3), a reasonable way of establishing an acceptable level of risk for healthy research volunteers is to define this risk in terms of a comparable socially acceptable activity. Although one might argue that social acceptability is not a sufficient basis for making ethical decisions about risks because it is a relative standard, it does have the virtue that is consistent with a public trust ethos. That is, research which is perceived to be riskier than comparable activities is more likely to undermine public trust than research which is perceived to have risks which are on par with comparable social activities (Resnik 2012).

If our argument is sound up to this point, the next step is to establish a comparable activity. London (2006) argues that volunteer firefighting is a comparable activity because it (1) involves significant risk; and (2) is socially valuable. Volunteer firefighters have a morality rate of about 3.4 deaths per 100,000 full-time equivalent (FTE) jobs per year (U.S. Bureau of Labor Statistics 2022a). Miller and Joffe (2009) argue that kidney donation is a comparable activity for the same reasons that London thinks that volunteer firefighting is a comparable activity. Kidney donation has a fatality rate of about 30 to 40 deaths per 100,000 donors (Taliercio et al. 2011).

These authors make a crucial mistake, however: for most healthy volunteers, research is a form of paid labor, not a purely altruistic act, like serving as a volunteer firefighter or donating a kidney to a relative (Fisher 2020; Walker et al. 2022). Thus, we should equate the acceptable upper bounds of healthy volunteer research with a comparable form of risky, paid labor. Table 7.3 includes data on fatal occupational injuries for selected U.S. occupations.

As one can see from Table 7.3, there is significant variation in occupational risks, which range from loggers with 100.7 deaths per 100,000 full-time equivalent (FTE) workers to computer scientists, with 0.2 deaths per 100,000 FTE workers. Emergency medical technicians have the U.S. average (mean) of 3.7 deaths per 100,000 FTE workers, with warehouse workers (4.2) just above and local government (3.0) just below.

What would be a comparable activity for acceptable risk in healthy volunteer research? Logging work seems way too risky, but computer scientist seems to be too safe. Somewhere in between these two extremes would seem to be a reasonable place to set risk limits. But where? Setting the bar limit lower than the mean of 3.7 deaths per 100,000 FTE workers seems too low because the public has come to accept that participating in research involves a certain amount of risk, especially when the research has considerable social value. For example, there was substantial public support for the COVID-19 vaccine challenge studies that took place in the U.K. even though these involved significant risks, including the possibility of hospitalization (22 per 100,000 participants) or death (2.5 per 100,000 participants) (Manheim et al. 2021; Barker et al. 2022; see Chap. 1). I would argue therefore, for setting the mortality risk **at 7.4 deaths per 100,000 FTE workers or double the average occupational risk.** For the typical Phase I study, which involves about an

Table 7.3 Job-related mortality risks for selected U.S. occupations (U.S. Bureau of Labor Statistics 2022a, b)

Occupation	Fatal occupational injury per full-time equivalent 100,000 workers at 40 hours per week and 50 weeks of work per year or 2000 hours per year
Logging workers	100.7
Roofers	57.5
Fishing and hunting workers	50.9
Aircraft pilots and flight engineers	35.9
Structural iron and steel workers	21.3
Landscapers	18.1
Mining (except oil and gas)	16.3
Construction workers	15.6
Police officers	14.3
Shuttle drivers, taxi drivers, chauffeurs	11.2
Automotive services technicians and mechanics	7.5
Machinery manufacturing	5.4
Warehouse workers	4.2
Emergency medical technicians	3.7
Firefighters[a]	3.4
Local government	3.0
Grocery stores	2.3
Retail sales	1.9
Sales/office occupations	1.1
Registered nurses	0.7
Lawyers	0.6
Scientists and lab technicians	0.6
Educators and librarians	0.4
Computers scientists and software developers	0.2
	U.S. average = 3.7

[a]This excludes data from September 11, 2001 terrorists attacks, in which 343 firefighters died trying to save people who were trapped inside the burning World Trade Center buildings

entire month of time (720 hours) (Pfizer.com 2024), or about 0.36 FTE worker years, this would be 2.6 deaths per 100,000 volunteers in Phase I studies or roughly the risks estimated for the COVID-19 challenge studies.[16]

[16] In the first edition of this book, I argued for a much higher limit of 65 deaths per 100,000 FTE workers. This number was based arrived at by splitting the difference between the highest at lowest rates for the data available at that time. I now think this number may have been too high (Resnik 2018b).

I will not argue for establishing a risk limit for SAEs, because these risks are very difficult to estimate and depend a great deal on the type of study under consideration. For some oncology clinical trials that enroll cancer patients (Phase II or Phase III), there may be 10–12 SAEs per fatal AE (FAE) (Wang et al. 2019). For other studies, that ratio may be much different. For example, for research involving a bronchoscopy, the risk of death may be close to 0.0, so that ratio of SAEs to FAEs would very high (>10 million to 1). So, while one might expect that for a study involving a bronchoscopy as the main risk there will be 500 SAEs per 100,000 subjects, there may be 0 deaths.

Now, one might object that all of this seems incredibly arbitrary. Why not set the risk level higher or lower? I agree that the choosing an exact number is somewhat arbitrary, but it seems that any number we pick could be regarded as arbitrary. At least the number I have picked has the virtue that it is closely linked to social acceptability and public trust, since it is based on occupational risk data. If we set the level much higher, then more subjects could suffer significant harms in research, which would have negative repercussions for the injured parties and the scientific enterprise. If we set the maximum level much lower, then we would prohibit investigators from conducting important research that benefits society.

Another objection is that the risk level I have set is relative to a particular society, i.e., the U.S. In other societies that accept higher risks for workers, the acceptable risk level for healthy volunteers might be much higher. I agree that the standard I have chosen is relative and not absolute. It might be higher or lower in other countries. But, barring a robust moral theory that can provide use with universal standards for the acceptability of risks, I don't see a reasonable way of avoiding this problem.

One might also object that the maximum level is too rigid. Situations might arise in which it would be desirable to exceed this risk level to conduct research with important benefits for society, such as the COVID-19 vaccine challenge trials (Eyal 2020). I agree that the maximum risk level should not be viewed as rigid: it should be a guideline, not an absolute rule. An IRB should be able to approve a study that exceeds this risk level if it determines that the research addresses a compelling societal need. An expected fatality rate of 7.4 deaths per 100,000 FTE workers should be a default position that can be exceeded for compelling moral reasons.

7.12 Phase I Trials on Seriously Ill Patients

Risk/benefit balancing in research involving patients is very different from risk/benefit balancing in research involving healthy volunteers because patients may benefit medically from their participation. We allow studies to be performed on patients that we would consider to be too risky to perform on healthy volunteers because patients often receive medical benefits from their participation, but healthy volunteers do not (Wendler and Miller 2008). If someone has a 90% chance of dying from terminal cancer in the next three months when treated with conventional therapy, then it may be reasonable for them to participate in clinical trial of an experimental drug with a 50% chance of killing them but also a 50% chance of curing them, since a 50% chance at a life is better than a 10% chance.

Although clinical research studies may offer patients direct benefits, their main purpose is to develop new knowledge. Clinical trials are designed to evaluate medical treatments, not to provide individualized care for patients. It is important for investigators to help their patients understand the difference between research and therapy and not take advantage of their desperate situation (see discussion of the therapeutic misconception in Chap. 4). Patients with a terminal or incurable illness may be willing to try anything that gives them a chance to obtain effective treatment.

Although Phase I clinical trials are usually performed on healthy volunteers, some Phase I studies are performed on patients with terminal or incurable diseases (such as some types of cancer) (Bittlinger et al. 2022). The rationale for conducting Phase I studies on these patients is that the treatments are so dangerous that it would be unethical to ask healthy volunteers to receive them. Patients with terminal or incurable illnesses have nothing to lose and much to gain by trying experimental treatments in Phase I studies (Agrawal and Emanuel 2003, 2008). Jesse Gelsinger (see discussion in Chap. 2) was one of these patients.

Some have argued that most Phase I oncology trials are unethical because the risks are not reasonable in relation to expected benefits since the patients' chances of benefit are remote (Miller 2000). Phase I trials are designed to evaluate safety and toxicity, not efficacy. These studies usually systematically escalate dosing to determine the maximum tolerable dose and the effective dosing window. Patients may therefore not receive an effective dose; or, if they do, they may only receive it for a short period of time. Phase I studies usually last 30 days or less, which may not be enough time for patients to derive significant or long-lasting benefits (Miller 2000).

Others have argued that the risks of Phase I trials involving seriously ill patients are justifiable if there is a reasonable expectation of direct benefit for the patient (Agrawal and Emanuel 2003, 2008). Studies have shown that the overall response rate to the treatment in Phase I oncology trials is 4.4% and that the mortality rate is about 0.5% or about 500 per 100,000 (Agrawal and Emanuel 2008).

In thinking about Phase I trials on sick patients from the perspective of trust, informed consent is a key consideration (King 2000). Patients with a terminal or incurable illness may be especially susceptible to the therapeutic misconception, since they may be willing to try anything that will give them a hope of a cure (Miller 2000). Investigators who enroll patients in Phase I trials should carefully explain to them the risks and benefits of the research as compared to standard treatment if any is available. Some cancer patients may prefer to forego treatment and live the remainder of their days with fewer adverse effects and symptoms than to take experimental treatments that do not significantly increase their lifespan but cause significant suffering (Bittlinger et al. 2022). Investigators who do not clearly discuss the different options (including the option of no treatment) with their oncology patients may betray their trust.

7.13 Randomized Controlled Trials

Randomized controlled trials (RCTs) are widely regarded as the "gold standard" for clinical research because they use methods, such as control groups (e.g., experimental vs. standard therapy or placebo), random assignment of patients to different treatment groups, blinding (taking steps to prevent patients and researchers from learning who is getting the experimental treatment), and clearly defined protocols to control bias and enhance rigor and reproducibility (Sackett et al. 1997). The SUPPORT study and the HIV prevention trials (discussed in Chap. 1) were RCTs. The first controlled trial in medical research, James Lind's (1716–1794) study of different treatments for scurvy, took place in 1747, but the first RCT did not occur until 1946, when the Medical Research Council of the U.K. conducted an experiment to test the efficacy of streptomycin in treating tuberculosis (Bhatt 2010). Some types of RCTs include:

- Placebo-controlled RCTs (discussed in Sect. 7.14). Subjects are randomized to the experimental treatment group or placebo-control group.
- Active-controlled RCTs. Subjects are randomized to experimental treatment group or a group that receives a standard therapy.
- Crossover study: All subjects are randomized to the treatment or placebo group, then, after an appropriate period all time, subjects switch groups (Prasad and Grady 2014).
- Pragmatic RCTs: the purpose of the trial is to compare treatments under real-world conditions, such as in an outpatient clinic or intensive care unit (Ford and Norrie 2016; Sugarman and Calliff 2018).
- Cluster RCTs: the treatments are randomized to among different groups of subjects instead of among individual subjects (Weijer et al. 2011). The Johns Hopkins catheter infection study (discussed in Chap. 1) was a cluster RCT.
- Adaptive RCTs: The design of the study is modified during the trial in response to incoming data. For example, a 50/50 allocation of experimental vs. placebo could be changed to 60/40 or the sample size could be decreased if the experimental treatment is proving to be much more effective than the placebo (van der Graaf et al. 2012).

In this chapter, we will focus on two ethical issues related to RCTs: randomization and the choice of control groups.

In 1970s and 1980s, physicians and ethicists began to question the ethics randomization because it seemed to conflict with a physician's obligation to provide the best treatment for their patients (Fried 1974; Marquis 1983; Hellman and Hellman 1991; Rothman and Michels 1994). In clinical practice, physicians recommend treatment based on their medical knowledge, the patient's condition, and the patient's preferences. For example, if a woman has breast cancer, a physician could recommend various treatment options, such as mastectomy, lumpectomy, different types of chemotherapy, and radiation. The woman's treatment decision should not be made at random but should depend on various factors, such as the size and stage of the tumor, her ability to tolerate different types of chemotherapy, her concerns about disfiguring surgery, etc.

So, if a physician should not randomize their treatment recommendations when practicing medicine, what justifies randomization in medical research? Charles Fried (1974) and Benjamin Freedman (1987) developed the concept of equipoise to justify random assignment to different treatment modalities in clinical research and other authors have refined this idea (Miller and Weijer 2003; London 2007; Joffe and Truog 2008; Nix and Weijer 2021; London and Seymour 2023). Equipoise refers to a situation in which the medical community is uncertain about which among two or more treatments (including no treatment at all) is better. The purpose of conducting an RCT is to compare current treatments to resolve clinical uncertainty and determine if one treatment is better than the other (s) (Freedman 1987; Miller 2008).[17] Randomization is justified because it is needed to control bias and produce a valid answer to this clinical question.[18] Moreover, clinical investigators may initiate an RCT <u>only</u> when equipoise exists (Miller and Weijer 2003). Randomly deciding a patient's treatment when one treatment is recognized as better than the others is unethical because it violates the physician's fiduciary duty to provide patients with the best treatment (Miller and Weijer 2003).

There are at least two major critiques of this position. First, some have argued that the notion of equipoise is inherently vague and subject to multiple interpretations because it is not clear what is meant by "uncertain" (is uncertainty the same as lack of evidence from rigorous studies, lack of professional agreement, both or neither?) and "better" (better in terms of what—cost, convenience, mortality, morbidity, disease biomarkers?) (Gifford 2000; Miller 2008; Miller and Brody 2007; Miller

[17] Different types of RCTs aim to achieve different goals. For example, for two treatments, A and B, a superiority trial attempts to determine whether $A > B$, a non-inferiority trials attempts to determine whether $A \geq B$, and an equivalence trial attempts to determine whether $A = B$. The null hypothesis for a superiority trial would be—($A \ngtr B$) and for a non-inferiority trial it would ($A \nless B$). Equivalence trials are, statistically speaking, more difficult to conduct than the other trials because to prove that $A = B$, one must have enough evidence to reject two null hypotheses, i.e., $A > B$ and $B > A$ (Christensen 2007).

[18] There is an interesting debate among philosophers of science and statisticians concerning the validity of randomization as a method of controlling bias. The original rationale for randomization was developed by the pioneering statistician R.A. Fisher (1890–1962), who argued that randomly assigning research subjects to different treatment groups reduces biasing effects due to (1) unknown confounding variables; (2) the experimenter's preferences (Fisher 1925; Hall 2007). Some have challenged viewpoint by arguing that randomization does not guarantee unbiased assignment, especially in small samples, and that other methods of assignment that an account for all the relevant variables work better (Urbach 1985). However, I believe that randomization is a valid method for assigning research subjects to different treatment groups because I am not convinced that other methods of assignment can fully account for unknown, confounding variables or the experimenter's preferences, which may subconsciously affect behavior (Mayo 1987). Additionally, there is moral argument for randomization: randomization is procedurally fair (like drawing lots) since a random process and not the investigator (who be prejudiced for or against certain patients) decides the assignment. It is important to note this is tangential to controversies about equipoise because equipoise is used justify scientific methods of treatment assignment, whether occur through randomization or some other method.

and Joffe 2011). Although defenders of equipoise have tried to clarify what is meant by this concept, a recent study by Dewar et al. 2023 involving semi-structured interviews with 15 clinical researchers, 15 research ethics board chairs, and 15 philosophers of science/bioethicists found that while most found the concept of equipoise to be useful, they had difficulty defining it.

Second, Frank Miller (Miller and Brody 2002, 2003, 2007; Litton and Miller 2005; Miller and Joffe 2011) have challenged the equipoise requirement by arguing that the ethics of clinical research is different from the ethics of clinical medicine. The ethical requirements are different in these settings because the goal of clinical research is to develop generalizable knowledge, while the goal of clinical medicine is to promote the best interests of patient. Consequently, clinical research differs from clinical medicine in numerous ways (see Table 7.4) (Miller and Brody 2002; Resnik 2009). Patients may benefit from participating in clinical research, but studies are designed to develop medical knowledge, not necessarily to benefit individual patients (National Commission 1979). When patients consent to participate in clinical research, they are told how their treatment will differ from standard care, and that the goal of the study is to gain new knowledge. **What justifies participation in an RCT is that the patient has consented and the RCT is designed to answer an important biomedical question.** It may be the case that the goal of the study is to determine whether a particular treatment is better than another, but this need not be the case. Equipoise may be an aspect of the research design, but it is not an essential condition for initiating an RCT.

Additional support for this position comes from the observation that in other, non-research contexts, patients can consent to suboptimal therapy (Chiong 2006;

Table 7.4 Clinical medicine vs. clinical research

Clinical medicine	Clinical research
Goal is to promote the best medical interests of the patient in accordance with their preferences and values.	Goal is to develop generalize knowledge.
Treatment can be individualized to deal with the patient's unique circumstances.	Treatment must follow a rigid protocol to reduce confounding variables and enhance rigor.
Test, procedures, interventions, and examinations are performed only to address medical questions relevant to the patient's care.	Test, procedures, interventions, and examinations may be performed for scientific reasons.
Treatment decisions are made by the patient and doctor.	Treatment decisions can be made by scientific procedures, such as randomization.
The doctor informs the patient about the treatment they are getting.	The patient and doctor may be blinded with respect to treatment information to control for the placebo effect.
Patients do not receive placebos, except in some special cases where placebos are used to diagnose certain psychogenic conditions.	The patient may receive a placebo.

Resnik 2009). For example, the best treatment for a disease may be an expensive medication that is still under patent. A patient may decide to take a medication that is less effective and expensive than the new medication because they cannot afford the new one. It would be acceptable for the physician to prescribe the less effective medication in this situation to accommodate the patient's financial needs, if the physician informs the patient about the option of taking the expensive medication. Additionally, patients receiving care in teaching hospitals or medical centers may agree to allow students or trainees to perform procedures on them under the supervision of a health professional. Although the best care would be provided by someone who already has the appropriate training and experience, most of us would agree that it is ethical to allow students and trainees to practice on patients because this helps them to learn, which is an important social goal (Chiong 2006). Thus, if a patient can consent to receiving suboptimal medical care in these situations, a patient may also consent to having his or her treatment randomized, regardless of whether equipoise exists, provided that the physician informs the patient of his or her options, including the option of not participating in research and obtaining treatment outside of a study, if it is available (Chiong 2006; Resnik 2009).

The dispute concerning the equipoise requirement involves a conflict between promoting the welfare of patients/subjects and advancing scientific knowledge (Resnik 2009). Proponents of the equipoise requirement argue that patients' best interests must always take precedence over scientific objectives, while critics of the requirement argue that patients' best interests need not always take precedence if patients agree to place their well-being at risk to help answer important scientific questions.

I agree with Miller's perspective that equipoise is not a necessary requirement for initiating an RCT. However, I do not view equipoise as an utterly useless concept since it can play an important role in framing research questions and protecting subjects from exploitation (London 2007). Also, the moral legitimacy of Miller's position very much depends on the patient/subject's consent, since this is what justifies involving them in clinical research activities that may not be in their best medical interests. As we saw in Chap. 5, patients often succumb to the therapeutic misconception by mistakenly believing that clinical studies are designed to benefit them. Because this misconception is difficult to overcome, physician/scientists must make a special effort to help their patients/subjects understand that they will be participating research and that the study is designed to develop new knowledge and not necessarily to benefit them. Patients/subjects should also understand methods, procedures, and tests that are part of the research (as opposed to their clinical care), such as randomization, the use of placebos, blinding, and medically unnecessary tests or procedures. Taking the time to help patients/subjects understand this information conveyed during the consent process will help to avoid misunderstanding and will promote trust.

7.14 Placebo-Controlled Trials

In a placebo-controlled RCT, subjects in at least one of the treatment groups receive an inactive treatment (or placebo) such as sugar pill or saline injection (Emanuel and Miller 2001). In an active-controlled RCT, all subjects receive an active treatment with some known efficacy. There are various types of placebo-controlled RCT designs. Some common types include:

- Experimental treatment vs placebo.
- Experimental vs. standard treatment vs. placebo.
- Two types of standard treatment vs. standard treatment and placebo (placebo add-on).

The purpose of including a placebo control group in an RCT is to control for bias due to the placebo effect, a well-documented phenomenon in which a person's belief that they are receiving an effective treatment can influence their response to treatment (Temple and Ellenberg 2000; Pardo-Cabello et al. 2022). The placebo effect is especially powerful in the treatment of pain and some psychiatric disorders, such a depression (Laughren 2001; Kirsch et al. 2008; Marchant 2015). Because the placebo effect is so strong, double-blinding (i.e., blinding the patient and researcher) is standard practice in placebo-controlled RCTs (Emanuel and Miller 2001).

Few people raise ethical objections to using placebo control groups when there is no effective treatment for a disease or condition because patients who receive a placebo are not being denied anything they would have received as part of standard therapy. From the perspective or equipoise, an RCT is acceptable because there is clinical uncertainty about whether the experimental treatment is better than what the patient would otherwise receive, i.e., no treatment (Miller and Weijer 2003). The use of placebo-control groups in RCTs raises ethical concerns, however, if there already is an effective treatment, because this study design would seem to violate a physician's duty to provide effective treatment for their patients (Emanuel and Miller 2001).

Some ethicists and physicians hold that placebo-controlled trials are unethical when there is an effective treatment because this violates the doctor's duty to provide their patients with the best medical therapy (World Medical Association 1996; Miller and Weijer 2003; Joffe and Truog 2008). If an effective treatment exists, then an RCT should incorporate an active-control study design (for example, one that compares the experimental treatment to standard therapy). Critics of the HIV prevention trials (discussed in Chap. 1) argued that these studies were unethical because some subjects received placebos even though a therapy for preventing mother-child transmission of HIV had been proven effective (Lurie and Wolfe 1997).[19]

[19] It is worth noting that this argument assumes that subjects who receive a placebo will be worse-off than those who receive an experimental treatment but that is not always the case. As noted above, placebos can be highly effective; also, experimental treatments can be ineffective and dangerous.

Others hold that these studies can be ethical, provided that (1) a placebo control group is needed to meet standards of scientific rigor; (2) subjects understand that they may receive a placebo; (3) subjects are not likely to suffer any serious harm from receiving a placebo, and (4) the research serves an important biomedical or public health goal (Varmus and Satcher 1997; Resnik 1998). The 2024 revision of Helsinki Declaration[20] reflects this approach to placebo-controlled RCTs:

> The benefits, risks, burdens and effectiveness of a new intervention must be tested against those of the best proven intervention(s), except in the following circumstances:
>
> Where no proven intervention exists, the use of placebo, or no intervention, is acceptable; or
> Where for compelling and scientifically sound methodological reasons the use of any intervention less effective than the best proven one, the use of placebo, or no intervention is necessary to determine the efficacy or safety of an intervention and the patients who receive any intervention less effective than the best proven one, placebo, or no intervention will not be subject to additional risks of serious or irreversible harm as a result of not receiving the best proven intervention.
> Extreme care must be taken to avoid abuse of this option (World Medical Association 2024).

The CIOMS guidelines take a similar stance toward placebos in RCTs:

> Placebo may be used as a comparator when there is no established effective intervention for the condition under study, or when placebo is added on to an established effective intervention.
> When there is an established effective intervention, placebo may be used as a comparator without providing the established effective intervention to participants only if:
>
> there are compelling scientific reasons for using placebo; and
> delaying or withholding the established effective intervention will result in no more than a minor increase above minimal risk to the participant and risks are minimized, including through the use of effective mitigation procedures (CIOMS 2016: 15).

While there seems to be an emerging consensus that placebo control groups are ethically acceptable in RCTs, provided that certain conditions are met to protect the rights and wellbeing of patients (Emanuel and Miller 2001), there are still some remaining issues that researchers and IRBs need to be mindful of:

- **Is a placebo control group needed to meet standards of scientific rigor?** This question is not always easy to answer. When there is no effective therapy a placebo control is a clearly necessary. In other situations, it may or may not be necessary, depending on the power of the placebo effect in a particular area of medicine and the effectiveness of existing therapies as compared to placebo (Laughren 2001). In any case, researchers should provide the IRB with a scientific justification for the use of placebo control group in their protocol.
- **Will the subjects have any problems understanding that they may receive a placebo?** Since, studies have shown that subjects often do not have a good understanding of what is meant by a placebo (Tam et al. 2015), researchers must take due care to ensure that they understand that they may receive a placebo. This requirement is also important for ensuring the placebo-controlled trials are

[20] Early versions of the Declaration only allowed placebos to be used in RCTs when there are no effective therapies. See World Medical Association (1996).

trustworthy because subjects who initially do not understand that they may receive a placebo may feel like they've been deceived if they later discover that they have received a placebo.

- **Are the subjects expected to suffer any serious harm from receiving a placebo?** In most cases, the harms that occur will be due to not receiving an effective therapy; for example, if a person stops taking medication for Type 2 (insulin-resistant) diabetes while participating in study (Aronoff et al. 2000). In some cases, the placebo itself may cause harm. For example, sham surgery is a type of placebo treatment that includes everything that a person would receive to have genuine surgery (e.g., analgesia, anesthesia, and an incision) except the experimental surgery. Because sham surgery can cause pain and discomfort and may involve significant risks, it is ethically controversial (Miller 2003). However, it may be appropriate in some cases where the efficacy of surgical intervention is unknown, such as some types of orthopedic surgery (Sihvonen et al. 2013). Establishing a maximum acceptable risk level can be difficult. In the first edition of this book, I argued that the maximum risk level for subjects who receive a placebo should be roughly the same as for healthy volunteers, because both types of subjects do not receive medical benefits (Resnik 2018b). I no longer agree with this position because subjects in the placebo control group may receive medical benefits, such as diagnosis and treatment related to study participation, and when they consented that had a chance of not receiving a placebo. The CIOMS guidelines state the maximum risk should be no more than a "minor increase over minimal risk," but what is meant by this phrase is unclear, because it its interpretation is linked to the concept of "minimal risk," which, as we shall see below, is also unclear.

7.15 Alternatives to RCTs in Clinical Research

Although RCTs are widely recognized as the "gold standard" for scientific rigor in clinical research, investigators may decide to conduct studies that deviate from the RCT paradigm for ethical or practical reasons (Saccà 2010). Some of these types of studies include:

- Phase I drug dosing studies in which all subjects receive the drug and there is no control group (discussed above).
- Human challenge studies in which all subjects receive the intervention (such as exposure to a pathogen) and there is no control group (discussed above).
- Phase IV (post-marketing) studies in which subjects in a cohort who are taking a medication are followed long-term.
- Studies involving the development of new methods of measurement, such as new methods of MRI-imaging or biomarker detection.

- Controlled trials in which randomization is not feasible for ethical or other limitations; for example, studies of drugs taken during pregnancy, which will be discussed in Chap. 8 (World Health Organization 2024).

One of the ethical reasons sometimes given for using trial designs that deviate from the RCT paradigm is to make experimental treatments more widely available during public health emergencies, such as the COVID-19 pandemic, in which initially there were no proven therapies (Caplan et al. 2015; Raus 2016). In the U.S., it takes ten years, on average, for a new drug to be approved by the FDA, including nine years of clinical trials (Phase I, II, and III) and six to ten months for the agency to review the data (Wysocki 2021). The typical new drug is tested on only a few thousand patients during clinical trials (Duijnhoven et al. 2013) but there may be hundreds of thousands of people who desperately need some form of treatment for the condition being studied and who may die before a medication has been approved.

One of the ways of making an unproven therapy more widely available is to conduct non-randomized and/or uncontrolled trials with the medication on a large number of patients (tens of thousands or more). For example, during the COVID-19 pandemic, in which more than 7 million people died, doctors tried many different unproven therapies on their patients, such as hydroxychloroquine and ivermectin (Taccone et al. 2022). The demand for unproven therapies was also very high during the 2014–2016 West African Ebola epidemic, in which more than 11,000 people died (Calain 2018) and during the AIDS pandemic of the 1990s and 2000s, in which about almost 2 million people died per year (Aizenman 2019).

Some commentators, such as London and Kimmelman (2020) and Kalil (2020), contend that clinical investigators and IRBs should not lower standards of scientific rigor during public health emergencies. Several arguments support this position. First, trials that deviate from the RCT paradigm are more likely to yield biased results, which can not only mislead the public and clinicians but also make it difficult for scientists to interpret evidence from different studies, which impedes progress toward discovering an effective treatment for the disease (Cox et al. 2014; Caplan and Bateman-House 2015; Lipworth et al. 2023). Second, allowing large numbers of patients to have access to unproven treatments can undermine recruitment for RCTs, which also can also impede progress toward discovering an effective treatment (London and Kimmelman 2020). Third, because unproven treatments can be dangerous, making them widely available before RCTs have been completed successfully could harm many patients (Kalil 2020). The best way to promote public health and safety, according to this view, is to maintain standards of scientific rigor so patients, clinicians and regulators will have good evidence for making medical and policy decisions (London and Kimmelman 2020).

While I agree that it is important to maintain high standards of rigor in clinical research, I am more of pragmatist than an idealist when it comes to these matters (Adebamowo et al. 2014; Caplan et al. 2015). When thousands or even hundreds of thousands of people are dying from a disease and there are no proven therapies to treat them with, there is enormous social pressure to do something—anything—to help. As became apparent during the AIDS crisis of the 1980s and 1990s, ignoring the impassioned pleas from people who are sick and dying can lead to substantial backlash and undermine the public's trust in the medicine and the research enterprise (Dresser 2001). The U.S. federal regulations, the Helsinki Declaration, and other guidelines do not require that clinical research be perfectly designed; they only require that it be appropriately designed to yield important knowledge. Thus, to promote maintain standards of rigor and promote access to unproven treatments during epidemics and pandemics, I recommend the following:

1. Patients who are not enrolled in RCTs may participate in prospective cohort (natural history) studies of off-label uses of medications, provided that these are appropriately designed to minimize, identify, and control bias and yield useful results (Song and Chung 2010). To produce useful results, prospective cohort studies must be sufficiently powered to enable statistical analysis of data that corrects for potential biases and confounders due to non-random assignments to treatment groups, lack of blinding, variation in medical care, and other factors. These may therefore need to be very large (thousands or tens of thousands of patients) and may need to collect many different types of data (dozens or even hundreds of variables). In the past, the types of studies I am suggesting would not have been feasible, due to the complexities of the analysis, but advances in statistics, mathematics, and computer science have greatly improved our ability process and analyze large and heterogenous data sets (Batko and Ślęzak 2022).
2. RCTs conducted during epidemics or pandemics could use adaptive designs to increase speed and efficiency without sacrificing rigor (see van der Graaf et al. 2012; Caplan et al. 2015).
3. To promote safety and advance scientific knowledge, physicians should be encouraged to recruit patients into research studies (RCTs or prospective cohorts) rather than prescribing treatments on experimental or *ad hoc* basis. Opportunities to collect data that can promote public health should not be squandered.
4. Patients that do not enroll in studies may be able to get access to medications through expanded access (or "compassionate use") programs, which allow patients to receive experimental treatments that have successfully completed Phase I testing but have not yet been approved for general use (Darrow et al. 2015)
5. The regulatory approval process for drugs, biologics, and medical devices should be accelerated. During the COVID-19 pandemic, for example, the FDA allowed treatments, tests, and vaccines to be available to the public under emergency use authorizations (or EUAs; Food and Drug Administration 2024). The FDA may grant an EUA if it determines, after reviewing the available evidence, that they product is likely to be proven safe and effective when studies are completed. However, considerable care should be used when granting an EUA and the agency should vigilantly monitor the data. The agency should be prepared to revoke an EUA if the data are trending away from safety and efficacy.[21]

[21] However, the agency may face significant pressures from sponsors and patient advocacy groups not to revoke on EUA.

7.16 Minimal Risk

The concept of minimal risk plays a pivotal role in the federal research regulations. The idea of a minimal risk level emerged from a debate between Catholic theologian/ethicist, Paul Ramsay and Protestant theologian/ethicist Richard McCormick, which was hosted by the National Commission for Protection of Human Subjects in Biomedical and Behavioral Research and published in the *Hastings Center Report.* Ramsay (1976) argued that children should be exposed to no risks in research that does not benefit them medically because parents and investigators have an obligation to promote children's best interests and children do not have an interest in participating in these no-benefit studies. Children can only be exposed to risks in research when it is in their best interests, such as participating in a Phase II or III oncology clinical trial. McCormick (1976) argued that, while it is important to protect children, children can participate in activities that benefit society but expose them to some risks. For example, the boy scouts do community service projects that involve some risk, such as picking up litter, collecting donations for food banks, planting trees, and shoveling snow from sidewalks (Boy Scouts of America 2024). Participation in service activities benefits society and can benefit children by contributing to their moral and emotional development. McCormick prevailed in this debate and the compromise position that the National Commission accepted is that children can participate minimal risk research that does not benefit them medically. Thus, the concept of minimal risk emerged from this debate as a threshold for a morally (and publicly)[22] acceptable level of risk (Kopelman 1981, 1997, 2000; Freedman et al. 1993). In 1991, minimal risk was incorporated into the federal regulations for research with children (45 CFR 45 Subpart D) and extended to pregnant women and fetuses (Subpart B) and prisoners (Subpart C). Minimal risk functions as a key factor in other aspects of IRB decision-making, such as using expedited review to approve a study (45 CFR 46.110) and waiving and altering consent requirements (45 CFR 46.116f; 46.117c).

In Chap. 8, I will discuss specific issues related to minimal risk research involving vulnerable populations in more depth. For now, I would like to consider important definitional issues related to minimal risk. As we saw in Chap. 2, the definition includes two ways of determining whether the risks of research qualify as minimal: (1) comparing these risks to the risks of daily life or (2) comparing these risks to the risk of routine physical or psychological examinations or tests. In some situations, minimal risk determinations will be straightforward because the research only involves routine physical or psychological examinations or tests, such as collecting a small amount of blood (e.g., 50 ml), performing a physical exam, or asking questions about a person's medical and social history. However, in other situations the research may include procedures, examinations, or tests that that are not part of a routine medical check-up, such as skin or muscle biopsies or administration of

[22] A survey by Schupmann et al. (2022) of 1658 U.S. adults suggests there is also strong public support for minimal risk pediatric research.

medications. In such situations, the IRB can determine the level of risk by comparing the risks of research to the risks of daily life (Resnik 2005; Wendler et al. 2005). But what are the risks of daily life? And how can we define and measure them?

Most of the controversies about minimal risk arise when IRBs must use the risks of daily life standard to make determinations about minimal risk. As discussed in Sect. 7.11 of this chapter, Shah et al. (2004) found that there is significant variation in how IRB chairs apply the minimal risk concept to different study procedures. An earlier study by Janofsky and Starfield (1981) also found significant variation in how pediatricians think about minimal risk in children, and a study by White et al. (2021) found significant variation in how IRB members apply the concept of minimal risk to research involving pregnant women. As a member of different IRBs for more than 25 years and as chair for 11 years, I have also witnessed significant disagreements about whether particular studies or procedures should be classified as minimal risk, including disputes about biopsies, cytapheresis, and vaginal examinations in girls younger than 10-years-old. In some cases, our decisions seemed arbitrary. For example, we decided that a skin biopsy of 2 mm or less in diameter would a minimal risk procedure but a biopsy of more than 2 mm would be more than minimal risk.

There are two different ways of interpreting the risks of daily life standard: the relative standard and the absolute standard (Kopelman 1997, 2000; Resnik 2005; Wendler et al. 2005). According to the relative standard, minimal risk should be equated with the risks of daily life for the population being studied. The main argument for this position is that since minimal risk represents a standard of morally acceptable risk, it should reflect the local community's judgments (Freedman et al. 1993). One problem with this interpretation is that it might permit unequal and unfair treatment of human research subjects because the risks of daily life could differ dramatically between different locations (Kopelman 1997, 2000; National Bioethics Advisory Commission 2001; Resnik 2005; CIOMS 2016). For example, the relative standard implies that a child living in impoverished neighborhood in Baltimore, MD could be exposed to significantly greater risks in non-beneficial research than a child living in rich neighborhood in Boston, MA which seems patently unfair.

The absolute standard avoids the unfairness problem by equating the risks of daily life with the general risks of daily life, not the risks that any particular population faces. Wendler et al. (2005), for example, try to quantify minimal risk for U.S. children by collecting and analyzing data on the risks that U.S. children ordinarily children face, such as risks associated with riding in automobiles, bathing, and participating in sports. They use this data to estimate the cumulative daily life risks of U.S. children for different age ranges (see Table 7.5). While this approach might work for U.S. children, what about other countries (Binik 2014)? Suppose the NIH funds a pediatric study in Ghana, which will be reviewed by an IRB in Ghana. Should minimal risk be interpreted in terms of the risks faced by the typical child living in the U.S. or in Ghana? To resolve conflicts related to international research collaborations, we would need to use a standard that transcends any particular country or geographic region, such as risks typically encountered by the average healthy child or adult living on the planet Earth. But it is difficult to conceive of how we would quantify these risks, since we would need to collect and analyze data from all over the world to produce a fair and reasonable estimate. Thus, it would seem that we may be stuck with some type of relative standard benchmarked to a particular country.

At a more fundamental level, we can see that questions about how to interpret "minimal risk" involve conflicts between protecting human subjects from harm and advancing scientific research (Freedman et al. 1993; Kopelman 2000). The absolute standard of minimal risk provides greater protection for human subjects but possibly at the expense of preventing socially valuable research, whereas the relative standard offers more flexibility (Binik 2014). Members of the National Commission (1979) recognized this conflict and developed the concept of minimal risk to deal with it. But, as we have seen, the concept of minimal risk does not completely resolve the underlying moral conflict (Binik 2014) (Table 7.6).

If we think minimal risk from the perspective of trust, we should ask ourselves which approach to interpreting minimal risk is most likely to promote trust. Although studies involving only minimal risk are not likely to cause serious harm to research subjects, harm—and especially the perception of harm— can still occur in this research. For example, the Johns Hopkins IRB classified the Kennedy Krieger study (discussed in Chap. 2) as minimal risk. The parents of two of the children in the study sued the institution and the investigators because they believed that the research exposed their children to the unsafe levels of lead, even though they were already living in homes with lead paint. Likewise, three IRBs classified the CHEERS study (discussed Chap. 2) as minimal risk. Although the study was never implemented, critics argued that the study would expose children to the risks of pesticide exposure, although the protocol did not require parents to use pesticide. Additionally, SUPPORT study investigators (see discussion in Chap. 1) argued that their research was minimal risk. But critics claimed that the study was more than minimal risk and that parents were not properly informed of the risks.

There are several lessons one can draw from these cases. First, investigators, institutions, and IRBs should carefully consider how to best protect vulnerable subjects in minimal risk research. If there is a dispute about whether a study is minimal risk, investigators and IRBs should explore ways of further reducing risks to resolve uncertainty concerning the study's risk classification. Second, subjects (or their representatives) may sometimes perceive that harm has occurred even when the evidence for harm is weak. When it comes to promoting trust, perceptions can matter a great deal, since trust is often based on the perception of fidelity or integrity.

Table 7.5 Cumulative daily life risks for U.S. children per million children per day by age group

Type of Harm	<1 year	1–4 years	5–9 years	10–14 years	15–19 years
Death	1.5	1.5	1.4	1.4	10
Hospitalization	1.0	1.3	1.7	2.1	6
Emergency room visit	6.4	16.4	26.0	36.1	64

Based on Wendler et al. (2005)

Table 7.6 Selected U.S. accidental deaths risks

Causes	Deaths per 100,000
Drug poisoning including drug overdose	30.9
Falls	14
Motor vehicle	13.4
Medical error	6.9
Drowning	1.3
Fire	1.3
Food poisoning	0.93
Extreme heat	0.69
Allergic reaction	0.25

Based on National Center for Health Statistics (2024), CDC (2024), Department of Health and Human Services (2024), Statista. (2024) U.S. Fire Administration (2023), Rodwin et al. (2020), and Neugut et al. (2001)

Although it is not realistic to expect that investigators, institutions, or IRBs can prevent adverse perceptions from occurring, they should nevertheless take steps to minimize them by gauging the community's attitudes toward risk. For example, community members who serve on the IRB could share their knowledge of the community's attitudes toward risk with the board when it is trying to determine whether a study should be classified as minimal risk. If the board understands "daily life risks" to mean the daily life risks faced by a typical adult or child in the U.S., it could adjust this standard to accommodate local sensibilities.

7.17 Compensation for Research-Related Injuries

As we have seen in this book, human subjects are sometimes injured as a result of participating in research. Injuries range from relatively minor problems, such as infections, benign allergic reactions, or short-term toxicity, to catastrophic events, such as stroke, cardiac arrest, permanent disability, or death (Steinbrook 2006; Pike 2014). The federal research regulations do not require investigators, institutions, or sponsors to compensate human subjects who are injured in research, although they do require investigators to inform subjects who are participating in more than minimal risk research about "whether any compensation and an explanation as to whether any medical treatments are available if injury occurs and, if so, what they consist of, or where further information may be obtained (45 CFR 46.116b6)." However, the federal regulations appear to prohibit institutions from asking subjects to waive their right to compensation, since they prohibit informed consent documents or discussions from including "any exculpatory language through which the subject or the representative is made to waive or appear to waive any of the subject's legal rights, or releases or appears to release the investigator, the sponsor, the institution or its agents from liability for negligence (45 CFR 46.116a6)."

Since the 1970s, numerous government commissions and commentators have recommended that the U.S. require that institutions or sponsors compensate human subjects for research-related injuries (Childress 1976; National Bioethics Advisory Commission 2001; Presidential Commission for the Study of Bioethical Issues. 2011; Pike 2014). Although the U.S. does not mandate that research subjects receive compensation for injuries, many European, Asian, and African countries do (Pike 2012; Avilds 2014). The CIOMS (2016) guidelines and Helsinki Declaration (World Medical Association 2024) both recommend the sponsors and researchers compensate subjects for harms resulting from research participation.

If an institution or sponsor refuses to compensate a research subject in the U.S. for their injuries, the subject can still sue for damages. However, relying on the legal system to rectify harms is neither an effective nor a fair method of compensation since injured research subjects may have difficulty prevailing in court or obtaining a satisfactory settlement, and some may be barred from obtaining compensation, due to legal defenses, such as government immunity or assumption of risk. Moreover, it may take many years for injured research subjects who win their cases to receive compensation (Resnik 2006; Pike 2014).

There are several ethical arguments for providing injured research subjects with some form of compensation beyond what they could obtain by means of litigation (Resnik 2006; Pike 2012). First, compensatory justice requires that investigators, institutions, or sponsors compensate human subjects for their injuries, since responsible parties should rectify the damages that they have caused. Second, distributive justice requires that research subjects receive compensation for their injuries, since people who are harmed in research may bear an unfair burden of the financial and non-financial costs of research. To ensure that the costs are distributed fairly, injured subjects should receive some form of compensation, i.e. money, medical treatment, or both. Third, compensating injured research subjects can help prevent additional harms that may occur if they do not receive treatment because they lack health insurance and cannot afford to pay for their own care (Pike 2012). For example, if a subject does not receive treatment for an infection related to the placement of a catheter in his arm, his infection could spread and threaten his health or life. Finally, the principle of trust also implies that institutions or sponsors should compensate human subjects for their injuries, since failing to provide subjects with medical treatment or some of other form of compensation when they are injured would undermine their trust in institutions and sponsors and by adding insult to injury (so to speak!).

Most U.S. research institutions do not have policies requiring that subjects be compensated for research-related injuries. Resnik et al. (2014) conducted a survey in 2012 of the compensation for injury policies of the top 169 U.S. academic institutions ranked by total research funding and compared the data they obtained to data collected by Parasidis in 2000 (Parasidis 2001). Resnik et al. (2014) found that 51.2% of institutions offered no compensation in 2012, compared to 56.1% in 2000;

36.9% offered compensation if certain conditions are met, compared to 31.8% in 2000; 8.1% offered compensation at the discretion of the institution or sponsor, compared to 11.2% in 2000, and 3.8% offered compensation without any conditions, compared to 0.9% in 2000. The only statistically significant difference between data from 2012 and 2000 emerged when the analysis focused on institutions included in both studies. Of these 57 institutions, 9.1% changed from no compensation in 2000 to discretionary compensation in 2012 ($p < 0.001$) (Resnik et al. 2014). The authors noted that some institutions may provide medical treatment or some other form of compensation to injured research subjects even though they do not have an official policy that requires them to do so (Resnik et al. 2014).

The study by Resnik et al. (2014) indicates that substantial changes to compensation for injury policies probably will not occur without a U.S. government mandate (Resnik et al. 2014). When federal agencies published their "Advance notice of proposed rulemaking" for revisions to the Common Rule in 2011 and "Notice of proposed rulemaking" in 2015 (Department of Health and Human Services, and Food and Drug Administration. 2011; Department of Homeland Security et al. 2015), there was an opportunity to include compensation for research-related injury requirements in federal regulations. Several commentators (e.g., Dresser 2012; Resnik 2018b) argued that the revisions to the Common Rule should include such requirements, but these recommendations were not acted upon.

It is worth noting that some institutions have exemplary compensation for injury policies. The NIH's Clinical Center, the Department of Defense, and the Department of Veterans Affairs provide injured research subjects with free access to short-term medical care (Pike 2012). Since 1979, the University of Washington has a self-funded compensation plan that provides injured research subjects with up to $250,000 in medical treatment (Pike 2012). Medicare also covers the cost of treatment for research subjects injured in qualifying clinical trials (Pike 2012).

Scott (2003), Resnik (2006) and Pike (2012) have proposed no-fault compensation systems for research injuries. Resnik's proposal, which is modelled after the U.S. National Vaccine Injury Compensation Program, includes four elements. First, subjects should be eligible to receive compensation regardless of whether they are partly responsible for their injuries. Second, panels composed of doctors, scientists, and other experts would review compensation claims filed by research subjects to determine whether they have been injured and what form compensation should take. Third, immediate medical care would be provided to subjects who require it. Fourth, institutions and sponsors would contribute money to a national insurance pool to cover the costs of compensation, including short-term and long-term medical care, if necessary (Resnik 2006). Of course, sponsors and institutions might object that compensation programs are too expensive. While compensation programs are likely to increase the short-term costs of conducting research, they may also save money in the long run by helping sponsors and institutions reduce litigation expenses (Resnik 2006). Another objection to compensation programs is that they may be cumbersome implement, but this problem can be overcome through careful planning and ample funding for administrative costs.

7.18 Non-Research-Related Benefits

In the last four sections of this chapter, I will discuss a cluster of issues related to providing benefits to research subjects and communities that are not directly related to the research design. When a cancer patient receives treatment as part of Phase II clinical trial, this is a medical benefit that is directly related to the research design.[23] However, the patient might also receive some benefits that are not directly related to the research, such as genetic test results or treatment for a headache. The Common Rule does not require the research subjects receive these types of benefits; it only requires that the risks of research be reasonable in relation to the benefits to the subject and the value of the knowledge expected to be gained. However, as we shall see below, there are strong moral arguments for providing subjects—and communities—with benefits that not directly related to research design.

7.19 Sharing Individualized Research Results with Participants

Researchers frequently collect data that may be useful to human participants, such as information about them obtained from physical examinations (e.g. blood pressure, heart rate, swollen lymph nodes), laboratory tests (e.g. blood sugar levels, environmental exposures, genetic/genomic variants), interviews (e.g. depression screening), and imaging studies (e.g. abnormal findings detected by X-rays, ultrasounds, or MRIs) (Brody et al. 2007; Wolf et al. 2008, 2012; Beskow and Burke 2010; Green et al. 2013; Jarvik et al. 2014; Darnell et al. 2016). The information (or individualized results) that researchers disclose to participants may be useful for making medical or reproductive decisions or managing hazards in the home or workplace. For example, in the Kennedy Krieger Institute study (discussed in Chap. 2) the investigators collected information on children's exposures to lead in the home. Parents of two children in the study sued the investigators and KKI on the grounds that they failed to disclose lead exposure information in a timely fashion. The parents argued that the researchers had a duty to disclose this information this information because the parents could use it to protect their children from lead-related hazards. Although disclosing individualized research can benefit human subjects, it can also cause harm if the subjects misunderstand or adversely react to the information. Hence, disclosure involves a potential conflict between benefits and risks (Resnik 2011; National Academies of Sciences, Engineering, and Medicine 2018).

[23] Because clinical trials provide participants with medical benefits, some commentators refer to them as "therapeutic research" but use of this term rests on a conceptual mistake because it confuses that the goals of research with the goals of medicine and reinforces the therapeutic misconception (Levine 1988).

Federal research regulations do not require researchers to disclose individualized results, but they do require them to inform subjects about any disclosure plans. The Common Rule requires that informed consent include "A statement regarding whether clinically relevant research results, including individual research results, will be disclosed to subjects, and if so, under what conditions (45 CFR 46.116c8)." Also, if there are plans to return individualized results, the research cannot be considered exempt under 45 CFR 46.104d.

In thinking about the issues, the following distinctions have proven to be useful (The Presidential Commission for the Study of Bioethical Issues 2013; National Academies of Sciences, Engineering, and Medicine 2018):

- **Aggregate vs. individualized results.**[24] Aggregate results are the results for the whole study; for example, whether a clinical trial found that a treatment was more effective than placebo; individualized results are the results for a specific participant, such as urine or blood tests.
- **Primary vs. secondary results.** Primary results are results investigators seek to collect to achieve the aims of study; secondary results are other results collected as part of a study. For example, pulmonary function tests results might be the primary results of an asthma study while liver function test results might be secondary results.
- **Incidental vs. targeted results.** Incidental results are results that happen to be discovered while conducting research but are not primary or secondary results. For example, an investigator who is studying fibroid tumors in women might discover an abnormal growth when conducting an ultrasound examination of a woman's uterus. An abnormal growth would be an incidental finding, whereas a fibroid tumor would be a targeted finding.
- **Anticipated vs. unanticipated results.** Anticipated results are results that one expects, given the tests one is using; for example, detecting a uterine fibroid in a study of uterine fibroids would be an anticipated result. Unanticipated results are results one does not expect, given one's understanding of the field and the tests one is using; for example, in a study of the environmental factors related to Parkinson's disease, discovering that nicotine exposure reduces the risk of the disease could be an unanticipated result.

Most commentators agree that it is appropriate for investigators to share aggregate results with participants because they contributed to the research and may want to know how the study turned out. Reciprocity requires at least this much. Most of the ethical debate has focused on the obligation to disclose individualized results (Wolf et al. 2008).

There are several arguments for disclosing individualized results to participants. The first is that disclosure of results is implied by the obligation to benefit participants (Fernandez et al. 2003; Wolf et al. 2008; Miller et al. 2008; Beskow and Burke

[24] Other terms that are commonly used instead of 'results' are 'findings' and 'outcomes.' To avoid confusion, I will just use one term, 'results.'

2010). For example, an investigator who discovers that a subject has abnormally high blood pressure can share this information with the participant to help them avoid additional health problems. The investigator could also recommend that the participant see a physician and provide a referral, if necessary. The obligation to benefit participants may be grounded in the principle of beneficence, moral theories that imply duties to help others, or in the fiduciary obligations arising out of relationships between researchers and participants (Richardson 2008; National Academies of Science, Engineering, and Medicine 2018).

The second argument is that disclosure helps to promote the subject's autonomy (Fernandez et al. 2003; Shalowitz and Miller 2005; Brody et al. 2007). For example, a research subject who learns that they have a genetic mutation that triples their lifetime risk of developing breast cancer can use this information to make decisions concerning lifestyle changes they can implement to reduce their risk of developing the disease, and a subject who learns that their home contains high levels mold allergens may decide to have their home treated for mold.

Studies indicate that most research subjects want to receive individualized results (Shalowitz and Miller 2008). For example, Fong et al. (2004) found that 90% of 429 survey respondents would want to individual results from research on stored biological samples and 80% would want these sent to their physicians. Hoeyer et al. (2004) found that 55% of 600 survey respondents would want to receive tests results concerning genetic predispositions for disease if treatment is available. Brody et al. (2007) found that 97% of 120 participants in a household environmental exposure study wanted to receive individual and aggregate results. Bollinger et al. (2012) found that most participants from ten focus groups wanted to receive individualized results from genetic research.

These two arguments apply to all the types of results. It makes little ethical difference whether individualized results are primary, secondary, targeted, incidental, anticipated, or unanticipated since individuals can benefit from—and may want to receive—all of these types of results. The main differences between types of results are logistical: for example, researchers may be well-prepared to deal with targeted results, but they may not be well-prepared to deal with incidental ones, since they are not actively looking for them.

While the principles of beneficence and autonomy favor disclosure of individual research results, other ethical considerations may caution against it. First, disclosure may cause harm in some cases. For example, a false HIV positive test result could cause a research subject needless stress and anxiety, whereas a false negative result might lull a subject into an erroneous sense of security (Wolf et al. 2008; Resnik 2011). To minimize these risks, results should be validated before they are shared with research subjects. Researchers should send biological samples to laboratories that follow testing procedures designed to enhance accuracy, precision, and reliability, as evidence by CLIA (Clinical Laboratory Improvement Amendments) certification or some other acceptable measure of quality (Ravitsky and Wilfond 2006; The Presidential Commission for the Study of Bioethical Issues 2013; National Academies of Sciences, Engineering, and Medicine 2018). If a test result pertains to a clinically significant outcome (such as a positive test for HIV antibodies), the subject should be retested to ensure that the result is accurate. MRI, x-ray, and other diagnostic images should be read by a qualified clinician.

Harm may also occur when the results are valid but have uncertain clinical or practical implications (Jarvik et al. 2014). For example, suppose that researchers discover that a participant has a genomic variation that increases the risk of developing Alzheimer's dementia at age 70 by 10%. Suppose also there is also no effective method of preventing or treating this disease. Sharing this finding with the participant might cause them considerable worry without offering them substantial benefits (Renegar et al. 2006). Moreover, they might misunderstand the implications of the finding and take drastic measures, such as deciding to spend all of their money or even committing suicide before reaching the age of 70. In some cases, subjects might undergo expensive and painful medical tests, such as invasive biopsies, due to receiving misleading findings or results (Fenton et al. 2018). Ethicists and researchers who are concerned about the harmful effects of disclosing research results argue that investigators should only disclose results which have a definite clinical or practical value (Wolf et al. 2008; Beskow and Burke 2010; Resnik 2011; Presidential Commission for the Study of Bioethical Issues 2013).

Empirical research on the psychological impact of uncertain results or findings is mixed (Oliveri et al. 2018; Carlsson et al. 2022). While some studies indicate that people have can have a difficult time dealing with uncertain results (Dorval et al. 2005; O'Neill et al. 2006; van Dijk et al. 2006), others suggest most adverse psychological reactions to inconclusive test results are minor or dissipate over time (Brody et al. 2007; Hamilton et al. 2009). Until we have a better understanding of the impact of informing research subjects about uncertain results, precaution may be the best policy (Resnik 2011).

Cost is another factor to consider when deciding whether to share individual research results with human subjects, since disclosing individual results might interfere with the aims of the research by drawing financial resources away from other study activities (Illes et al. 2006; Resnik 2011). Investigators who are conducting brain imaging studies on healthy volunteers, for example, may need to hire radiologists to examine images for incidental findings, such as possible tumors. Investigators who are conducting genetic or genomic research may need to hire genetic counselors to discuss results with participants (Jarvik et al. 2014; Darnell et al. 2016). Investigators may also need to spend considerable time and effort locating research subjects to inform them of individual results, especially if they have moved and not provided the research team with their new contact information. Actively searching for incidental findings, for example, by performing a battery of genetic tests on biospecimens, may also inflate the costs of research (Presidential Commission for the Study of Bioethical Issues 2013). One might argue that disclosure is not necessary when the costs are significant and the benefits to the subjects are marginal (Affleck 2009). However, if the benefits of disclosure are high and the costs are low, then cost should not be a factor (Resnik 2011). For example, it costs very little to inform a subject that they have dangerously high blood pressure.

Some commentators have pointed out that promising to disclose individualized research results to participants may contribute to the therapeutic misconception (see discussion in Chap. 5) by causing them to think that the purpose of the research is to provide them with medical benefits (Ravitsky and Wilfond 2006; Clayton and McGuire 2012). Others have argued, however, that disclosure of individualized results is not likely to contribute significantly to the therapeutic misconception, provided that investigators carefully explain the purposes of research to participants (Shalowitz and Miller 2008).

The principle of trust implies that individualized results should be shared with participants in some situations but not others. As discussed in Chap. 4, trust-based relationships involve shared expectations and reciprocity. When human subjects agree to allow researchers to collect, analyze and use their data and biospecimens, the expect researchers to handle them in a compassionate and respectful way. A researcher who learns about an important result, such as dangerous exposure to lead (see discussion of the Kennedy Krieger Institute case in Chap.2), but does not share it with the participant, betrays their trust (Richardson 2008; Richardson and Cho 2012). On the other side of the ledger, sharing results with clinical and practical value in a timely fashion can foster trust, since subjects are likely to feel that investigators care about them and want to promote their well-being. For example, a subject who learns for the first time that she has high blood sugar may be grateful to the research team for sharing this information with her and helping her get treatment. Sharing research results with uncertain clinical or practical implications could positively impact trust if subjects want to receive this information and the investigators help them understand its implications. For example, Brody et al. (2007) gave the subjects in their study information about their in-home chemical exposures, including a chart that compared their exposure levels to exposure levels for the average home. They also informed subjects about any dangerous exposures and steps they could take to reduce their exposures. Sharing uncertain results could negatively impact trust if subjects do not want to receive them, or they want to receive them, but the investigators do not provide them with adequate counseling or advice. Subjects who receive their individualized research results but get little counseling from investigators may feel overwhelmed or abandoned. Of course, subjects may feel overwhelmed or abandoned if they do not receive adequate counseling or advice related to research results with definite clinical or practical implications, but this problem is more likely to occur when they receive results which have uncertain clinical or practical value.

Considering the pros and cons of sharing individualized research results with participants, I offer the following recommendations[25]:

1. Researchers should clearly describe plans and procedures concerning the disclosure of individualized research results in the consent document and protocol.
2. Results should be disclosed only if they have been appropriately validated and are likely to be clinically or practically useful.

[25] This is similar to the guidance provided by the National Academies of Science, Engineering, and Medicine (2018).

3. The default position should be that validated and useful results will be disclosed unless there is scientific or practical justification for non-disclosure; for example, the researchers are planning to anonymize the data and samples, the tests that will be performed have not been validated, and results may not be known for several years.
4. Participants should have the opportunity to decide (yes or no) whether they will receive results and their wishes should be respected.[26] If the participant was enrolled in research as a minor and their parents consented to receiving results, they should have the opportunity to opt-out of receiving results when they become an adult.
5. Appropriate counseling and education should be offered to participants when results are disclosed.
6. Researchers should frequently update contact information for participants in case results need to be shared with them.
7. Research budgets should allocate sufficient funding disclosure of individualized results.

7.20 Ancillary Care

Ancillary care is another type of benefit that investigators might offer to research subjects. We already discussed this topic in Chap. 4 when examining Richardson and Belsky's partial entrustment view, but we will say a bit more about it here. Ancillary care, as the reader will recall, is medical care provided to human subjects beyond what is required to (a) address the study's research questions, (b) protect subjects from harm related to their participation in the study, or (c) redress harms that have occurred because of participation (Belsky and Richardson 2004). Ancillary care may include medical testing, diagnosis, treatment, or prevention (Belsky and Richardson 2004). Although ancillary care is major concern when conducting clinical research in low-income countries in which most people cannot afford basic health care, it can also be an issue in high-income countries, such as the U.S., when a research subject arrives at a clinic with a symptom or problem (such as high blood pressure) that requires medical attention.

Several moral arguments support a duty to provide ancillary care to research subjects in some circumstances. First, because ancillary care is a benefit, duties of beneficence support the provision of ancillary care for participants. As was previously mentioned, such duties could be grounded in the principle of beneficence, moral theories that imply duties to help others, or in fiduciary obligations arising out of relationships between clinical researchers (e.g., doctors or nurse) and participants

[26] This option may be limited in some situations. For example, in the U.S., health care professionals are legally obligated to report the results of HIV tests to patients and public health agencies. Patients cannot refuse to receive their HIV test results. They only way to not receive HIV test results is to opt-out of HIV testing.

(i.e., patients) (Richardson and Belsky 2004; National Academies of Science, Engineering, and Medicine 2018).

Second, one could argue that the principle of justice implies a duty to provide ancillary care to socioeconomically disadvantaged populations which may not have access to care outside of the study, as is often the case in research conducted in low-income countries (Dickert and Wendler 2009; Hooper 2010; Pratt et al. 2013; London 2022). Providing ancillary care to disadvantaged research subjects can improve their health and therefore help to address socioeconomic inequalities. It is also a way of sharing benefits with the community (see discussion in Sect. 7.22, below).

Third, one could argue that the goal of promoting trust supports the provision of ancillary care because providing ancillary care helps to assure participants that researchers have their best interests in mind and will take care of them.

However, there are objections to these arguments.

Concerning obligations to help others, one might object that these have limitations (Resnik 1998). If you see someone drowning in a pool but you are a poor swimmer, most people would say that while you have an obligation to take reasonable measures to help this person, such as calling a lifeguard or throwing them a life preserver, you do not have an obligation to dive in the pool and risk your own life to save theirs (McNamara 1996; Miller 2021). Duties of beneficence derived from doctor-patient relationships also have limitations. For example, doctors are not obligated to provide medical care that is outside the scope of their expertise or the standard of care (Patrone 2009). Additionally, claiming that researchers should provide subjects with ancillary care may contribute to the therapeutic misconception (see discussion in Chap. 5) by encouraging subjects to think that the main purpose of a study is to provide them with medical treatment (Resnik 2009). Some research subjects may enroll in a study because they want to receive ancillary care benefits.

Concerning justice, one might argue that claiming that justice implies extensive ancillary care obligations conflates the goals of research with the goals of social and economic development. Although helping people who have significant medical needs is a worthy social goal (London 2005), that is not the main purpose of research. Research is conducted to generate knowledge which can help people. Local governments, health agencies, or international relief/development organizations, should be responsible for providing health care to socioeconomically disadvantaged people and addressing issues of global justice (Participants in 2006 the Georgetown University Workshop on Ancillary-Care Obligations of Medical Researchers Working in Developing Countries 2008).

Concerning trust, while the goal of promoting trust might support some ancillary care obligations, it probably does not support unlimited obligations. For example, while a patient might expect a clinical researcher to treat a minor cut that they received on their way to the clinic, they probably would not (and should not) expect them to treat their fractured tooth. Trust-based obligations also have limitations (Richardson and Belsky 2004).

So, if there are limits to the obligation to provide ancillary care, what might these be based on?

First, cost is an important consideration. As noted in Chap. 4, providing ancillary care may significantly increase the costs of research (Richardson and Belsky 2004; Dickert and Wendler 2009). Although some ancillary treatments are relatively inexpensive (for example, providing a vaccination) others, such as dental care or chemotherapy, could be very expensive. If a sponsor is unwilling to cover the costs of ancillary care, they may decide to not fund the study (Wertheimer 2011), or they may decide to shift funds from research activities to cover the costs of ancillary care, which could undermine the scientific and social value of the study (Richardson et al. 2017). Investigators and institutions who ask for sponsors to pay for ancillary care may face a conflict between benefitting individual research subjects by providing them with ancillary care and benefitting the local population or the larger society by conducting a study that is adequately funded to achieve its goals.

Protecting the integrity and quality of research is another important consideration. As noted in Chap. 4, providing ancillary care may bias data and results in some cases, especially in observational studies which are collecting information on the health of a cohort (Hyder and Merritt 2009; Richardson et al. 2017). For example, suppose that researchers are studying the health impacts of coal miners' exposure to occupational hazards. Treating the miners for diseases related to their work may bias the study results by giving the miners access to health care which they would not have had if they were not in the study.

Roles, responsibilities, and expertise may also limit ancillary obligations. For example, a clinical researcher might be obligated to help negotiate an agreement with a sponsor to fund ancillary but not obligated to pay for it themselves or even provide it if they have designated this responsibility to other members of the research team. A sponsor's obligation to pay for ancillary care may be limited by the availability of care in the community. For example, a sponsor may be under no obligation to pay for ancillary care that is already widely available and affordable in the community (Participants in the 2006 Georgetown University Workshop on Ancillary-Care Obligations of Medical Researchers Working in Developing Countries 2008; Hyder and Merritt 2009).

The nature of the relationship between participations and researchers may also limit ancillary care obligations. For example, these obligations may be very weak if the researcher is someone who has no prior relationship with the subjects or community and is only on-site to collect biospecimens for a genetic study. However, a physician/researcher who cares for patients in the study may have stronger obligations based on the nature of their relationship with their patients (Richardson and Belsky 2004; Resnik 2009).

Given all of these all of these questions and concerns about providing ancillary care, I offer the following recommendations[27]:

[27] These are similar to some of the recommendations included in the CIOMS (2016) guidelines and made by Participants in the 2006 Georgetown University Workshop on Ancillary-Care Obligations of Medical Researchers Working in Developing Countries (2008).

1. For studies that are taking place in areas where ancillary care is likely to be a concern (for example in low-income countries), researchers, institutions, sponsors, local governments, and other stakeholders (such as humanitarian organizations) should work together and reach written agreements concerning the provision of ancillary care. These agreements should clearly delineate the (a) type of care that will be provided, (b) who will provide it, and (c) who will pay for it.
2. Given the ethical, logistical, and financial complexities of providing ancillary care in research, the most reasonable approach is to make decisions about the type of ancillary care that will be provided on a case-by-case basis, taking various factors into account, such as costs, the health needs of the population, local health care resources, the goals and design of the study, the expertise and skill of members of the research team, and the nature of the relationship between the researchers and the research subjects.
3. The type of ancillary care that will be provided and who will provide it and pay for it should be described in the research protocol and consent form.
4. The provision of ancillary care should not interfere with achieving the goals of the study, the researchers' willingness to conduct the study, or the sponsors' willingness to fund it.
5. Ancillary care agreements should not prevent clinicians from providing emergency care to patients; for example, if a patient arrives at clinic severely dehydrated, a clinician should be able to begin hydration therapy without worrying about whether this treatment is covered by the ancillary care agreement. Formal agreements should not interfere with medical commonsense.
6. To reduce the risk of therapeutic misconception, study advertisements and recruitment materials should not highlight or overemphasize the possibility of receiving ancillary care.

7.21 Post-trial Access to Medications

Post-trial access to medications emerged as controversial topic during the late 1990s, when investigators were conducting HIV/AIDS prevention or treatment clinical trials in low-income countries (Glantz et al. 1998). Patients who had access to experimental drugs used to combat HIV could not obtain these or other medications when studies ended, due to limited local health care resources. Many commentators argued that it was cruel and unfair for investigators and sponsors to provide patients with treatment during the trial and then abandon them once the trial ended (Grady 2005). An international consensus soon emerged that investigators and sponsors have an ethical obligation to help secure post-trial access to treatment in low-income countries (Lavery 2008). A statement concerning post-trial access was added to the 2000 revision of the Helsinki Declaration:

> At the conclusion of the study, every patient entered into the study should be assured of access to the best proven prophylactic, diagnostic and therapeutic methods identified by the study (World Medical Association 2000).

The 2024 revision includes the above statements as well as this additional one:

> In advance of a clinical trial, post-trial provisions must be arranged by sponsors and researchers to be provided by themselves, healthcare systems, or governments for all participants who still need an intervention identified as beneficial and reasonably safe in the trial. Exceptions to this requirement must be approved by a research ethics committee. Specific information about post-trial provisions must be disclosed to participants as part of informed consent (World Medical Association 2024).

The 2016 of the revision of the CIOMS guidelines detailed guidance on post-trial access:

> Especially in the context of clinical trials, researchers and sponsors must make adequate provisions for addressing participants' health needs during research and, if necessary, for the transition of participants to care when the research is concluded. The obligation to care for participants' health needs is influenced, among other things, by the extent to which participants need assistance and established effective care is available locally...Addressing participants' health needs requires at least that researchers and sponsors make plans for:
>
> ...
>
> - transitioning participants who continue to need care or preventive measures after the research to appropriate health services;
> - providing continued access to study interventions that have demonstrated significant benefit; and
> - consulting with other relevant stakeholders, if any, to determine everyone's responsibilities
> - and the conditions under which participants will receive continued access to a study intervention, such as an investigational drug, that has demonstrated significant benefit in the study (Council for the International Organizations of Medical Sciences 2016:21)

Sofaer and Strech (2012) have conducted a systematic review of the bioethics literature on post-trial obligations and identified 36 types of arguments for or against providing post-trial access to treatments. Many of these arguments are similar to those that appear in the bioethics literature on ancillary care. I will review a condensed list of arguments here, focusing on the obligation to provide post-trial antiretroviral treatment (ART) in HIV/AIDS research, since this issue has generated the most debate.

Despite the existence of a broad consensus concerning the idea that patients in clinical trials conducted in low-income countries should have post-trial access to treatments, many questions remain unsettled, such as: what is the justification for providing post-trial treatment? What is the scope of and extent of post-trial treatment obligations? Who is responsible for providing and funding post-trial treatment? Should sponsors be required to pay for post-trial treatment when it is not otherwise available? Should there be a time limit on this obligation? (Zong 2008; Millum 2011). The support for post-trial access to ART appears to stem from a common emotional reaction to a compelling human need rather than a careful reflection on the ethical issues (Sachs 2011). Not surprisingly, most clinical trial participants in resource-poor settings believe that they deserve access to medications when the study ends (Pace et al. 2006; Andanda and Wathuta 2018).

The first argument for post-trial access to ART is that patients/subjects can benefit from continuing to stay on the treatment (Sofaer and Strech 2012). Duties of beneficence may be grounded in moral theories that imply obligations to help others, the principle of beneficence, fiduciary obligations arising out of relationships between patients and clinical researchers, and even private companies' social responsibilities (Resnik 2001; London 2022). However, as we have already seen, duties to help others are not absolute and can be limited by financial, logistical, and other considerations (McNamara 1996; Resnik 2001). If private sponsors are required to cover the lifetime costs of ART for patients enrolled in HIV/AIDS clinical trial, many of them will decide to not conduct these studies to avoid onerous financial burdens (Wertheimer 2011).

The second argument for post-trial access to ART is that the principle of nonmaleficence ("do not harm") implies obligations to provide this treatment to patients since they will be harmed if they discontinue their treatment (Sofaer and Strech 2012). To evaluate this argument, we need to distinguish between two types of alleged harm:

Case 1: Patient A has HIV prior to enrolling in a clinical trial testing an ART drug. A's health improves during the trial but then worsens when it ends because she no longer has access to ART.

Case 2: Patient B does not have HIV prior to enrolling in clinical trial of an HIV vaccine. B develops HIV during the trial. After the trial ends, an analysis of the data indicates that subjects receiving the vaccine were more likely to develop HIV than subjects receiving a placebo.

One might argue that failing to provide A with access to ART after the trial ends does not harm them because they are no worse-off than before the trial started (Wertheimer 2011). Harm involves making someone worse-off: if you encounter a starving person and provide them with a meal in exchange for their labor, you do not harm him if you stop feeding them when their work is finished because they are no worse-off than before you met them. Proponents of post-trial access obligations could respond to this critique by arguing that participants in Case 1 may experience psychological harm as a result of their disappointment and distress at no longer having access to medications. Critics of post-trial access obligations could this type of harm occurs all the time when people are disappointed and is not significant enough to warrant payment for ART.

Case 2 is different from Case 1 because it appears that some subjects were made worse-off because of their participation, since they developed HIV during the study. In this situation, one could argue that subjects who received the vaccine were injured because of their participation and that sponsors should compensate them for this injury (see discussion in Sect. 7.16 of this chapter). Compensation could include free medical treatment, including ART, after the study ends (Lo et al. 2007). It is important to note, however, that this argument does not imply that sponsors have a duty to pay for ART medications for all subjects in the study when it ends; it only supports paying for medications for subjects who can prove that they were injured in this study.

According to the non-exploitation argument, sponsors should pay for post-trial access to ART to avoid taking unfair advantage of research subjects (Sofaer and Stench 2012). In Chap. 5, I argued that underpayment could be exploitative when it denies human subjects a fair share of the benefits of research (Wertheimer 1999). To determine whether failing to pay for post-trial ART treatment is exploitative, we need to consider whether this would deny research subjects or host communities a fair share of the benefits of research (Resnik 2003; Hawkins 2008; Wertheimer 2011). The answer to this question depends on how the benefits of research are distributed among the various stakeholders, i.e., human subjects, host communities, investigators, institutions, and sponsors.

Let's first focus on potential exploitation of human subjects. If the sponsor is a private company that expects to reap a huge profit from the study, and the subjects only benefit from treatments received during the study, then one could argue that fairness demands that the sponsor should provide the subjects with additional benefits, such as post-trial access to medications (Hawkins 2008). However, if the sponsor is a government agency or a private charity that is conducting research to increase human knowledge and benefit the local population, then its decision to not pay for post-trial access to medications would probably not be exploitative, because the sponsor would not be acquiring an unfair share of the benefits, since the sponsor would not be profiting from the research and would be benefitting the community in other ways.

Turning to the potential exploitation of communities, it is worth noting that local populations often derive significant economic benefits from research. Clinical trials can create jobs and infuse money into the local economy. Although these economic benefits are significant, proponents of the non-exploitation argument for ART would claim that fairness demands that the sponsor provide additional benefits to the host community, such as funding post-trial ART for research participants (Hawkins 2008). However, this argument for providing post-trial treatment has limited force, because there are many ways that the sponsor or investigator could benefit the host community besides providing post-trial ART, such as investing in the community's health care or educational infrastructure, developing medical products or services that will be available to the community, or providing low cost or free medications to community members (Participants in the 2001 Conference on Ethical Aspects of Research in Developing Countries 2002). One could argue that community leaders, sponsors, and investigators should decide how to share benefits fairly and that the agreements they reach may or may not include post-trial access to medications (see discussion below). Exploitation of the community would be less likely to occur if the sponsor is a government agency or private charity because the sponsor would not profit from the research.

According to the reciprocity argument, sponsors should fund post-trial ART for research subjects as *quid pro quo* payment for their contributions to research (Sofaer and Stench 2012). The reciprocity argument is similar to the non-exploitation argument in that it holds that human subjects should receive a fair share of the benefits of research, which includes post-trial access to medications. One might object to this argument by claiming that reciprocity may or may not require that human subjects receive post-trial access to medications, since they might receive meaningful benefits during the conduct of the study, such as access to treatments, ancillary care, or payment for participation.

According to the global justice argument, sponsors should pay for post-trial ART for research subjects to promote justice and address historical injustice (Sofaer and Stench 2012; London 2005). Some of the most socially disadvantaged people in the world are HIV-infected patients who do not have access to ART due to limited local health care resources in the countries in which they live. Additionally, most of these countries have been victimized by colonialism, slavery, political and religious oppression, and theft of natural resources by industrialized nations. Sponsors should therefore pay for post-trial ART subjects in poor countries to atone for these injustices and promote global justice. However, one might object to this argument that requiring research sponsors to pay for post-trial access to ART conflates the goals of research with the goals of human development (see discussion of this issue in Sect. 7.20 of this chapter). Also, why should researchers and research sponsors be held responsible for past injustices they had little to do with? Governments are the most appropriate locus of responsibility, it would seem.

Turning to arguments against sponsors' obligations to pay for post-trial ART, the first type of argument appeals cost: requiring sponsors to pay for post-trial ART treatment would increase the cost of research in poor countries unreasonably (Sofaer and Stench 2012). The cost of ART varies across the globe, depending on market forces and government price controls. In African countries, the price of ART ranges from about \$290 to \$952 per person per year (Tagar et al. 2014).[28] If a person lives 40 years after a trial has ended, a sponsor might have to pay \$11,600 to \$38,000 for lifetime, post-trial access to ART. If the trial has 500 subjects the cost of lifetime access to ART would come to \$5.8 to \$19 million. A sponsor of a clinical trial might accommodate these higher costs by cutting back on some research activities, which could compromise the scientific and social value of the study; or, worse yet, the sponsor might decide to not fund the study, due to its excessive costs (Wertheimer 2011).

Cost would be a significant consideration regardless of the profit/non-profit status of the sponsor. The NIH, for example, does not fund post-trial access to ART but encourages investigators to work with authorities in host countries and other stakeholders to ensure continued access to care (National Institutes of Health 2005). The NIH's justification for this policy is that the agency has been legally authorized to fund biomedical research but not treatment (National Institutes of Health 2005). In other words, according to the NIH, funding post-trial access to ART conflates scientific and humanitarian goals. Proponents of post-trial treatment obligations could respond to this financial argument by suggesting that sponsors could cover the costs of ART medications until a government agency, private charity, or international aid organization picks up the costs. Sponsor-supported post-trial access to treatment would be a temporary bridge until other stakeholders assume this responsibility.

[28] This is based on \$208 to \$682 reported by Tagar et al. (2014), factoring in inflation.

A second argument against sponsors' post-trial treatment obligations is that informing human subjects that they are entitled to receive ART treatment after the study ends might contribute to the therapeutic misconception (see discussion in Chap. 5) by encouraging them to think that the purpose of the study is to provide them with treatment (Sofaer and Stench 2012). Proponents of post-trial access obligations could argue that investigators can help dispel the therapeutic misconception by making sure that subjects understand the purpose of the study is to develop scientific knowledge, not to provide them with treatment.

A third argument against sponsors' post-trial treatment obligations is that offering human subjects ART after the study ends might constitute undue inducement (see discussion in Chap. 5) to participate in research. HIV-infected participants in resource-poor countries might not be especially susceptible to this form of inducement. Proponents of post-trial treatment obligations could argue that offering treatment to subjects after the study ends would not constitute undue inducement provided that the subjects are able to make intelligent decisions concerning the benefits and risks of the study.

As one can see, there are compelling arguments on both sides of this issue. While most people would agree that researchers, institutions, and sponsors should make arrangements for post-trial access to ART in HIV/AIDS treatment or prevention studies in resource-poor settings, deciding who should cover the costs of ART is controversial. Approaching the perspective of trust, it appears that failing to make provisions for post-trial access to ART in poor countries may undermine subjects' and communities' trust in researchers, institutions, and sponsors, since most believe that they deserve post-trial access to treatment. HIV/AIDS patients who no longer have access to ART at the completion of a study may feel that they have been abandoned, exploited, and betrayed. Members of the communities in which these patients live may harbor similar feelings. Research sponsors, institutions, investigators, and IRBs should do their best to avoid this negative outcome. Ideally, sponsors should offer to temporarily pay for post-trial ART until local governments and other parties can cover these costs. Priority should be given to providing ART to subjects who have developed HIV because of their participation in the study. If sponsors do not have funds to pay for post trial ART, or paying for treatment outside of a study is not part of their mandate, then investigators and institutions should work with other stakeholders, such as private charities, to provide post-trial access to HIV care. They should also inform research subjects about the availability of treatment after the study ends, and they should consult with local community leaders concerning post-trial treatment issues.

7.22 Fair Benefit Sharing

A common theme in our discussion of returning individual results, ancillary care, post-trial access to treatment is the importance of providing benefits not only to human subjects but also to communities, populations, or nations. Many bioethicists agree that investigators, institutions, sponsors, and IRBs should take make special efforts to minimize the potential for exploitation when conducting, funding or overseeing research in developing nations (Hawkins 2008). Studies should be designed and implemented so that host communities or nations are likely to receive a fair share of the benefits of the research. The ethical obligation to share benefits fairly takes on added importance when considers the historical record of exploitation of developing nations by developed ones. Since people who live in developing nations are likely to be aware of this legacy, they may be especially sensitive to any interaction that appears to be unfair (Hawkins 2008).

Since the 1990s, pharmaceutical companies have outsourced clinical trials to developing nations to take advantage of cheap labor, willing volunteers, and lax regulatory environments (Hawkins 2008). Also, some developing nations, such as India, have courted pharmaceutical and biotechnology companies to obtain economic benefits from their research investments (The Economic Times 2024).[29] These trends have important implications for benefit-sharing, since private sponsors may reap significant profits from research conducted in developing nations.

In 2001, investigators, bioethicists, and IRB members from the U.S., Europe, and eight African nations met in Malawi to consider issues related to the sharing of benefits in research with human subjects. Members of this group developed a proposal known as the fair benefits framework. The framework consists of the following key points (Participants in the 2001 Conference on Ethical Aspects of Research in Developing Countries 2002):

- **Collaborative partnership**: investigators, institutions, and sponsors should work with community representatives to develop a plan for benefit sharing; benefits should address the community's needs and concerns.
- **Broad and flexible benefits**: the benefits to the community can take many different forms, such as: making treatments shown to be effective in a study reasonably available to members of the community; providing ancillary care or post-trial access to medications; providing health care information, services or medications to other community members; and developing local health care infrastructure.
- **Transparency**: an independent group, such as the World Health Organization, should collect and publish benefit sharing arrangements to enable host communities to have the information they need to negotiate with investigators, sponsors and institutions.

[29] The FDA will accept data from clinical trials conducted outside the U.S. that are submitted to the agency as part of an application for a new drug, biologic, or medical device, provided that the studies meet certain conditions, such as conformity with good clinical practices guidelines (Food and Drug Administration 2012). The good clinical practices guidelines (GCPs) are guidelines for conducting clinical research developed by the International Conference on Harmonization (2016, 2023), an industry group. They include ethical, scientific, and logistical guidance. The ethical guidance provided by the GCPs is similar to the guidance provided by other ethics codes, such as CIOMS, except GCP guidance does not include restrictions on the use of placebo control groups, which is one reason why the FDA endorses this document.

Although the fair benefits framework was developed to provide guidance for research conducted in developing nations, the key points of the proposal apply to other types of research with human subjects when exploitation might be an issue. Investigators conducting research in impoverished regions of the U.S. should also consider taking measures to share benefits with local populations or communities. For example, NIEHS investigators who were studying the health impacts of cleanup activities related to the Deepwater Horizon oil spill in the Gulf of Mexico in 2010 were cognizant of the need to provide tangible benefits to the participants, most of whom were socioeconomically disadvantaged people who were living in the area impacted by oil spill. Investigators shared individual research results with volunteers and made referrals to medical care, if appropriate. The investigators became aware of the importance of providing benefits to the subjects because of discussions with representatives from the host communities (Resnik et al. 2015).

7.23 Conclusion

Risks and benefits are key concerns in almost all ethical issues in research involving human subjects. This chapter has discussed a variety of topics related to the management of risks and benefits, including risk minimization, risk evaluation, social and other benefits, minimal risk, compensation for research-related injuries, sharing individualized results with participants, ancillary care, and risk/benefit considerations in different study designs. The next chapter will address risk/benefit, consent, and other ethical issues related to enrolling vulnerable populations in research.

References

Adebamowo, C., O. Bah-Sow, F. Binka, R. Bruzzone, A. Caplan, J.F. Delfraissy, D. Heymann, P. Horby, P. Kaleebu, J.J. Tamfum, P. Olliaro, P. Piot, A. Tejan-Cole, O. Tomori, A. Toure, E. Torreele, and J. Whitehead. 2014. Randomised controlled trials for Ebola: Practical and ethical issues. *Lancet* 384 (9952): 1423–14244.

Affleck, P. 2009. Is it ethical to deny genetic research participants individualised results? *Journal of Medical Ethics* 35 (4): 209–213.

Agrawal, M., and E.J. Emanuel. 2003. Ethics of phase I oncology studies: Reexamining the argument and data. *Journal of the American Medical Association* 290 (8): 1075–1082.

———. 2008. Phase I oncology research. In *The Oxford handbook of clinical research ethics*, ed. E.J. Emanuel, C. Grady, R.A. Crouch, R.K. Lie, F.G. Miller, and D. Wendler, 356–366. New York: Oxford University Press.

Aizenman, N. 2019. How to demand a medical breakthrough: Lessons from the AIDS fight. NPR, 9 February. Available at: https://www.npr.org/sections/health-shots/2019/02/09/689924838/how-to-demand-a-medical-breakthrough-lessons-from-the-aids-fight. Accessed 9 Aug 2024.

Andanda, P., and J. Wathuta. 2018. Human dignity as a basis for providing post-trial access to healthcare for research participants: A South African perspective. *Medicine, Health Care, and Philosophy* 21 (1): 139–155.

Anderson, E.E., J.M. DuBois, and JM. 2012. IRB decision-making with imperfect knowledge: A framework for evidence-based research ethics review. *Journal of Law, Medicine, and Ethics* 40 (4): 951–969.

Aronoff, S., S, Rosenblatt, S. Braithwaite, J.W. Egan, A.L. Mathisen, and R.L. Schneider. 2000. Pioglitazone hydrochloride monotherapy improves glycemic control in the treatment of patients with type 2 diabetes: A 6-month randomized placebo-controlled dose-response study. The Pioglitazone 001 Study Group. Diabetes Care 23 (11):1605–1611.

Avilds, M.R. 2014. Compensation of research-related injuries in the European Union. *European Journal of Health Law* 21 (5): 473–487.

Azad, M. 2014. Q&A: Barry Marshall. *Nature* 514: S6–S7.

Bambery, B., M. Selgelid, C. Weijer, J. Savulescu, and A.J. Pollard. 2016. Ethical criteria for human challenge studies in infectious diseases. *Public Health Ethics* 9 (1): 92–103.

Barker, C., K. Collet, D. Gbesemete, M. Piggin, D. Watson, P. Pristerà, W. Lawerence, E. Smith, M. Bahrami-Hessari, H. Johnson, K. Baker, A. Qavi, C. McGrath, C. Chiu, R.C. Read, and H. Ward. 2022. Public attitudes to a human challenge study with SARS-CoV-2: A mixed-methods study. *Wellcome Open Research* 7: 49.

Batko, K., and A. Ślęzak. 2022. The use of big data analytics in healthcare. *Journal of Big Data* 9 (1): 3.

Belsky, L., and H.S. Richardson. 2004. Medical researchers' ancillary clinical care responsibilities. *British Medical Journal* 328 (7454): 1494–1496.

Bernabe, R.D., G.J. van Thiel, J.A. Raaijmakers, and J.J. van Delden. 2012. The risk-benefit task of research ethics committees: An evaluation of current approaches and the need to incorporate decision studies methods. *BMC Medical Ethics* 13: 6.

Bernstein, D., M. Wanner, L. Borish, G.M. Liss, and Immunotherapy Committee; American Academy of Allergy, Asthma and Immunology. 2004. Twelve-year survey of fatal reactions to allergen injections and skin testing: 1990–2001. *Journal of Allergy and Clinical Immunology.* 113 (6): 1129–1136.

Beskow, L.M., and W. Burke. 2010. Offering individual genetic research results: Context matters. *Science: Translational Medicine* 2 (38): 38cm20.

Bhatt, A. 2010. Evolution of clinical research: a history before and beyond James Lind. *Perspectives in Clinical Research* 1 (1): 6–10.

Binik, A. 2014. On the minimal risk threshold in research with children. *American Journal of Bioethics* 14 (9): 3–12.

Binik, A., and S.P. Hey. 2019. A framework for assessing scientific merit in ethical review of clinical research. *Ethics & Human Research* 41 (2): 2–13.

Bittlinger, M., S. Bicer, J. Peppercorn, and J. Kimmelman. 2022. Ethical considerations for phase I trials in oncology. *Journal of Clinical Oncology* 40: 3474–3488.

Bollinger, J.M., J. Scott, R. Dvoskin, and D. Kaufman. 2012. Public preferences regarding the return of individual genetic research results: Findings from a qualitative focus group study. *Genetics in Medicine* 14 (4): 451–457.

Boy Scouts of America. 2024. Service projects. Available at: https://www.scouting.org/outdoor-programs/conservation-and-environment/youth-service/. Accessed 28 July 2024.

Brody, J.G., R. Morello-Frosch, P. Brown, R.A. Rudel, R.G. Altman, M. Frye, C.A. Osimo, C. Pérez, and L.M. Seryak. 2007. Improving disclosure and consent: "Is it safe?": New ethics for reporting personal exposures to environmental chemicals. *American Journal of Public Health* 97 (9): 1547–1554.

Bull, S., E. Jamrozik, A. Binik, and M.J. Parker. 2020. SARS-CoV-2 challenge studies: Ethics and risk minimisation. *Journal of Medical Ethics* 47 (12): e79.

Calain, P. 2018. The Ebola clinical trials: A precedent for research ethics in disasters. *Journal of Medical Ethics* 44: 3–8.

Caplan, A.L., and A. Bateman-House. 2015. Should patients in need be given access to experimental drugs? *Expert Opinion on Pharmacotherapy* 16 (9): 1275–1279.

Caplan, A.L., C. Plunkett, and B. Levin. 2015. Selecting the right tool for the job. *American Journal of Bioethics* 15 (4): 4–10.

Carlsson, L., E. Thain, B. Gillies, and K. Metcalfe. 2022. Psychological and health behaviour outcomes following multi-gene panel testing for hereditary breast and ovarian cancer risk: A mini-review of the literature. *Hereditary Cancer and Clinical Practice* 20 (1): 25.

Castellano-Hinojosa, A., M.J. Gallardo-Altamirano, J. González-López, and A. González-Martínez. 2023. Anticancer drugs in wastewater and natural environments: A review on their occurrence, environmental persistence, treatment, and ecological risks. *Journal of Hazardous Materials* 447: 130818.

Centers for Disease Control and Prevention (CDC). 2024. Estimates of foodborne illnesses in the United States. Available at: https://www.cdc.gov/foodborneburden/2011-foodborne-estimates.html. Accessed 30 July 2024.

Chan S. 2016. 6 hospitalized, one of them brain-dead, after drug trial in France. *New York Times*, January 15, 2016: B2.

Childress, J.F. 1976. Compensating injured research subjects: I. The moral argument. *Hastings Center Report* 6 (6): 21–27.

Chiong, W. 2006. The real problem with equipoise. *American Journal of Bioethics* 6 (4): 37–47.

Christensen, E. 2007. Methodology of superiority vs. equivalence trials and non-inferiority trials. *Journal of Hepatology* 46: 947–954.

Clayton, E.W., and A.L. McGuire. 2012. The legal risks of returning results of genomics research. *Genetics in Medicine* 14: 473–477.

Corbie-Smith, G., S.B. Thomas, and D.M. St George. 2002. Distrust, race, and research. *Archives of Internal Medicine* 162 (21): 2458–2463.

Council for the International Organizations of Medical Sciences. 2016. *International ethical guidelines for health-related research involving humans*. Geneva, Switzerland: Council for the International Organizations of Medical Sciences.

Cox, E., L. Borio, and R. Temple. 2014. Evaluating Ebola therapies—The case for RCTs. *New England Journal of Medicine* 371 (25): 2350–2351.

Crocco, A., and D. D'Elia. 2007. Adverse reactions during voluntary donation of blood and/or blood components. A statistical-epidemiological study. *Blood Transfusion* 5 (3): 143–152.

Darnell, A.J., H. Austin, D.A. Bluemke, R.O. Cannon 3rd, K. Fischbeck, W. Gahl, D. Goldman, C. Grady, M.H. Greene, S.M. Holland, S.C. Hull, F.D. Porter, D. Resnik, W.S. Rubinstein, and L.G. Biesecker. 2016. A clinical service to support the return of secondary genomic findings in human research. *American Journal of Human Genetics* 98 (3): 435–441.

Darrow, J.J., A. Sarpatwari, J. Avorn, and A.S. Kesselheim. 2015. Practical, legal, and ethical issues in expanded access to investigational drugs. *New England Journal of Medicine* 372 (3): 279–286.

Davis, J.K. 2003. Self-experimentation. *Accountability in Research* 10 (3): 175–187.

Department of Health and Human Services. 2024. Extreme heat deaths. Available at: https://www.hhs.gov/climate-change-health-equity-environmental-justice/climate-change-health-equity/climate-health-outlook/extreme-heat/index.html#:~:text=Heat%2Drelated%20deaths%20have%20been,2022%2C%20and%202%2C302%20in%202023. Accessed 30 July 2024.

Department of Health and Human Services, and Food and Drug Administration. 2011. Advance notice of proposed rulemaking. Human subjects research protections: Enhancing protections for research subjects and reducing burden, delay, and ambiguity for investigators. *Federal Register* 76 (143): 44512–44531.

Department of Homeland Security, Department of Agriculture, Department of Energy, National Aeronautics and Space Administration, Department of Commerce, Social Security Administration, Agency for International Development, Department of Justice, Department of Labor, Department of Defense, Department of Education, Department of Veterans Affairs, Environmental Protection Agency, Department of Health and Human Services, National Science Foundation, and Department of Transportation. 2015. Notice of proposed rulemaking. Federal policy for the protection of human subjects. *Federal Register* 80 (173): 53933–54061.

Dewar, B., S. Chevrier, J. De Meulemeester, M. Fedyk, R. Rodriguez, S. Kitto, R. Saginur, and M. Shamy. 2023. What do we talk about when we talk about "equipoise"? Stakeholder interviews assessing the use of equipoise in clinical research ethics. *Trials* 24 (1): 203.

Dickert, N.W., and D. Wendler. 2009. Ancillary care obligations of medical researchers. *Journal of the American Medical Association* 302 (4): 424–428.

Dorval, M., G. Gauthier, E. Maunsell, M.J. Dugas, I. Rouleau, J. Chiquette, M. Plante, R. Laframboise, M. Gaudet, P.J. Bridge, and J. Simard. 2005. No evidence of false reassurance among women with an inconclusive BRCA1/2 genetic test result. *Cancer Epidemiology Biomarkers and Prevention* 14: 2862–2867.

Dresser, R. 2001. *When science offers salvation: Patient advocacy and research ethics*. New York, NY: Oxford University Press.

———. 2012. Research ethics: Aligning regulations and ethics in human research. *Science* 337 (6094): 527–528.

Duijnhoven, R.G., S.M. Straus, J.M. Raine, A. de Boer, A.W. Hoes, and M.L. De Bruin. 2013. Number of patients studied prior to approval of new medicines: A database analysis. *PLoS Medicine* 10 (3): e1001407.

DuVal, G. 2004. Ethics in psychiatric research: Study design issues. *Canadian Journal of Psychiatry* 49 (1): 55–59.

Earman, J. 1992. *Bayes of bust? A critical examination of Bayesian confirmation theory*. Cambridge, MA: M.I.T. Press.

Eder, M.M. 2023. Aligning clinical research ethics with community-engaged and participatory research in the United States. *Frontiers in Public Health* 11: 1122479.

Edwards, S.J.L., R. Ashcroft, and S. Kirchin. 2004. Research ethics committees: Differences and moral judgement. *Bioethics* 18 (5): 408–427.

Emanuel, E., and F.G. Miller. 2001. The ethics of placebo-controlled trials: A middle ground. *New England Journal of Medicine* 345: 915–919.

Emanuel, E.J., D. Wendler, and C. Grady. 2000. What makes clinical research ethical? *Journal of the American Medical Association* 283 (20): 2701–2711.

Emanuel, E.J., G. Bedarida, K. Macci, N.B. Gabler, A. Rid, and D. Wendler. 2015. Quantifying the risks of non-oncology phase I research in healthy volunteers: Meta-analysis of phase I studies. *British Medical Journal* 350: h3271.

Eyal, N. 2020. Is there an ethical upper limit on risks to study participants? *Public Health Ethics* 13 (2): 143–156.

Eyal, N., M. Lipsitch, and P.G. Smith. 2020. Human challenge studies to accelerate coronavirus vaccine licensure. *Journal of Infectious Disease* 221 (11): 1752–1756.

Fenton, J.J., M.S. Weyrich, S. Durbin, Y. Liu, H. Bang, and J. Melnikow. 2018. Prostate-specific antigen-based screening for prostate cancer: Evidence report and systematic review for the US Preventive Services Task Force. *Journal of the American Medical Association* 319 (18): 1914–1931.

Fernandez, C.V., E. Kodish, and C. Weijer. 2003. Informing study participants of research results: An ethical imperative. *IRB* 25 (3): 12–19.

Fisher, R.A. 1925. *Statistical methods for research workers*. London, UK: Oliver and Boyd.

Fisher, J.A. 2020. *Adverse events: Race, inequality, and the testing of new pharmaceuticals*. New York, NY: NYU Press.

Fong, M., K.L. Braun, and R.M. Chang. 2004. Native Hawaiian preferences for informed consent and disclosure of results from research using stored biological specimens. *Pacific Health Dialogue* 11 (2): 154–159.

Food and Drug Administration. 2012. Guidance for industry and FDA Staff: FDA acceptance of foreign clinical studies not conducted under an IND. Frequently asked questions. Available at: https://www.fda.gov/regulatory-information/search-fda-guidance-documents/fda-acceptance-foreign-clinical-studies-not-conducted-under-ind-frequently-asked-questions; https://www.fda.gov/. Accessed 4 Aug 2024.

———. 2018. Payment and reimbursement to research subjects: Guidance for institutional review boards and clinical investigators. Available at: https://www.fda.gov/regulatory-information/search-fda-guidance-documents/payment-and-reimbursement-research-subjects. Accessed 30 June 2024.

———. 2019. Clinical lactation studies: Considerations for study design guidance for industry. https://www.fda.gov/media/124749/download.

———. 2023. What is a serious adverse event? Available at: https://www.fda.gov/safety/reporting-serious-problems-fda/what-serious-adverse-event. Accessed 16 July 2024.

———. 2024. Emergency use authorization. Available at: https://www.fda.gov/emergency-preparedness-and-response/mcm-legal-regulatory-and-policy-framework/emergency-use-authorization. Accessed 9 Aug 2024.

Ford, I., and J. Norrie. 2016. Pragmatic trials. *New England Journal of Medicine* 375: 454–463.

Freedman, B. 1987. Equipoise and the ethics of clinical research. *New England Journal of Medicine* 317 (3): 141–145.

Freedman, B., A. Fuks, and C. Weijer. 1993. In loco parentis. Minimal risk as an ethical threshold for research upon children. *Hastings Center Report* 23 (2): 13–19.

Fried, C. 1974. *Medical experimentation: Personal integrity and social policy*. New York, NY: Elsevier Publishing.

Friedman, L.W., and E.B. Schron. 2008. Data and safety monitoring boards. In *The Oxford handbook of clinical research ethics*, ed. E.J. Emanuel, C. Grady, R.A. Crouch, R.K. Lie, F.G. Miller, and D. Wendler, 569–576. New York, NY: Oxford University Press.

Friesen, P., A.N.M. Yusof, and M. Sheehan. 2019. Should the decisions of institutional review boards be consistent? *Ethics and Human Research* 41 (4): 2–14.

Gallin, J.I., and F.P. Ognibene. 2012. *Principles and practice of clinical research*. 3rd ed. Amsterdam, The Netherlands: Elsevier.

Gibbons, L., S.N. Blair, H.W. Kohl, and K. Cooper. 1989. The safety of maximal exercise testing. *Circulation* 80 (4): 846–852.

Gifford, F. 2000. Freedman's 'clinical equipoise' and sliding-scale all-dimensions-considered equipoise'. *Journal of Medicine and Philosophy* 25 (4): 399–426.

Glantz, L.H., G.J. Annas, M.A. Grodin, and W.K. Mariner. 1998. Research in developing countries: Taking "benefit" seriously. *Hastings Center Report* 28: 38–42.

Goodyear, M. 2006. Learning from the TGN1412 trial. *British Medical Journal* 332 (7543): 677–688.

Gordon, B.G., J.S. Brown, C.J. Kratochvil, T.L. Schonfeld, and E.D. Prentice. 2006. Paying research subjects. In *Institutional review board: Management and function*, ed. E.A. Bankert and R.J. Amdur, 154–159. Sudbury, MA: Jones and Bartlett.

Grady, C. 2005. The challenge of assuring continued post-trial access to beneficial treatment. *Yale Journal of Health Policy, Law, and Ethics* 5: 425–435.

Grady, C., and C. Denny. 2008. Research involving women. In *The Oxford handbook of clinical research ethics*, ed. E.J. Emanuel, C. Grady, R.A. Crouch, R.K. Lie, F.G. Miller, and D. Wendler, 407–422. New York: Oxford University Press.

Green, L.A., J.C. Lowery, C.P. Kowalski, and L. Wyszewianski. 2006. Impact of institutional review board practice variation on observational health services research. *Health Services Research* 41 (1): 214–230.

Green, R.C., J.S. Berg, W.W. Grody, S.S. Kalia, B.R. Korf, C.L. Martin, A.L. McGuire, R.L. Nussbaum, J.M. O'Daniel, K.E. Ormond, H.L. Rehm, M.S. Watson, M.S. Williams, L.G. Biesecker, and American College of Medical Genetics and Genomics. 2013. ACMG recommendations for reporting of incidental findings in clinical exome and genome sequencing. *Genetics in Medicine* 15 (7): 565–574.

Gunsalus, C.K., E. Bruner, N.C. Burbules, L. Dash, M. Finkin, J.P. Goldber, W.T. Greenough, G.A. Miller, M.G. Pratt, and D. Aronson. 2005. The Illinois white paper: Improving the system for protecting human subjects: Counteracting IRB 'Mission Creep'. *Qualitative Inquiry* 13 (5): 617–649.

Hall, N.S. 2007. R. A. Fisher and his advocacy of randomization. *Journal of the History of Biology* 40 (2): 295–325.
Hamilton, R., J.K. Williams, H. Skirton, and B.J. Bowers. 2009. Living with genetic test results for hereditary breast and ovarian cancer. *Journal of Nursing Scholarship* 41 (3): 276–283.
Hanser, M. 2020. Bystander risk and informed consent. *Bioethics* 34: 912–917.
Hawkins, J.S. 2008. Research ethics, developing countries, exploitation: A primer. In *Exploitation and developing countries: The ethics of clinical research*, ed. J.S. Hawkins and E.J. Emanuel, 21–54. Princeton: Princeton University Press.
Hellman, S., and D. Hellman. 1991. Of mice but not men: Problems of the randomized clinical trial. *New England Journal of Medicine* 324 (2): 185–189.
Hirshon, J.M., S.D. Krugman, M.D. Witting, J.P. Furuno, M.R. Limcangco, A.R. Perisse, and E.K. Rasch. 2002. Variability in institutional review board assessment of minimal-risk research. *Academic Emergency Medicine* 9 (12): 1417–1420.
Hoeyer, K., B.O. Olofsson, T. Mjörndal, and N. Lynöe. 2004. Informed consent and biobanks: A population-based study of attitudes towards tissue donation for genetic research. *Scandinavian Journal of Public Health* 32 (3): 224–229.
Hooper, C.R. 2010. Ancillary care duties: The demands of justice. *Journal of Medical Ethics* 36 (11): 708–711.
Hyder, A.A., and M.W. Merritt. 2009. Ancillary care for public health research in developing countries. *Journal of the American Medical Association* 302 (4): 429–431.
Illes, J., M.P. Kirschen, E. Edwards, L.R. Stanford, P. Bandettini, M.K. Cho, P.J. Ford, G.H. Glover, J. Kulynych, R. Macklin, D.B. Michael, and S.M. Wolf. 2006. Incidental findings in brain imaging research. *Science* 311: 783–784.
International Conference on Harmonization. 2016. Good clinical practice. Available at: https://database.ich.org/sites/default/files/E6_R2_Addendum.pdf. Accessed 10 Aug 2024.
———. 2023. Good clinical practice, draft version. Available at: https://database.ich.org/sites/default/files/E6_R2_Addendum.pdf. Accessed 10 Aug 2024.
Jamrozik, E., K. Littler, S. Bull, C. Emerson, G. Kang, M. Kapulu, E. Rey, C. Saenz, S. Shah, P.G. Smith, R. Upshur, C. Weijer, and M.J. Selgelid; WHO Working Group for Guidance on Human Challenge Studies in COVID-19. 2021. Key criteria for the ethical acceptability of COVID-19 human challenge studies: Report of a WHO Working Group. *Vaccine* 39(4):633–640.
Janofsky, J., and B. Starfield. 1981. Assessment of risk in research on children. *Journal of Pediatrics* 98 (5): 842–846.
Jarvik, G.P., L.M. Amendola, J.S. Berg, K. Brothers, E.W. Clayton, W. Chung, B.J. Evans, J.P. Evans, S.M. Fullerton, C.J. Gallego, N.A. Garrison, S.W. Gray, I.A. Holm, I.J. Kullo, L.S. Lehmann, C. McCarty, C.A. Prows, H.L. Rehm, R.R. Sharp, J. Salama, S. Sanderson, S.L. Van Driest, M.S. Williams, S.M. Wolf, W.A. Wolf, eMERGE Act-ROR Committee and CERC Committee, CSER Act-ROR Working Group, and W. Burke. 2014. Return of genomic results to research participants: The floor, the ceiling, and the choices in between. *American Journal of Human Genetics* 94 (6): 818–826.
Joffe, S., and R. Truog. 2008. Equipoise and randomization. In *The Oxford textbook of clinical research ethics*, ed. E. Emanuel, C. Grady, R. Crouch, R. Lie, F. Miller, and D. Wendler, 245–260. New York, NY: Oxford University Press.
Jordan, S.R., and P.W. Gray. 2017. Clarifying the concept of the "social" in risk assessments for human subjects research. *Accountability in Research* 25 (1): 1–20.
Kalil, A.C. 2020. Treating COVID-19—Off-label drug use, compassionate use, and randomized clinical trials during pandemics. *Journal of the American Medical Association* 323 (19): 1897–1898.
Kane, C., K. Takechi, M. Chuma, H. Nokihara, T. Takagai, and H. Yanagawa. 2019. Perspectives of non-specialists on the potential to serve as ethics committee members. *Journal of International Medical Research* 47 (5): 1868–1876.
Kenter, M.J., and A.F. Cohen. 2006. Establishing risk of human experimentation with drugs: Lessons from TGN1412. *Lancet* 368 (9544): 1387–1391.

Keulartz, J., and H. van den Belt. 2016. DIY-bio – Economic, epistemological and ethical implications and ambivalences. *Life Sciences and Social Policy* 12 (1): 7.
Khodyakov, D., L. Mikesell, and E. Bromley. 2017. Trust and the ethical conduct of community-engaged research. *European Journal for Person-Centered Healthcare* 5 (4): 522–526.
Kimmelman, J. 2004. Valuing risk: The ethical review of clinical trial safety. *Kennedy Institute of Ethics Journal* 14 (3): 369–393.
———. 2005. Medical research, risk, and bystanders. *IRB* 27 (4): 1–6.
———. 2008. The ethics of human gene transfer. *Nature Reviews Genetics* 9: 239–244.
———. 2020. Why IRBs should protect bystanders in human subjects research. *Bioethics* 34: 933–936.
Kimmelman, J., and V. Henderson. 2016. Assessing risk/benefits trials for using preclinical evidence: A proposal. *Journal of Medical Ethics* 42 (1): 50–53.
Kimmelman, J., D.B. Resnik, J. Peppercorn, and M.J. Ratain. 2017. Burdensome research procedures in trials: Why less is more. *Journal of National Cancer Institute* 109 (4): djw315.
King, N.M. 2000. Defining and describing benefit appropriately in clinical trials. *Journal of Law, Medicine & Ethics* 28 (4): 332–343.
Kionka, E. 2020. *Torts in a nutshell*. St. Paul, MN: Westlaw.
Kirsch, I., B.J. Deacon, T.B. Huedo-Medina, A. Scoboria, T.J. Moore, and B.T. Johnson. 2008. Initial severity and antidepressant benefits: A meta-analysis of data submitted to the Food and Drug Administration. *PLoS Medicine* 2: e45.
Klitzman, R.L. 2012. Institutional review board community members: Who are they, what do they do, and whom do they represent? *Academic Medicine* 87 (7): 975–881.
———. 2013. How IRBs view and make decisions about social risks. *Journal of Empirical Research in Human Research Ethics* 8 (3): 58–65.
Klitzman, R., E. Pivovarova, A. Murray, P.S. Appelbaum, D.F. Stiles, and C.W. Lidz. 2019. Local knowledge and single IRBs for multisite studies: Challenges and solutions. *Ethics and Huma Research* 41 (1): 22–31.
Kolodziejczyk, B. 2017. Do-it-yourself biology shows safety risks of an open innovation movement. Brookings Institute, 9 October. Available at: https://www.brookings.edu/articles/do-it-yourself-biology-shows-safety-risks-of-an-open-innovation-movement/. Accessed 22 July 2024.
Kopelman, L.M. 1981. Estimating risk in human research. *Clinical Research* 29 (1): 1–8.
———. 1997. The best-interests standard as threshold, ideal, and standard of reasonableness. *Journal of Medicine and Philosophy* 22 (3): 271–289.
———. 2000. Children as research subjects: A dilemma. *Journal of Medicine and Philosophy* 25 (6): 745–764.
Kritsaneepaiboon, S., A. Jutiyon, and A. Krisanachinda. 2018. Cumulative radiation exposure and estimated lifetime cancer risk in multiple-injury adult patients undergoing repeated or multiple CTs. *European Journal of Trauma Emergency Surgery* 44 (1): 19–27.
Laughren, T.P. 2001. The scientific and ethical basis for placebo-controlled trials in depression and schizophrenia: An FDA perspective. *European Psychiatry* 16 (7): 418–423.
Lavery, J.V. 2008. The obligation to ensure access to beneficial treatments for research participants at the conclusion of clinical trials. In *The Oxford handbook of clinical research ethics*, ed. E.J. Emanuel, C. Grady, R.A. Crouch, R.K. Lie, F.G. Miller, and D. Wendler, 261–272. New York, NY: Oxford University Press. 697–710.
Levine, R.J. 1988. *Ethics and the regulation of clinical research*. 2nd ed. New Haven: Yale University Press.
Lipworth, W., I. Kerridge, C. Stewart, D. Silva, and R. Upshur. 2023. The fragility of scientific rigour and integrity in "sped up science": Research misconduct, bias, and hype and in the COVID-19 pandemic. *Journal of Bioethical Inquiry* 20 (4): 607–616.
Litton, P., and F.G. Miller. 2005. A normative justification for distinguishing the ethics of clinical research from the ethics of medical care. *Journal of Law, Medicine & Ethics* 33 (3): 566–574.
Lo, B., N. Padian, and M. Barnes. 2007. The obligation to provide antiretroviral treatment in HIV prevention trials. *AIDS* 21 (10): 1229–1231.

London, A.J. 2005. Justice and the human development approach to international research. *Hastings Center Report* 35 (1): 24–37.
———. 2006. Reasonable risks in clinical research: A critique and a proposal for the integrative approach. *Statistics in Medicine* 25: 2869–2885.
———. 2007. Clinical equipoise: Foundational requirement or fundamental error. In *The Oxford handbook of bioethics*, ed. B. Steinbock, 571–596. Oxford, UK: Oxford University Press.
———. 2022. *Philosophical foundations of research ethics*. New York, NY: Oxford University Press.
London, A.J., and J. Kimmelman. 2020. Against pandemic research exceptionalism. *Science* 368 (6490): 476–477.
London, A.J., and C.W. Seymour. 2023. The ethics of clinical research: Managing persistent uncertainty. *Journal of the American Medical Association* 329 (11): 884–885.
Lurie, P., and S.M. Wolfe. 1997. Unethical trials of interventions to reduce perinatal transmission of the human immunodeficiency virus in developing countries. *New England Journal of Medicine* 337 (12): 853–856.
Lux, A.L., S.W. Edwards, and J.P. Osborne. 2000. Responses of local research ethics committees to a study with approval from a multicentre research ethics committee. *BMJ* 320 (7243): 1182–1183.
Manheim, D., W. Więcek, V. Schmit, J. Morrison, and 1Day Sooner Research Team. 2021. Exploring risks of human challenge trials for COVID-19. *Risk Analysis* 41 (5): 710–720.
Marchant, J. 2015. Strong placebo response thwarts painkiller trials. Nature, 6 October. Available at: https://doi.org/10.1038/nature.2015.18511. Accessed 27 July 2024.
Marquis, D. 1983. Leaving therapy to chance. *Hastings Center Report* 13 (4): 40–47.
Martin, D.K., E.M. Meslin, N. Kohut, and P.A. Singer. 1995. The incommensurability of research risks and benefits: Practical help for research ethics committees. *IRB* 17 (2): 8–10.
Mayo, D. 1987. Comments on "randomization and the design of experiments" by P. Urbach. *Philosophy of Science* 54 (4): 592–596.
Mayo Clinic. 2024. Bronchoscopy. Available at: https://www.mayoclinic.org/tests-procedures/bronchoscopy/about/pac-20384746. Accessed 15 July 2024.
McCormick, R.A. 1976. Experimentation in children: Sharing sociality. *Hastings Center Report* 6 (6): 41–46.
McNamara, P. 1996. Making room for going beyond the call. *Mind* 105: 415–450.
Melo, H., C. Brandao, G. Rego, and R. Nunes. 2001. Ethical and legal issues in xenotransplantation. *Bioethics* 15 (5–6): 427–442.
Miller, M. 2000. Phase I cancer trials: A collusion of misunderstanding. *Hastings Center Report* 30 (4): 34–43.
Miller, F.G. 2003. Sham surgery: An ethical analysis. *American Journal of Bioethics* 3 (4): 41–48.
———. 2008. The ethics of placebo-controlled trials. In *The Oxford handbook of clinical research ethics*, ed. E.J. Emanuel, C. Grady, R.A. Crouch, R.K. Lie, F.G. Miller, and D. Wendler, 261–272. New York, NY: Oxford University Press.
Miller, D. 2021. Responsibility and the duty to rescue. *Journal of Applied Philosophy* 39 (2): 313–326.
Miller, F.G., and H. Brody. 2002. What makes placebo-controlled trials unethical? *American Journal of Bioethics* 2 (2): 3–9.
———. 2003. A critique of clinical equipoise. Therapeutic misconception in the ethics of clinical trials. *Hastings Center Report* 33 (3): 9–28.
———. 2007. Clinical equipoise and the incoherence of research ethics. *Journal of Medicine and Philosophy* 32 (2): 151–165.
Miller, F.G., and C. Grady. 2001. The ethical challenge of infection-inducing challenge experiments. *Clinical Infectious Diseases* 33 (7): 1028–1033.
Miller, F.G., and S. Joffe. 2009. Limits to research risks. *Journal of Medical Ethics* 35 (7): 445–449.
———. 2011. Equipoise and the dilemma of randomized clinical trials. *New England Journal of Medicine* 364 (5): 476–480.

Miller, P.B., and C. Weijer. 2003. Rehabilitating equipoise. *Kennedy Institute of Ethics Journal* 13 (2): 93–118.

Miller, F.G., and A. Wertheimer. 2007. Facing up to paternalism in research ethics. *Hastings Center Report* 37 (3): 24–34.

Miller, F.G., M.M. Mello, and S. Joffe. 2008. Incidental findings in human subjects research: What do investigators owe research participants? *The Journal of Law, Medicine & Ethics* 36: 271–279.

Millum, J. 2011. Post-trial access to antiretrovirals: Who owes what to whom? *Bioethics* 25 (3): 145–154.

Moreno, J., A.L. Caplan, P.R. Wolpe, and the Members of the Project on Informed Consent, Human Research Ethics Group. 1998. Updating protections for human subjects involved in research. *Journal of the American Medical Association* 280 (22): 1951–1958.

Morse, M.A., R.M. Califf, and J. Sugarman. 2001. Monitoring and ensuring safety during clinical research. *Journal of the American Medical Association* 285 (9): 1201–1205.

National Academies of Sciences, Engineering, and Medicine. 2018. *Returning individual research results to participants: Guidance for a new research paradigm*. Washington, DC: National Academies Press (US). 2018 Jul 10.

National Bioethics Advisory Commission. 2001. *Ethical and policy issues in research involving human participants*. Washington, DC: National Bioethics Advisory Commission.

National Center for Health Statistics. 2024. Accidents or unintentional injuries. Available at: https://www.cdc.gov/nchs/fastats/accidental-injury.htm. Accessed 30 July 2024.

National Commission for the Protection of Human Subjects of Biomedical or Behavioral Research. 1979. *The Belmont report: Ethical principles and guidelines for the protection of human subjects of research*. Washington, DC: Department of Health, Education, and Welfare.

National Institutes of Health. 2005. Guidance for addressing the provision of antiretroviral treatment for trial participants following their completion of NIH-funded HIV antiretroviral treatment trials in developing countries. NOT-OD-05-038. Available at: https://grants.nih.gov/grants/guide/notice-files/NOT-OD-05-038.html. Accessed 4 Aug 2024.

———. 2023. IRP Scientific Review Process of Human Subjects Research. Available at: https://clinicalcenter.nih.gov/chief_scientific_officer/scientific_review_process/sop_srp.html. Accessed 15 July 2024.

Nature. 2006. Can super-antibody drugs be tamed? *Nature* 440: 855–856.

Neugut, A.I., A.T. Ghatak, and R.L. Miller. 2001. Anaphylaxis in the United States: An investigation into its epidemiology. *Archives of Internal Medicine* 161 (1): 15–21.

Nix, H.P., and C. Weijer. 2021. Uses of equipoise in discussions of the ethics of randomized controlled trials of COVID-19 therapies. *BMC Medical Ethics* 22: 143.

Nuremberg Code. 1949. *Trials of war criminals before the Nuremberg military tribunals under Control Council Law 2 (10): 181–182*. Washington, DC: U.S. Government Printing Office. Available at: https://ori.hhs.gov/content/chapter-3-The-Protection-of-Human-Subjects-nuremberg-code-directives-human-experimentation. Accessed 7 May 2024.

O'Fallon, L.R., and A. Dearry. 2002. Community-based participatory research as a tool to advance environmental health sciences. *Environmental Health Perspectives* 110 Suppl 2 (Suppl 2): 155–159.

O'Neill, S.C., T. DeMarco, B.N. Peshkin, S. Rogers, J. Rispoli, K. Brown, H. Valdimarsdottir, and M.D. Schwartz. 2006. Tolerance for uncertainty and perceived risk among women receiving uninformative BRCA1/2 test results. *American Journal of Medical Genetics C: Seminars in Medical Genetics* 142C (4): 251–259.

Office of Human Research Protections. 1998. Expedited review: Categories of research that may be reviewed through an expedited review procedure. Available at: https://www.hhs.gov/ohrp/regulations-and-policy/guidance/categories-of-research-expedited-review-procedure-1998/index.html. Accessed 16 July 2024.

Oliveri, S., F. Ferrari, A. Manfrinati, and G. Pravettoni. 2018. A systematic review of the psychological implications of genetic testing: A comparative analysis among cardiovascular, neurodegenerative and cancer diseases. *Frontiers in Genetics* 9: 624.

Pace, C., C. Grady, D. Wendler, J.D. Bebchuk, J.A. Tavel, L.A. McNay, H.P. Forster, J. Killen, and E.J. Emanuel. 2006. Post-trial access to tested interventions: The views of IRB/REC chair, investigators, and research participants in a multinational HIV/AIDS study. *AIDS Research and Human Retroviruses* 22 (9): 837–841.

Parasidis, E. 2001. Compensation for research-related injuries involving human participants. *Harvard Journal on Medical Ethics* 2: 26–33.

Pardo-Cabello, A.J., V. Manzano-Gamero, and E. Puche-Cañas. 2022. Placebo: A brief updated review. *Naunyn Schmiedebergs Archives of Pharmacology* 395 (11): 1343–1356.

Participants in 2006 the Georgetown University Workshop on Ancillary-Care Obligations of Medical Researchers Working in Developing Countries. 2008. The ancillary care obligations of medical researchers working in developing countries. *PLoS Medicine* 5 (5): e90.

Participants in the 2001 Conference on Ethical Aspects of Research in Developing Countries. 2002. Ethics. Fair benefits for research in developing countries. *Science* 298 (5601): 2133–2134.

Patrone, D. 2009. Disfigured anatomies and imperfect analogies: Body integrity identity disorder and the supposed right to self-demanded amputation of healthy body parts. *Journal of Medical Ethics* 35 (9): 541–545.

Pfizer.com. 2024. How clinical trials work. Available at: https://www.pfizerclinicaltrials.com/about/how-clinical-trials-work. Accessed 7 July 2024.

Pike, E.R. 2012. Recovering from research: A no-fault proposal to compensate injured research participants. *American Journal of Law and Medicine* 38 (1): 7–62.

———. 2014. In need of remedy: US policy for compensating injured research participants. *Journal of Medical Ethics* 40 (3): 182–185.

Prasad, V., and C. Grady. 2014. The misguided ethics of crossover trials. *Contemporary Clinical Trials* 37 (2): 167–169.

Pratt, B. 2024. Defending and defining environmental responsibilities for the health research sector. *Science and Engineering Ethics* 30: 25.

Pratt, B., D. Zion, K.M. Lwin, P.Y. Cheah, F. Nosten, and B. Loff. 2013. Ancillary care: From theory to practice in international clinical research. *Public Health Ethics* 6 (2): 154–169.

Presidential Commission for the Study of Bioethical Issues. 2011. *Moral science: Protecting participants in human subjects research*. Washington, DC: Presidential Commission.

———. 2013. *Anticipate and communicate: Ethical management of incidental and secondary findings in the clinical, research, and direct to consumer contexts*. Washington, DC: Presidential Commission.

Pritchard, I.A. 2011. How do IRB members make decisions? A review and research agenda. *Journal of Empirical Research in Human Research Ethics* 6 (2): 31–46.

Pue, C.A., and E.R. Pacht. 1995. Complications of fiberoptic bronchoscopy at a university hospital. *Chest* 107 (2): 430–432.

Ramsay, P. 1976. The enforcement of morals: Nontherapeutic research on children. *Hastings Center Report* 6 (4): 21–30.

Raus, K. 2016. An analysis of common ethical justifications for compassionate use programs for experimental drugs. *BMC Medical Ethics* 17: 60.

Ravitsky, V., and B.S. Wilfond. 2006. Disclosing individual genetic results to research participants. *American Journal of Bioethics* 6 (6): 8–17.

Renegar, G., C.J. Webster, S. Stuerzebecher, L. Harty, S.E. Ide, B. Balkite, T.A. Rogalski-Salter, N. Cohen, B.B. Spear, D.M. Barnes, and C. Brazell. 2006. Returning genetic research results to individuals: Points-to-consider. *Bioethics* 20 (1): 24–36.

Resnik, D.B. 1998. The ethics of HIV research in developing nations. *Bioethics* 12 (4): 286–306.

———. 2001. Developing drugs for the developing world: An economic, legal, moral, and political dilemma. *Developing World Bioethics* 1 (1): 11–32.

———. 2003. Exploitation in biomedical research. *Theoretical Medicine and Bioethics* 24 (3): 233–259.

———. 2005. Eliminating the daily life risks standard from the definition of minimal risk. *Journal of Medical Ethics* 31 (1): 35–38.

———. 2006. Compensation for research-related injuries. Ethical and legal issues. *Journal of Legal Medicine* 27 (3): 263–287.

———. 2007. Intentional exposure studies of environmental agents on human subjects: Assessing benefits and risks. *Accountability in Research* 14 (1): 35–55.

———. 2009. The clinical investigator-subject relationship: A contextual approach. *Philosophy, Ethics, and Humanities in Medicine* 4: 16.

———. 2011. Disclosure of individualized research results: A precautionary approach. *Accountability in Research* 18 (6): 382–397.

———. 2012. Limits on risks for healthy volunteers in biomedical research. *Theoretical Medicine and Bioethics* 33 (2): 137–149.

———. 2015a. Unequal treatment of human research subjects. *Medicine Health Care and Philosophy* 18 (1): 23–32.

———. 2015b. Paternalism and utilitarianism in research with human participants. *Health Care Analysis* 23 (1): 19–31.

———. 2017a. The role of intuition in risk/benefit decision-making in human subjects research. *Accountability in Research* 24 (1): 1–29.

———. 2017b. Risk communication in EPA's controlled inhalation exposure studies and in SUPPORT. *Perspectives in Biology and Medicine* 60 (1): 117–129.

———. 2018a. Examining the social benefits principle in research with human participants. *Health Care Analysis* 26 (1): 66–80.

———. 2018b. *The ethics of research with human subjects: Protecting people, advancing science, promoting trust*. Cham: Springer.

———. 2021a. Standards of evidence for institutional review board decision-making. *Accountability in Research* 28 (7): 428–455.

———. 2021b. *Precautionary reasoning in environmental and public health policy*. Cham: Springer.

Resnik, D.B., and K.C. Elliott. 2016. The ethical challenges of socially responsible science. *Accountability in Research* 23 (1): 31–46.

Resnik, D.B., and C.E. Kennedy. 2010. Balancing scientific and community interests in community-based participatory research. *Accountability in Research* 17 (4): 198–210.

Resnik, D.B., and G. Koski. 2011. A national registry for healthy volunteers in phase 1 clinical trials. *Journal of the American Medical Association* 305: 1236–1237.

Resnik, D.B., and R.R. Sharp. 2006. Protecting third parties in human subjects research. *IRB* 28 (4): 1–7.

Resnik, D.B., and S.S. Tinkle. 2007. Ethical issues in clinical trials involving nanomedicine. *Contemporary Clinical Trials* 28 (4): 433–441.

Resnik, D.B., and D.C. Zeldin. 2008. Environmental health research on hazards in the home and the duty to warn. *Bioethics* 22 (4): 209–217.

Resnik, D.B., E. Parasidis, K. Carroll, J.M. Evans, E.R. Pike, and G.E. Kissling. 2014. Research-related injury compensation policies of U.S. research institutions. *IRB* 36 (1): 12–19.

Resnik, D.B., A.K. Miller, R.K. Kwok, L.S. Engle, and D.P. Sandler. 2015. Ethical issues in environmental health research related to public health emergencies: Reflections in the GuLF study. *Environmental Health Perspectives* 123 (9): A227–A231.

Reynolds, T. 2002. For institutional review boards, decisions can be subjective. *Journal of the National Cancer Institute* 94 (20): 1518–1520.

Richardson, H.S. 2008. Incidental findings and ancillary-care obligations. *The Journal of Law, Medicine & Ethics* 36: 256–270.

Richardson, H.S., and L. Belsky. 2004. The ancillary-care responsibilities of medical researchers. *Hastings Center Report* 34 (1): 25–33.

Richardson, H.S., and M.K. Cho. 2012. Secondary researchers' duties to return incidental findings and individual research results: A partial-entrustment account. *Genetics in Medicine* 14 (4): 467–472.

Richardson, H.S., N. Eyal, J.I. Campbell, and J.E. Hargerer. 2017. When ancillary care clashes with study aims. *New England Journal of Medicine* 377: 1213–1215.

Rid, A., and D. Wendler. 2011. A framework for risk-benefit evaluations in biomedical research. *Kennedy Institute of Ethics Journal* 21 (2): 141–179.

Rid, A., E.J. Emanuel, and D. Wendler. 2010. Evaluating the risks of clinical research. *Journal of the American Medical Association* 304 (13): 1472–1479.

Rodwin, B.A., V.P. Bilan, N.B. Merchant, C.G. Steffens, A.A. Grimshaw, L.A. Bastian, and C.G. Gunderson. 2020. Rate of preventable mortality in hospitalized patients: A systematic review and meta-analysis. *Journal of General Internal Medicine* 35 (7): 2099–2106.

Rossi, J., and R.M. Nelson. 2012. Is there an objective way to compare research risks? *Journal of Medical Ethics* 38 (7): 423–427.

Rosenheck, M. 2022. Risk, benefit, and social value in Covid-19 human challenge studies: Pandemic decision making in historical context. *Monash Bioethics Review* 40 (2): 188–213.

Rothman, K., and K. Michels. 1994. The continuing unethical use of placebo controls. *New England Journal of Medicine* 331 (6): 394–398.

Rothstein, M.A., K.C. Zimmerer, P. Andanda, T. Arawi, F. Arzuaga, H. Chen, M. de Vries, E.S. Dove, M. Ghaly, R. Hatanaka, A.C. Hendriks, M.C. Hernández, C.W.L. Ho, Y. Joly, D. Krekora-Zając, W.B. Lee, T. Mattsson, F. Molnár-Gábor, K. Namalwa, P. Nicolás, J. Nielsen, O. Nnamuchi, M. Otlowski, N. Palmour, E. Rial-Sebbag, G. Siegal, J.M. Wathuta, M.H. Zawati, and B.M. Knoppers. 2024. International scope of biomedical research ethics review. *Science* 385 (6705): 145–147.

Różyńska, J. 2015. On the alleged right to participate in high-risk research. *Bioethics* 29 (7): 451–461.

Saccà, L. 2010. The uncontrolled clinical trial: Scientific, ethical, and practical reasons for being. *Internal and Emergency Medicine* 5 (3): 201–204.

Sachs, B. 2011. Going from principles to rules in research ethics. *Bioethics* 25 (1): 9–20.

Sackett, D., W. Richardson, W. Rosenberg, and R. Haynes. 1997. *Evidence-based medicine*. New York, NY: Churchill Livingstone.

Schiff, J.H., A. Welker, B. Fohr, A. Henn-Beilharz, U. Bothner, H. Van Aken, A. Schleppers, H.J. Baldering, and W. Heinrichs. 2014. Major incidents and complications in otherwise healthy patients undergoing elective procedures: Results based on 1.37 million anaesthetic procedures. *British Journal of Anaesthesia* 113 (1): 109–121.

Schmidt, M.H., J. Marshall, J. Downie, and M.R. Hadskis. 2011. Pediatric magnetic resonance research and the minimal-risk standard. *IRB* 33 (5): 1–6.

Schneider, C.E. 2015. *The Censor's hand: The misregulation of human-subject research*. Cambridge, MA: MIT Press.

Schott, S.L., A. Adams, R.J. Dougherty, T. Montgomery, F.C. Lapite, and F.E. Fletcher. 2023. Renewed calls for abortion-related research in the post-Roe era. *Frontiers in Public Health* 11: 1322299.

Schupmann, W., X. Li, and D. Wendler. 2022. Acceptable risks in pediatric research: Views of the US public. *Pediatrics* 149 (1): e2021052687.

Scott, L.D. 2003. Research-related injury: Problems and solutions. *Journal of Law, Medicine & Ethics* 31 (3): 419–428.

Secretary's Advisory Committee on Human Research Protections. 2022. The protection of non-subjects from research harm. Available at: https://www.hhs.gov/ohrp/sachrp-committee/recommendations/tab-c-the-protection-of-non-subjects-from-research-harm.html. Accessed 25 July 2024.

Serpico, K. 2024. Defining institutional review board application quality: Critical research gaps and future opportunities. *Research Ethics* 20 (1): 19–35.

Seykora, A., C. Coleman, S.J. Rosenfeld, B.E. Bierer, and H.F. Lynch. 2021. Steps toward a system of IRB precedent: Piloting approaches to summarizing IRB decisions for future use. *Ethics and Human Research* 43 (6): 2–18.

Shah, S., A. Whittle, B. Wilfond, G. Gensler, and D. Wendler. 2004. How do institutional review boards apply the federal risk and benefit standards for pediatric research? *Journal of the American Medical Association* 291 (4): 476–482.

Shah, S., J. Kimmelman, A.D. Lyerly, H.F. Lynch, F.G. Miller, R. Palacios, C.A. Pardo, and C. Zorrilla. 2018. Bystander risk, social value, and ethics of human research: Contentious risks demand a new approach. *Science* 360 (6385): 158–159.

Shalowitz, D.I., and F.G. Miller. 2005. Disclosing individual results of clinical research: Implications of respect for participants. *Journal of the American Medical Association* 294 (6): 737–740.

———. 2008. Communicating the results of clinical research to participants: Attitudes, practices, and future directions. *PLoS Medicine* 5 (5): e91.

Shamoo, A.E., and T.J. Keay. 1996. Ethical concerns about relapse studies. *Cambridge Quarterly of Healthcare Ethics* 5 (3): 373–386.

Shamoo, A.E., and D.B. Resnik. 2022. *Repsonsible conduct of research*, 4th ed. New York, NY: Oxford University Press.

Shrader-Frechette, K.S. 1991. *Risk and rationality: Philosophical foundations for populist reforms*. Berkeley: University of California Press.

Sihvonen, R., M. Paavola, A. Malmivaara, A. Itälä, A. Joukainen, H. Nurmi, J. Kalske, T.L. Järvinen, and Finnish Degenerative Meniscal Lesion Study (FIDELITY) Group. 2013. Arthroscopic partial meniscectomy versus sham surgery for a degenerative meniscal tear. *New England Journal of Medicine* 369 (26): 2515–2524.

Silberman, G., and K.L. Kahn. 2011. Burdens on research imposed by institutional review boards: The state of the evidence and its implications for regulatory reform. *Milbank Quarterly* 89 (4): 599–627.

Silverman, H., S.C. Hull, and J. Sugarman. 2001. Variability among institutional review boards' decisions within the context of a multicenter trial. *Critical Care Medicine* 29 (2): 235–241.

Sofaer, N., and Strech, D. 2012. The need for systematic reviews of reasons. *Bioethics* 26 (6):315–328.

Song, J.W., and K.C. Chung. 2010. Observational studies: Cohort and case-control studies. *Plastic Reconstructive Surgery* 126 (6): 2234–2242.

Statista. 2024. Rates of death due to drowning in the United States from 1905 to 1922. Available at: https://www.statista.com/statistics/527366/death-rate-due-to-drowning-in-the-us/. Accessed 30 July 2024.

Steinbrook, R. 2006. Compensation for injured research subjects. *New England Journal of Medicine* 354 (18): 1871–1873.

Stunkel, L., and C. Grady. 2011. More than the money: A review of the literature examining healthy volunteer motivations. *Contemporary Clinical Trials* 32 (3): 342–352.

Sugarman, J., and R. Calliff. 2018. Ethics and regulatory complexities for pragmatic clinical trials. *Journal of the American Medical Association* 311 (23): 2381–2382.

Tabery, J. 2015. Why is studying the genetics of intelligence so controversial? *Hastings Center Report* 45 (5 Suppl): S9–S14.

Taccone, F.S., M. Hites, and N. Dauby. 2022. From hydroxychloroquine to ivermectin: How unproven "cures" can go viral. *Clinical Microbiology and Infection* 28 (4): 472–474.

Tagar, E., M. Sundaram, K. Condliffe, B. Matatiyo, F. Chimbwandira, B. Chilima, R. Wanamanga, C. Moyo, B.M. Chitah, J.P. Nyemazi, Y. Assefa, Y. Pillay, S. Mayer, L. Shear, M. Dain, R. Hurley, R. Kumar, T. McCarthy, P. Batra, D. Gwinnell, S. Diamond, and M. Over. 2014. Multi-country analysis of treatment costs for HIV/AIDS (MATCH): Facility-level ART unit cost analysis in Ethiopia, Malawi, Rwanda, South Africa and Zambia. *PLoS ONE* 9 (11): e108304.

Taliercio, J., S. Nurko, and E. Poggio. 2011. Living donor kidney transplantation: N update on evaluation and medical implications of donation. *Minerva Urologica e Nefrologica* 63 (1): 73–87.

Tam, N.T., N.T. Huy, L.T.B. Thoa, N.P. Long, N.T. Trang, K. Hirayama, and J. Karbwang. 2015. Participants' understanding of informed consent in clinical trials over three decades: Systematic review and meta-analysis. *Bulletin of the World Health Organization* 93 (3): 186–98H.

Taylor, H.A., S. Ehrhardt, and A.M. Ervin. 2019. Public comments on the proposed common rule mandate for single-IRB review of multisite research. *Ethics Human Research* 41 (1): 15–21.

Temple, R., and S.S. Ellenberg. 2000. Placebo-controlled trials and active-control trials in the evaluation of new treatments. Part 1: Ethical and scientific issues. *Annals of Internal Medicine* 133 (6): 455–463.

The Economic Times. 2024. India becoming attractive for clinical trials, says pharma industry leaders. The Economic Times, 27 April. Available at: https://economictimes.indiatimes.com/industry/healthcare/biotech/pharmaceuticals/india-becoming-attractive-for-clinical-trials-says-pharma-industry-leaders/articleshow/109651758.cms. Accessed 4 Aug 2024.

Thomas, S.R., D.R. Jamieson, and K.W. Muir. 2000. Randomised controlled trial of atraumatic versus standard needles for diagnostic lumbar puncture. *British Medical Journal* 321 (7267): 986–990.

Tranter, E., G. Peters, M. Boyce, and S. Warrington. 2013. Giving monoclonal antibodies to healthy volunteers in phase 1 trials: Is it safe? *British Journal of Clinical Pharmacology* 76 (2): 164–172.

U.S. Bureau of Labor Statistics. 2022a. Challenges of calculating occupational fatality rates. Available at: https://www.bls.gov/opub/mlr/2022/article/challenges-in-calculating-occupational-fatality-rates.htm. Accessed 23 July 2024.

———. 2022b. Fatal occupational injuries by hours. 2022. Available at: https://www.bls.gov/iif/additional-resources/cfoi-hours-based-rates.htm. Accessed 22 July 2024.

U.S. Fire Administration. 2023. Fire death and injury risk. Available at: https://www.usfa.fema.gov/statistics/deaths-injuries/ Accessed 30 July 2024.

University of Michigan. 2022. Blood draw guidance. Available at: https://az.research.umich.edu/medschool/guidance/blood-draw-guidance#:~:text=Blood%20draws%20for%20healthy%20adults,maximum%20volume%20to%20be%20drawn. Accessed 5 Aug 2024.

Urbach, P. 1985. Randomization and the design of experiments. *Philosophy of Science* 52 (2): 256–273.

van der Graaf, R., K.C. Roes, and J.J. van Delden. 2012. Adaptive trials in clinical research: Scientific and ethical issues to consider. *Journal of the American Medical Association* 307 (22): 2379–2380.

van Dijk, S., D.R. Timmermans, H. Meijers-Heijboer, A. Tibben, C.J. van Asperen, and W. Otten. 2006. Clinical characteristics affect the impact of an uninformative DNA test result: The course of worry and distress experienced by women who apply for genetic testing for breast cancer. *Journal of Clinical Oncology* 24 (22): 3672–3677.

Varmus, H., and D. Satcher. 1997. Ethical complexities of conducting research in developing countries. *New England Journal of Medicine* 337 (14): 1003–1005.

Wald, M.L. 2010. When a cancer therapy puts others at risk. *New York Times*, October 24, 2010:A14.

Walker, R.L., D. MacKay, M. Waltz, D. Lyerly, and J.A. Fisher. 2022. Ethical criteria for improved human subject protections in phase I healthy volunteer trials. *Ethics and Human Research* 44 (5): 2–21.

Wang, Y., D. Simar, K. Anderberg, Y. Mavros, S. Kay, R. Zhao, B.T. Baune, A. O'Sullivan, N. Singh, and M.F. Singh. 2011. Muscle and adipose tissue biopsy in older adults with type 2 diabetes. *Journal of Diabetes Mellitus* 1 (3): 27–35.

Wang, Z., X. Yang, J. Xiao, S. Wang, X. Wang, M. Mao, Y. Li, W. Zhao, X. Qi Wang, and T. Wu. 2019. Risk of serious adverse event and fatal adverse event with molecular target anticancer drugs in cancer patients: A meta-analysis. *Journal of Cancer Research and Therapeutics* 15 (7): 1435–1449.

Weijer, C., J.J. Grimshaw, M. Taljaard, A. Binik, R. Boruch, J.C. Brehaut, A. Donner, M.P. Eccles, A. Gallo, A.D. McRae, R. Saginur, and M. Zwarenstein. 2011. Ethical issues posed by cluster randomized trials in health research. *Trials* 12: 100.

Weiss, P.A. 2011. *Introductory statistics*. 9th ed. Upper Saddles River: Pearson.

Wendler, D., and F.G. Miller. 2007. Assessing research risks systematically: The net risks test. *Journal of Medical Ethics* 33 (8): 481–486.

———. 2008. Risk-benefit analysis and the net risks test. In *The Oxford handbook of clinical research ethics*, ed. E.J. Emanuel, C. Grady, R.A. Crouch, R.K. Lie, F.G. Miller, and D. Wendler, 503–513. New York: Oxford University Press.

Wendler, D., and A. Rid. 2017. In defense of a social value requirement for clinical research. *Bioethics* 31 (2): 77–86.
Wendler, D., L. Belsky, K.M. Thompson, and E.J. Emanuel. 2005. Quantifying the federal minimal risk standard: Implications for pediatric research without a prospect of direct benefit. *Journal of the American Medical Association* 294 (7): 826–832.
Wertheimer, A. 1999. *Exploitation*. revised ed. Princeton: Princeton University Press.
———. 2011. *Rethinking the ethics of clinical research: Widening the lens*. New York: Oxford University Press.
———. 2013. Is payment a benefit? *Bioethics* 27 (2): 105–116.
———. 2015. The social value requirement reconsidered. *Bioethics* 29 (5): 301–208.
White, A., C. Grady, M. Little, K. Sullivan, K. Clark, M. Ngwu, and A.D. Lyerly. 2021. IRB decision-making about minimal risk research with pregnant participants. *Ethics and Human Research* 43 (5): 2–17.
Wolf, S.M., F.P. Lawrenz, C.A. Nelson, J.P. Kahn, M.K. Cho, E.W. Clayton, J.G. Fletcher, M.K. Georgieff, D. Hammerschmidt, K. Hudson, J. Illes, V. Kapur, M.A. Keane, B.A. Koenig, B.S. Leroy, E.G. McFarland, J. Paradise, L.S. Parker, S.F. Terry, B. Van Ness, and B.S. Wilfond. 2008. Managing incidental findings in human subjects research: Analysis and recommendations. *Journal of Law, Medicine and Ethics* 36 (2): 219–248.
Wolf, S.M., B.N. Crock, B. Van Ness, F. Lawrenz, J.P. Kahn, L.M. Beskow, M.K. Cho, M.F. Christman, R.C. Green, R. Hall, J. Illes, M. Keane, B.M. Knoppers, B.A. Koenig, I.S. Kohane, B. Leroy, K.J. Maschke, W. McGeveran, P. Ossorio, L.S. Parker, G.M. Petersen, H.S. Richardson, J.A. Scott, S.F. Terry, B.S. Wilfond, and W.A. Wolf. 2012. Managing incidental findings and research results in genomic research involving biobanks and archived datasets. *Genetics in Medicine* 14: 61–384.
Wolinetz, C.D., and F.S. Collins. 2017. Single-minded research review: The common rule and single IRB policy. *American Journal Bioethics* 17 (7): 34–36.
World Health Organization. 2024. Clinical trials and observational studies. Available at: https://www.who.int/tools/antiretrovirals-in-pregnancy-research-toolkit/clinical-trials-and-observational-studies. Accessed 9 Aug 2024.
World Medical Association. 1996. Declaration of Helsinki: Ethical principles for medical research involving human subjects (1996 revision). Available at: https://www.wma.net/wp-content/uploads/2018/07/DoH-Oct1996.pdf. Accessed 27 July 2024.
———. 2000. Declaration of Helsinki: Ethical principles for medical research involving human subjects (2000 revision). Available at: https://www.wma.net/wp-content/uploads/2018/07/DoH-Oct2000.pdf. Accessed 3 Aug 2024.
———. 2013. Declaration of Helsinki: Ethical Principles for Medical Research Involving Human Subjects (2013 revision). Available at: https://www.wma.net/wp-content/uploads/2024/10/DoH-Oct2013.pdf. Accessed 31 January 2025.
———. 2024. Declaration of Helsinki: Ethical principles for medical research involving human subjects (2023 revision). Available at: https://www.wma.net/policies-post/wma-declaration-of-helsinki-ethical-principles-for-medical-research-involving-human-subjects/ Accessed 22 Oct 2024.
Wysocki, E. 2021. What does it take to get a medication approved through the FDA? Nationwide Children's. Available at: https://www.nationwidechildrens.org/family-resources-education/700childrens/2018/03/what-does-it-take-to-get-a-drug-approved-through-the-fda. Accessed 8 Aug 2024.
Yarborough, M., and R.R. Sharp. 2009. Public trust and research a decade later: What have we learned since Jesse Gelsinger's death? *Molecular Genetics and Metabolism* 97 (1): 4–5.
Zong, Z. 2008. Should post-trial provision of beneficial experimental interventions be mandatory in developing countries? *Journal of Medical Ethics* 34 (3): 188–192.

Chapter 8
Vulnerable Subjects

Abstract This chapter explores a variety of issues related to the participation of vulnerable subjects in research including the definition of vulnerability; categories of vulnerable subjects; criticisms of classifying subjects or populations as vulnerable; arguments for providing additional protections for vulnerable subjects; arguments for including vulnerable subjects in research; evolving approaches to the protection vs. exclusion dilemma; the types of additional protections for vulnerable subjects; federal regulations and international guidance pertaining to vulnerable subjects; and ethical issues related to enrolling children, prisoners, pregnant women, students, employees, and adults with impaired decision-making abilities in research. The chapter also applies ethical principles, including the principles of trust, justice, beneficence, nonmaleficence, and respect for persons, to research with vulnerable subjects. The chapter concludes by arguing that there are no simple solutions to the protection vs. inclusion dilemma and that the best way forward is to carefully consider the different factors in favor of inclusion or exclusion of vulnerable participants for each study, such as the risks of the research, the ability of the subject to provide informed consent, and the importance of the research for promoting the health of the population of which the group is a member.

Keywords Ethics · Human subjects · Regulation · Policy · Vulnerable people · Children · Pregnant women · People with impaired decision-making abilities · Justice · Trust

In previous chapters, we have touched on some ethical issues pertaining to research involving human subjects considered to be vulnerable, such as the ethics of exposing children and fetuses to risks. In this chapter, we will focus on dilemmas faced by investigators, institutions, IRBs, and sponsors related to research involving

D. B. Resnik, *The Ethics of Research with Human Subjects*, The International Library of Bioethics 111, https://doi.org/10.1007/978-3-031-82757-0_8

vulnerable subjects.[1] In Chap. 2, we noted that the federal regulations include a general requirement for protecting vulnerable subjects:

> When some or all of the subjects are likely to be vulnerable to coercion or undue influence, such as children, prisoners, individuals with impaired decision-making capacity, or economically or educationally disadvantaged persons, [the IRB must determine that] additional safeguards have been included in the study to protect the rights and welfare of these subjects (45 CFR 46.111b).[2]

The Common Rule also includes specific requirements pertaining to pregnant[3] women, fetuses and neonates (Subpart B), prisoners (Subpart C), and children (Subpart D), which will discuss in greater depth in this chapter.

The Helsinki Declaration requires that "Those persons incapable of giving free and informed consent are in situations of particular vulnerability and are entitled to the corresponding safeguards (World Medical Association 2024)" and the CIOMS guidelines state that:

> When vulnerable individuals or groups are considered for recruitment in research, researchers and research ethics committees must ensure that specific protections are in place to safeguard the rights and welfare of these individuals and groups in the conduct of research (Council for the International Organizations of Medicals Sciences 2016:57).

The CIOMS guidelines also include sections which address research involving children, pregnant women, and adults with mental disabilities or diseases that affect decision-making (Council for the International Organizations of Medicals Sciences 2016).

---

[1] In this chapter, I will follow the Common Rule's lexicon and use the term "vulnerable subject." Some writers and institutions refer to "vulnerable populations," but I prefer not to speak in this manner because (1) it can be stigmatizing to speak of an entire population as vulnerable; (2) treating populations as policy units ignores significant variations within populations; and (3) it is people, not populations, that undergo research procedures and therefore face risks (National Bioethics Advisory Commission 2001).

[2] It is interesting to note that the pre-2018 version of the Common Rule included pregnant women on this list of vulnerable subjects, but this was changed because many commentators pointed out that pregnancy, by itself, does not make one especially susceptible to coercion or undue influence (Blehar et al. 2013; Little et al. 2024). Also, the 2018 version replaces "mentally disabled persons" with "impaired decision-making capacity" because this new term is more inclusive than the old and includes many different types of people who might have trouble making decisions.

[3] Some have argued that to respect people's gender identities we should use the inclusive terms "pregnant persons" or "pregnant people" instead of "pregnant women" because a person who is pregnant might self-identify as a man and not as a woman. I will stick with the old-fashioned "pregnant women" terminology in this book because (1) the federal regulations use this terminology; and (2) it is not currently possible to support a pregnancy without a functioning uterus and a vagina and a person who has a functioning uterus and a vagina is, biologically speaking, a woman (female sex) (Lewis 2021). Although gender is a social and psychological construct, sex is biologically real. Of course, a person who identifies as a man and is carrying a fetus in their uterus could be planning to have sex-change surgery so their body will no longer have female characteristics. But their plans to do this in the future do not change the fact that when they are carrying a fetus, they are a biologically a woman.

Before we consider specific policies related to vulnerable subjects, we first need to address some foundational questions:

- What makes a human research subject vulnerable?
- Why is it important to protect vulnerable subjects?
- What are some potential problems with classifying groups of subjects as vulnerable?

8.1 What Makes a Human Research Subject Vulnerable?

Webster's Dictionary of the English language (2024) defines vulnerable as: "capable of being physically or emotionally wounded…open to attack or damage." While this definition is a good starting place, it is too general because every human being is capable of being wounded. Perhaps a better way of phrasing it would be, "more susceptible to being wounded, attacked, or damaged than the average human adult." Thus, vulnerability would be equated with **susceptibility to physical or psychological harm**. Clearly, many types of research subjects are regarded as vulnerable because they are susceptible to harm. These include fetuses, children, the elderly, people with physical or cognitive disabilities, gravely ill patients, and victims of natural disasters (Sach and Cassel 1990; Kipnis 2003; Rosenstein and Miller 2008; Menikoff 2009; Levine 2008; Schonfeld 2013; Packenham et al. 2017; DeCormier Plosky et al. 2022; Ouelette 2024). We will call this type of vulnerability **physical or psychological vulnerability** (Coleman 2009).

However, not all people that are regarded as vulnerable are especially susceptible to harm. For example, prisoners, students, employees, and military personnel are not especially susceptible to harm, but they are susceptible to **manipulation,**[4] **coercion, or undue influence** which is this sense of vulnerability found in (45 CFR 46.111b). This way of thinking about vulnerability links it to problems with decision-making and consent (see discussion in Chap. 5). This definition of vulnerability would seem to apply to many types of subjects widely considered to be vulnerable, such as children, prisoners, people who are illiterate or have other language deficits, and adults with disabilities, diseases, or other conditions that impair DMC (Sachs and Cassel 1990; Hornblum 1997, 1998; Moskop 1998; McManus et al. 2005; Resnik and Jones 2006; Bonham and Moreno 2008; Lo and Garan 2008; Wendler 2010; Rogers and Lange 2013; Resnik 2016; Dickert and Largent 2024; Christopher 2024). We will call this type of vulnerability **decisional vulnerability** (Coleman 2009).

---

[4] Manipulation involves using a various unethical tactics to control a person's decision-making, such as deception, coercion, undue influence, bribery, and emotional extortion. Manipulation is unethical because it treats the person as a mere means to a particular end or goal and does not respect their autonomy (Noggle 2022).

Another condition that might make one vulnerable is susceptibility to exploitation. Macklin (2003) argues that subjects in extreme poverty, such as the participants in the Tuskegee Syphilis Study (see Chap. 2) or people in low-income countries (Pratt 2024, see Chap. 7), should be considered vulnerable because they are **susceptible to exploitation** (Coleman 2009). These subjects may be easily exploited in research because they are socioeconomically disadvantaged and/or lack access to health care (see discussions of exploitation in Chaps. 5 and 7).[5] Exploitation can also occur because of one's social position; for example, students are susceptible to being exploited by their teachers and employees are susceptible to being exploited by their employers (Resnik 2024). Other types of subjects who may be susceptible to exploitation include indigenous peoples and racial, ethnic, and gender minorities (King 1998; Lo and Garan 2008; Rogers and Lange 2013; Kelley et al. 2013; Brunger et al. 2024). We will call this type of vulnerability **social or economic vulnerability.**

Putting this all together, we can say that an individual is a vulnerable research subject if and only if they have increased susceptibility to psychological or physical harm, manipulation, coercion, undue influence, or exploitation, due to their age, institutionalization, health, dependency, social position, economic status, or other factors (Coleman 2009; Boldt 2019). In some cases, a person may be vulnerable in all of these senses: for example, most children are susceptible, harm, manipulation/coercion/undue influence, and exploitation (Kipnis 2003).

8.2 Why Is Important to Protect Vulnerable Subjects?

Once we understand that vulnerability can be equated with increased susceptibility to harm, manipulation, coercion, undue influence, or exploitation, it is easy to see why it is important to protect vulnerable subjects. It is important to protect vulnerable subjects from these transgressions because they are less capable of protecting themselves from them than other people (Goodin 1986; Rogers et al. 2012; Porro 2022). For example, children are less able to defend themselves against coercion, manipulation, or exploitation than adults. Pearl Buck (1892–1973), who won the Nobel Prize in literature in 1938, spoke about the obligation to care for society's helpless members: "Our society must make it right and possible for old people not to fear the young or be deserted by them, for the test of a civilization is the way that it cares for its helpless members (Buck, quoted in Atkins 2018)." Former Vice President Hubert Humphrey (1911–1978) echoed these sentiments in an address at

[5] However, Denny and Grady (2007) and Levine (2008) critique the notion that subjects in extreme poverty should be considered vulnerable. They argue that these subjects often can make sound decisions and they can refuse to participate in studies that distribute benefits and risks unfairly. Classifying such subjects as vulnerable could exclude them from research that offers benefits to the participants or their communities.

the Democratic National Convention in New York City on July 13, 1976: "The ultimate moral test of any government is the way it treats three groups of its citizens. First, those in the dawn of life — our children. Second, those in the shadows of life — our needy, our sick, our handicapped. Third, those in the twilight of life — our elderly (Humphrey quoted by Atkins 2018)."[6] In sum, obligations to protect vulnerable subject are supported by the principles of beneficence and justice, discussed in in Chap. 3.

The authors of the Belmont Report had this to say about protecting vulnerable subjects:

> One special instance of injustice results from the involvement of vulnerable subjects. Certain groups, such as racial minorities, the economically disadvantaged, the very sick, and the institutionalized may continually be sought as research subjects, owing to their ready availability in settings where research is conducted. Given their dependent status and their frequently compromised capacity for free consent, **they should be protected against the danger of being involved in research** solely for administrative convenience, or because they are easy to manipulate as a result of their illness or socioeconomic condition (National Commission 1979: 7, emphasis added).

8.3 Some Types of Additional Protections for Vulnerable Subjects

Some of the types of additional protections for vulnerable subjects found in regulations and ethical guidelines include:

- **Scientific justification for including or excluding vulnerable subjects in research.** The Helsinki Declaration and Belmont Report recommend that a scientific justification be provided for including vulnerable subjects in research, but the CIOMS guidelines take the opposite approach and state that certain groups of subjects, such as including children, women, and adults with impaired or uncertain decision-making capacity (DMC) must be included unless there is a scientific justification for exclusion (see CIOMS 2016: 61, 64, 69). The federal research regulations require that the selection of research subjects be equitable (45 CFR 46.111a3), which implies that vulnerable subjects should not be unfairly included in research (Mastroianni and Kahn 2001).
- **Limits on research risks.** Federal regulations place limits on the risks that fetuses, neonates, children, pregnant women, and prisoners can be exposed to in research that does not offer them direct benefits (45 CFR 46, Subparts B, C, D). The CIOMS guidelines also include limits on risks that children and pregnant women can be exposed to in research (CIOMS 2016: 65,71).

[6]These quotes are similar to a quote misattributed to Mahatma Gandhi (1869–1948): "The greatness of a nation can be judged by how it treats its weakest members." However, Gandhi never said this, and Buck and Humphrey are the most likely originator (Atkins 2018).

- **Rules for IRB Composition.** Federal regulations require that an IRB which reviews research on prisoners include a prisoner or prisoner representative as a member (45 CFR 46.304). Commentators have argued that IRBs which review research on people with impaired decision-making capacity (DMC), children, and other vulnerable subjects also should include members with the appropriate background and experience concerning the study population or that they should consult with outside experts (National Bioethics Advisory Commission 1998; Institute of Medicine 2004).
- **Special procedures for obtaining consent and assent; assessing DMC; and using an LAR.** Federal regulations specify procedures for obtaining consent from the parent(s) and assent from the child in pediatric research (45 CFR 46 Subpart D). The National Bioethics Advisory Commission (1998), National Institutes of Health (2009), and Secretary's Advisory Committee on Human Research Protections (SACHRP 2009) recommend that investigators follow procedures for assessing DMC in adults with cognitive disabilities or other conditions that may impair DMC and identifying legally authorized representatives (LARs) for subjects who are incapable providing consent. In some cases, an IRB might require independent monitoring of the consent process to protect vulnerable subjects.
- **Additional safeguards for privacy/confidentiality.** Additional protections for privacy and confidentiality might be appropriate for research involving students or employees to prevent their professors or employers from having access to private information that could be used against them (Bonham and Moreno 2008; Resnik 2016).

8.4 Critiques of Protecting Vulnerable Subjects

As we have seen, regulations, ethical guidelines, and moral principles all require additional protections for vulnerable research subjects. However, ethicists, researchers, and patient advocates have criticized some of these protections as well the concept of vulnerability itself. Let's consider each of these critiques.

The Concept of Vulnerability Is Too Vague Levine et al. (2004) argue that the concept of vulnerability is not clearly defined in the U.S. federal regulations, Helsinki Declaration, and other important documents and has therefore been applied to so many categories of research subjects that is has lost its utility (see Table 8.1). I agree that commentators and organization have used—and overused—the concept of vulnerability without clearly defining it. However, I think this problem can be overcome if the concept is clearly defined, which I have tried to do in Sect. 8.1. Other writers have also proposed definitions that help clarify the concept of vulnerability (Kipnis 2003; Schroeder and Gefenas 2009; Coleman 2009).

Table 8.1 Some categories of vulnerable subjects[a]

Type of subject	Type of vulnerability
Older children (i.e., capable of assent)	Physical, psychological, social, economic, decisional
Fetuses,[b] infants, neonates, and young children	Physical, psychological
Adults with compromised DMC[c]	Physical, psychological, social, economic, decisional
Seriously ill patients	Physical, psychological, decisional
Institutionalized people, including prisoners	Physical, psychological, social, economic, decisional
Elderly people	Physical, psychological, social, economic, decisional
Employees (when participating in research conducted by their employer)	Social, economic, decisional
Students	Social, economic, decisional
Military personnel	Decisional, social
Socioeconomically disadvantaged people	Social, economic
Disaster victims	Physical, psychological
Indigenous peoples, including native Americans	Social, economic
Individuals who are illiterate or lack language proficiency	Decisional
Individuals with physical disabilities, including blindness and deafness	Physical, psychological, social, economic, decisional
Racial, ethnic, or gender minorities	Social, economic

[a]The categories in this table are based on data from Jonathan et al. (2024). Two of the authors on this paper, Irene Jonathan and Eliza Akers, worked with me as interns at NIEHS from May to August 2024. I appreciate the hard work they did in collecting and coding this data
[b]As discussed in Footnote 1, it is the fetus not the pregnant woman who is vulnerable
[c]As will be discussed in more depth below, this is a highly diverse group which includes people with cognitive disabilities (such as Down's syndrome), mental illnesses (such as psychosis or depression), dementia, and people suffering from emotional trauma or temporary medical conditions, such as delirium or shock (National Bioethics Advisory Commission 1998; Rosenstein and Miller 2008).

The Concept of Vulnerability Stigmatizes People Levine et al. (2004) and Dubois et al. (2012) argue that the concept of vulnerability stigmatizes people. The term 'vulnerable' is associated with weakness, helplessness, cowardliness, dependency, and other pejorative terms (Boldt 2019; Porro 2022). I agree that the term 'vulnerability' can be stigmatizing but I am not sure that this is a good argument for expunging it from our discourse on research ethics, since we need some term to flag research subjects who may need additional protections. If we use a different term, such as "special population," it would probably also acquire negative connotations. For an instructive precedent, consider that words we use to refer to people who have disabilities. Prior to the 1980s, disabled people were often referred to as 'crippled' or 'handicapped.' Disability rights advocates argued that these terms demeaned people and should be replaced with 'disabled.' Then some people argued that 'disabled' was demeaning and should be replaced with 'differently abled.' Now, many people have changed their minds about this idea and argue that calling someone 'differently abled' can be condescending and that 'disabled' is therefore preferable (Stanford University 2024). The main point here is that attempting to change beliefs and attitudes by insisting that people use "politically correct" language often does not work. The National Bioethics Commission (2001) argues that one way of avoiding stigmas associated with vulnerability is to apply the term to individuals and not populations.

Additional Protections for Vulnerable Subjects Undermine Autonomy Dubois et al. (2012) argue that special protections for vulnerable subjects undermine autonomy by denying people the opportunity to make decisions related to research participation. For example, Subpart C of the Common Rule prevents prisoners from participating in more than minimal risk research that does not offer the direct benefits. Subpart B of the Common Rule prevents pregnant women from participating in more than minimal risk research that does not benefit the woman or her fetus. This critique is largely correct, of course, but the bigger issue is whether these restrictions on autonomy are morally justified. Recall the discussion of strong vs. weak paternalism in Chap. 2. Strong paternalism is more controversial than weak because it involves restricting the choices of fully autonomous people for their own good, whereas weak paternalism involves restricting the choices of people who have nascent, impaired, or compromised autonomy, such as children, people with cognitive disabilities, or prisoners. Because rules and guidelines play a crucial role in research ethics and oversight, it might be the case that to protect a class of people we might need to enact a rule or policy that seems to be unfair in some situations (Miller and Wertheimer 2007). Consider research with prisoners, for example. It might be the case that some prisoners are being kept under ideal circumstances for making free and informed choices to participate in risky research that offers them no benefits, such as Phase I drug trials. Nevertheless, one might argue that we need a rule that prohibits this type of research to prevent abuses of prisoners who do not live under these ideal conditions (Hornblum 1997). However, one might argue that a more flexible approach, which allows participation based on the IRBs determination the risks are reasonable and the assumption that investigators are trustworthy, better respects the rights of prisoners (Gostin et al. 2007).

Additional Protections for Vulnerable Subjects May Deny People Access to Research and Therefore Adversely Impact Their Health of the Populations to Which They Belong The fourth, and by far the most important, critique of additional protections for vulnerable research subjects is that these protections may deny people access to research and therefore negatively impact the health of the populations to which they belong (Mastroianni and Kahn 2001; Lyerly et al. 2008; Meltzer and Childress 2008; Wendler 2010; Dubois et al. 2012; DeCormier Plosky et al. 2022; Friesen et al. 2023). This critique of vulnerable participant policies has its origins in HIV/AIDS crisis of the 1980s and 1990s, in which activists argued for increased access to experimental medications (Dresser 2001; see discussion in Chap. 7, Sec. 16). Since then, advocates for different groups of people have argued for increased access to research.

For example, pregnant and non-pregnant women have been unfairly excluded from clinical research and this has negatively impacted their health by creating knowledge gaps (Dresser 1992; Lyerly et al. 2008; Schonfeld 2013; Little et al. 2024). Prior the 1990s, drug dosing was based on research conducted on men and extrapolated to women. Women—not just pregnant women—were excluded from clinical trials because researchers believed that certain aspects of female biology, such as the menstrual cycle, were confounding factors that could interfere with data interpretation. They were also concerned that women might become pregnant during the study, or they might have an undetected pregnancy, which could result in harm to the fetus (Dresser 1992; Grady and Denny 2008). However, basing women's drug dosing on studies conducted on men was a dangerous practice because there are physiological, metabolic, and hormonal difference between men and women that affect pharmacokinetics (Mastroianni and Henry 2024). Heart disease is another area where extrapolation from men to women has negatively impacted women's health (Wenger et al. 2022). Myocardial infarction, for example, can produce very different signs and symptoms in women than in men. With regard to pregnant women, evidence indicates most medications are prescribed in pregnancy on an off-label basis due to lack of research on pregnant women (Herring et al. 2010; Little et al. 2024).[7]

Pediatric researchers and bioethicists have argued that children have been unfairly excluded from research (Ross 2008; Wendler 2010). Most pediatric drugs are also prescribed an off-label basis due to lack of research on children (Bazzano et al. 2009). However, this practice is also dangerous, due to metabolic and physiologic differences between children and adults (Ross 2008).

Finally, disability rights advocates have argued that people with disabilities have been unfairly excluded (St. John et al. 2022; Ouelette 2024). Several studies provide evidence for this claim. Camanni et al. (2021) found that, among 2809 studies registered in ClinicalTrials.gov between June 2010 and June 2020, 17.4% excluded individuals with cognitive impairment, 21.9%. McDonald et al. (2022) found that, among 248 Phase II, III, or IV NIH-funded clinical trials listed in ClinicalTrials.gov

[7] Off-label means prescribing in a manner not approved by the FDA. When the FDA approves a new drug (or biologic), it approves it with a labelling that describes the approved use and dosage. Once the drug is one the market, physicians are allowed to prescribe it according to their best medical judgement consistent with the standard of care. The FDA is not authorized to regulate off-label prescribing. Medical societies have the authority to control off-label prescribing by sanctioning doctors but rarely do unless a physician's prescribing practices pose a significant threat to patient safety. While off-label prescribing can play an important role in medical innovation, some have argued that it should be monitored more closely to protect patients and prevent ineffective and wasteful medical practice (Radley et al. 2006).

between February 2018 and February 2021, 74.6% of the studies had eligibility criteria that directly or indirectly exclude adults with cognitive disabilities. DeCormier-Plosky et al. (2022) also studied clinical trials registered in ClinicalTrials.gov and found that 68% excluded people with psychiatric conditions and 42% excluded people with cognitive disabilities.

In the U.S., federal agencies have taken some steps to promote access to clinical research for different populations. For example, in 1993, the U.S. Congress passed the NIH Revitalization Act, which required the NIH clinical trials to include adequate numbers of women and minorities and that Phase III trials include enough women to allow for valid statistical comparisons between men and women (Mastroianni and Henry 2024). In 1994 the NIH published guidelines for the inclusion of women and minorities in research, and in 2001 the agency strengthened these guidelines (National Institutes of Health 2001; Grady and Denny 2008). To encourage private companies to invest in R & D related to children's health, the U.S. Congress passed the Best Pharmaceuticals for Children Act in 1997. The law, which became effective in 2002 and has been amended several times, grants companies an additional 6 months of market exclusivity if they test a drug on children and allows the FDA to require pediatric testing in some cases (National Institutes of Health 2024). In May 2024, DHHS announced regulations that implement Section 504 of the Rehabilitation Act, a law that prohibits recipients of federal funds from discriminating against disabled people. According to DHHS, Section 504 applies to clinical research and implies that "study eligibility criteria [must] be written in a way that does not unnecessarily screen out people with disabilities whose research participation would not alter the intended purpose of the program of clinical research being undertaken (DHHS 2024: 40089)."

I consider this fourth critique to be a compelling reason to carefully examine protections for vulnerable subjects to make sure that they strike a reasonable and fair balance between protection and inclusion (Friesen et al. 2023). I think few people would disagree with the idea that there needs to be a reasonable balance between protection and inclusion. However, acknowledging this abstract idea does not really solve the problem of inclusion vs. access: the devil, as they say, is in the details. To make some headway on this problem, we need to consider some specific policies regarding vulnerable subjects to see if they strike a reasonable balance.

8.5 Research with Children

Children are the first type of vulnerable subject we will discuss in this chapter. In Chap. 7, we considered some issues related to research involving children when we discussed the topic of minimal risk. In Chap. 2, we mentioned that that Common Rule includes special protections for children under Subpart D. Under the Subpart D, an IRB may approve three types of research with children:

46.404 Research not involving greater than minimal risk. HHS will conduct or fund research in which the IRB finds that no greater than minimal risk to children is presented, only if the IRB finds that adequate provisions are made for soliciting the assent of the children and the permission of their parents or guardians, as set forth in §46.408.

46.405 Research involving greater than minimal risk but presenting the prospect of direct benefit to the individual subjects. HHS will conduct or fund research in which the IRB finds that more than minimal risk to children is presented by an intervention or procedure that holds out the prospect of direct benefit for the individual subject, or by a monitoring procedure that is likely to contribute to the subject's well-being, only if the IRB finds that:

(a) The risk is justified by the anticipated benefit to the subjects;
(b) The relation of the anticipated benefit to the risk is at least as favorable to the subjects as that presented by available alternative approaches; and.
(c) Adequate provisions are made for soliciting the assent of the children and permission of their parents or guardians, as set forth in §46.408.

46.406 Research involving greater than minimal risk and no prospect of direct benefit to individual subjects, but likely to yield generalizable knowledge about the subject's disorder or condition. HHS will conduct or fund research in which the IRB finds that more than minimal risk to children is presented by an intervention or procedure that does not hold out the prospect of direct benefit for the individual subject, or by a monitoring procedure which is not likely to contribute to the well-being of the subject, only if the IRB finds that:

(a) The risk represents a minor increase over minimal risk;
(b) The intervention or procedure presents experiences to subjects that are reasonably commensurate with those inherent in their actual or expected medical, dental, psychological, social, or educational situations;
(c) The intervention or procedure is likely to yield generalizable knowledge about the subjects' disorder or condition which is of vital importance for the understanding or amelioration of the subjects' disorder or condition; and.
(d) Adequate provisions are made for soliciting assent of the children and permission of their parents or guardians, as set forth in §46.408.

Pediatric research which does not fit into these three categories can only be approved by a special panel convened by DHHS (45 CFR 46.407). The FDA has adopted Subpart D for FDA-regulated research involving children (21 CFR 50.5-50.56). As noted in Chap. 2, the EPA has its own regulations pertaining to research involving children. The EPA does not allow funded-investigators to conduct <u>any</u> research that intentionally exposes children to environmental agents (Resnik 2007). The Department of Education (DOE) also has rules for enrolling children in research conducted in public schools that receive DOE funds, but I will not consider those in this book (see Hicks 2006).

Most commentators agree that it is ethical to enroll children in studies, such as clinical trials of new medications, that offer participants direct medical benefits

(46.405), but there is considerable controversy about enrolling children in non-beneficial research that may expose them to risks (46.406 and 46. 407) (Institute of Medicine 2004). We allow competent adults to enroll in risky, non-beneficial research (such as Phase I trials on healthy volunteers, discussed in Chap. 7, Sec. 12) because they can make their own decisions concerning their risk exposure. But since children cannot make such decisions, it is incumbent on their parents or guardians to make these decisions for them.

As discussed in Chap. 7, the National Commission (and later the Common Rule) adopted the concept of minimal risk as a threshold for a morally acceptable level of risk for children, and this concept was later applied to other categories of vulnerable subjects, such as prisoners and fetuses (Wendler 2010). However, debates continue about how to interpret and apply the concept of minimal risk (Institute of Medicine 2004; Wendler et al. 2005).

If we accept that the idea that children may participate in non-beneficial, minimal risk research as part of their contribution to society (Wendler 2010, 2012), the next question we need to address is whether children should ever be able to participate in non-beneficial research that poses more than minimal risk (e.g., research covered by 46.406 and 46.407).

To focus our discussion, it may be useful to consider an actual case involving pediatric more than minimal risk research.

On November 17, 2005, the FDA's Pediatric Advisory Committee met to consider whether to approve a pediatric study titled "Gonadotropin Releasing Hormone Agonist Test in Disorders of Puberty" proposed by University of Chicago pediatrician Robert Rosenfield (Food and Drug Administration 2005). The purpose of the study was to test the effectiveness of a leuprolide injection at diagnosing disorders of puberty, such as premature onset and delayed onset puberty. Other methods of diagnosing puberty disorders were not reliable or too expensive to be reimbursed by insurers. The study included a group of children with puberty disorders and a group of normal controls (age 7–18). Normal controls would be paid $150 (about $240 in 2024 dollars) but the diseased children would not be paid because they would be receiving treatment. Study procedures included (Food and Drug Administration 2005):

- 36-h admission to the hospital's clinical research unit;
- Placement of venous catheter for obtaining blood samples totaling 150–240 ml (5–8 ounces), depending on the weight of the child;
- Injection of a single dose of leuprolide (10 micrograms per kilogram of body weight);
- X-rays to measure bone age;
- DNA testing;
- Administration of oral iron to prevent anemia.

Enrolling the subjects with puberty disorders was not in dispute since they would undergo the same tests and procedures even if they were not in the study. For them, the study would be considered beneficial research approvable under 21 CFR 50.52.

Enrolling the normal controls presented a problem, since they would be exposed to more than minimal risks with no compensating medical benefits. The study could not be approved under 21 CFR 50.52 or 21 CFR 50.53, because the normal volunteers did not have a disease or condition. So, the study could only be approved by a special panel convened by the FDA (similar to a 46.407 panel for DHHS). To approve the study under 21 CFR 50.54 the federal panel had to determine that:

(i) the clinical investigation presents a reasonable opportunity to further the understanding, prevention, or alleviation of a serious problem affecting the health or welfare of children;
(ii) the clinical investigation will be conducted in accordance with sound ethical principles;
(iii) adequate provisions are made for soliciting the assent of children and the permission of their parents or guardians (21 CFR 50.54; same as 45 CFR 46.407, except "clinical investigation" is substituted for "research").

The panel assessed the risks of administration of leuprolide, the insertion of an indwelling catheter, the loss of blood, and the potential psychological harm of a 36-h hospital admission for the children. Panel members all agreed that the risks were more than minimal (Food and Drug Administration 2005). Public comments reviewed by the committee expressed concerns about the risks of a leuprolide injection and problems with the consent process. Several professional societies endorsed the study and said that it was necessary to obtain data from healthy controls to evaluate the effectiveness of the leuprolide test. The committee determined that the study could be approved if it was modified as follows:

A) The results of the leuprolide test should not be disclosed to the normal control subjects or their parents because these results would be clinically uncertain and might convey the false impression that the child was abnormal, which could cause psychological harm;
B) The consent and assent documents should provide more information about the risks of the leuprolide injection and more clearly state that the normal controls would not receive any medical benefits (Food and Drug Administration 2005).

Were the risks for the normal controls in this study reasonable in relation to the value of the knowledge expected to be gained? Were the risks excessive? Let's consider some of the most significant ones:

- **Risks of the leuprolide injection**. While prolonged use leuprolide use includes significant risks, including allergic reactions, depression, memory disorders, severe headache, high blood sugar, vomiting, and muscle, joint, or bone pain (Mayo Clinic 2024), a single dose of this medication involves very little risk other than pain or bleeding at the injection site (Cincinnati Children's Hospital 2024).
    - Risks of a venous catheter: infection, bleeding, and pain and discomfort (see Chap. 7).

- Risks of a 36-h hospital stay: risk of a hospital-acquired infection; psychological stress, discomfort, boredom.
- Risks of collecting 150–240 ml of blood: anemia, dizziness, fainting, shortness of breath.
- Risks of x-ray: negligible for the level of exposure to ionizing radiation in the study (see Chap. 7).

Although the risks of the study are clearly more than minimal, none of them stand out as egregiously risky. The study did not involve a significant risk of an SAE or death and included plans to control and minimize the medical risks (Food and Drug Administration 2005). One might argue that the most significant risk of the study was the 36-h hospital admission, which could be stressful for the younger children in the study (and their parents) and would expose them to hospital-acquired infections. However, parents could withdraw their children from the study if they were unable to tolerate the hospital stay.

To determine whether these risks were excessive, we can compare them to the risks of other community service activities that are considered to be socially acceptable for children to participate in (Wendler and Emanuel 2005). As discussed in Chap. 7, social acceptability is important to setting risk limits because its relationship to public trust. Permitting non-beneficial research that exceeds socially acceptable risk thresholds can undermine trust when adverse events happen to volunteers. Some examples of socially acceptable volunteer activities for children that involve risks include:

- Picking up trash on the roadside (North Carolina Department of Transportation 2024, age restricted to 12+).
- Building ramps for disabled people to access their homes (Scouting Wire 2013).
- Helping to build houses for Habitat Humanity (Habitat for Humanity of the Charlotte Region 2024, age restricted to 16+ with parental consent).
- Working at food banks and soup kitchens (Second Harvest 2023, age restricted to 16+).
- Donating blood (American Red Cross 2024, age restricted to 16+, with parental consent).

All these activities are probably riskier than the leuprolide study. However, most have age restrictions. There are several likely reasons for these restrictions: (1) older children are less likely to be injured than younger children because they are better able to follow instructions, make decisions and protect themselves; (2) older children are better at making their own decisions concerning risk exposure[8] than younger children and in many cases are approaching the agent of consent; (3) older children may benefit more from being engaged in volunteer activities than younger children because they are better able to appreciate the value of the work.

These preceding ethical reflections support the following conclusion: **the maximum level of risk for children participating in non-beneficial research should**

[8] Parents of teenagers may disagree, of course!

not be greater than the risks children typically are exposed to in comparable age-appropriate community service activities. Under this standard, 17-year-olds would be permitted to face risks comparable to the risks of working for Habitat for Humanity, but 7-year-olds would not be allowed to face these risks. They could only face risk equivalent to the age-appropriate volunteer activity, such as soliciting contributions for a charitable organization like the Boys Scouts or Girl Scouts.

As mentioned above, the age range for the leuprolide study was 7–18 years old. Applying my age-based risk guideline to this study, one could argue that younger normal controls should have been excluded from the study, since they would have faced risks greater than those typically encountered by children in comparable age-appropriate community service activities. 12–18 years old would have been a more reasonable age range for normal controls in this study, assuming that there is not a compelling scientific justification for including younger children.

Some might object that my age-based risk guideline may prevent investigators from conducting some research that could have important benefits for children's health. I agree that this could happen, but keep in mind that the guideline is only a guideline, and there could be justifiable exceptions to this rule. However, the burden of proof should fall on those who propose non-beneficial studies that would involve risks exceeding those of comparable age-appropriate public service activities.

Before concluding this section, it is important to note that protectionist pediatric research policies may have had an overall negative impact on children's health. As noted above, due to lack of research, most drugs used to treat children are prescribed on an off-label basis (Bazzano et al. 2009). Off-label prescribing can pose serious problems for children's health, since drugs may affect children and adults differently, which means that dosing by weight can be unreliable. Also, some drugs have very different effects in children and adults. For example, some drugs used to treat attention deficit hyperactivity disorder (ADHD) help children to maintain focus and attention but may induce anxiety and agitation in adults (Bazzano et al. 2009).

There are a couple of reasons why drugs are infrequently tested in children. The main reason is economic: medications used to treat chronic illnesses that affect mostly adults usually generate higher profits for pharmaceutical companies than medications used to treat or prevent childhood diseases. An adult with hypertension may need to take a drug for the rest of their life, whereas a child with a bacterial ear infection may only need to take an antibiotic for a week, and only a few doses of vaccine may be needed to prevent a childhood disease. As noted above, the Best Pharmaceuticals for Children Act helps to increase economic incentives for including children in clinical trials.

Another important reason why drugs are infrequently tested in pediatric populations is that investigators and sponsors may be hesitant to propose pediatric clinical trials that include healthy controls due to the regulatory roadblocks for obtaining approval or these studies. As we saw above, pediatric studies with healthy controls who are exposed to more than minimal risks can only be approved by a special federal panel. Investigators may decide to forego conducting these types of studies, given the difficulties with obtaining approval. However, healthy control groups can be very important in establishing the safety of medications. One way of getting around this problem would be to allow IRBs to approve studies involving healthy controls in a clinical trial if the risk is no more than a minor increase over minimal risk (Iltis 2007). The justification for this would be that there is not a significant moral difference between a subject with a disorder or condition who does not benefit medically from the research and a subject who is healthy control. 45 CFR 46.406 and 21 CFR 50.53 could be amended to read "Research involving greater than minimal risk and no prospect of direct benefit to individual subjects, but likely to yield generalizable knowledge about the subject's disorder or condition or **necessary to ensure scientific validity of a controlled clinical trial**." To ensure that healthy controls are not exposed to excessive risk, it may be necessary to more clearly define "minor increase over minimal risk" (Wendler and Emanuel. 2005).

Before concluding this section, it is important discuss a couple of issues related to consent and assent with pediatric research.

First, sometimes children may object to their parents' decisions to enroll them in research. For example, suppose that a 15-year-old girl is dying of Hodgkin's lymphoma and her parents want to enroll her in a Phase II trial of an experimental medication to treat her disease. If she enrolls in the study, she has a 50% chance of surviving for another year. The experimental medication has significant side effects including fatigue, nausea, fever, rashes, depression, peripheral numbness, and high blood pressure. If she does not enroll in the study, she has a 90% chance of dying within 6 months because she has run out of other options for treatment. The girl does not want to enroll in the study and just wants to make the best of remaining days of life. Under state laws, her parents have the right to make the decision for her, provided that the decision is not a form of abuse or neglect, which it appears not to be (Krause 1995). Although the U.S. federal regulations require that the assent of the pediatric research subjects be obtained (45 CFR 46.404, 405, 406), they include some exceptions:

> If the IRB determines that the capability of some or all of the children is so limited that they cannot reasonably be consulted or that the intervention or procedure involved in the research holds out a prospect of direct benefit that is important to the health or well-being of the children and is available only in the context of the research, the assent of the children is not a necessary condition for proceeding with the research (45 CFR 46.408).

Thus, her parents would appear to have the legal right, under state and federal law, to enroll the girl in this study against her objections. But should they? That's the ethical question. This issue is complex, and there are many factors to consider, including the girl's emotional and intellectual maturity, other available treatment options, the potential benefits of participating in the study, religious or other beliefs, and the psychological impact of forcing her to do something against her wishes (Caplan 2007; Blake 2012). One could argue that the moral obligation to honor the child's wishes grows stronger as she approaches adulthood, because she is better to able to make reasonable decisions and exercise autonomy. It makes a difference in this type of situation, whether the child is 8 years old or almost 18 years old.

Second, questions may arise as to whether both parents need to give permission for their child to participate in research. According to 45 CFR 46.408, the consent of only one parent is needed for minimal risk research (45 CFR 46.404) and beneficial research (45 CFR 46.405) but two parents are needed to consent for more than minimal risk, non-beneficial research (45 CFR 46.406, 407) "unless one parent is deceased, unknown, incompetent, or not reasonably available, or when only one parent has legal responsibility for the care and custody of the child (45 CFR 46.408)." When both parents agree about participation, there shouldn't be major ethical concerns obtaining parental consent. However, suppose that both parents have custody, but they are separated or divorced and disagree their child's participation in research. Suppose, for example, that the child has cancer and that one parents wants to enroll the child in a clinical trial that holds out the prospect for a significant and direct benefit for the child but the other does not. Under 45 CFR 46.408, consent for only one parent is needed. Even if consent from only one parent is legally required, one might argue that, to avoid placing the researchers and institution in the middle of a family conflict, the child should be enrolled in the trial only if both parents agree. However, one might argue that if the trial holds out the prospect of a significant and direct benefit for the child, only one parent's permission is needed, because participating in the study is in the child's best interests. As one can see, situations like these can be complicated, messy, and emotional. As practical matter, an IRB might decide that consent of both parents is required for research participation if they are separated or divorced (Forysthe 2022).

A related issue is when two parents are needed for consent, but one parent is not reasonably available. Although 45 CFR 46.408 allows for the consent of only one parent in a situation like this, questions can arise as to what is meant by "reasonably available" (Nelson et al. 2013). Some form of electronic consent, such as through websites, email, teleconferences, videoconferences, or some combination thereof, may facilitate parental consent when one parent cannot come to the research site to provide consent (Klint 2013). The IRB would need to approve this type of electronic consent procedure, which should include all the required elements for informed consent (Office of Human Research Protections 2016).

8.6 Research Involving Pregnant Women

We will introduce our discussion of the ethical issues pertaining to research involving pregnant women by briefly discussing the thalidomide tragedy of the 1950s and 1960s, because this episode has had a profound impact on how researchers, institutions, and sponsors think about enrolling pregnant women in clinical trials (Grady and Denny 2008; Lyerly et al. 2009). In 1956, the German pharmaceutical company Chemi Grunenthal began marketing thalidomide as a sedative medication. After it was discovered that the thalidomide could relieve nausea, it was marketed in Europe, Japan, and Australia as a treatment for morning sickness during pregnancy. In 1961, William McBride and Widukind Lenz independently observed that thalidomide use in pregnancy was associated with severe congenital malformations, such deformed or missing limbs. Their findings were confirmed by several other researchers.

U.S. children were largely unaffected by the thalidomide tragedy because Frances Kathleen Oldham Kelsey (1914–2015), a drug reviewer for the FDA, refused to approve thalidomide since she was concerned about its safety. Soon, countries around the world banned the drug. In 1962, Kelsey was awarded the President's Gold Medal Award for Distinguished Federal Civilian Service. About 10,000 infants, mostly in West Germany, were born with severe birth defects, such deformed limbs, due to exposure to thalidomide in utero, and about 2000 children died from drug. Most of the defects occurred when fetuses were exposed to the drug between 34 and 49 days after the woman's last menstrual period (Kim and Scialli 2011; Rehman et al. 2011).

Investigators and sponsors are justifiably hesitant to include pregnant women in clinical trials, given the potential for serious harm to the fetus and future child. Numerous studies have demonstrated the potential adverse of effects of drugs, chemicals, and infections on the development of the fetus and subsequent health and development of the child, if the fetus survives (Gómez-Roig et al. 2021; Gunatilake and Patil 2023). Some well-studied teratogens include lead (Hu et al. 2006), alcohol (Popova et al. 2023), nicotine (Shea and Steiner 2008), and cocaine (Offidani et al. 1995). Research also indicates that the fetus has differential susceptibility to different teratogens at different times during pregnancy (Gómez-Roig 2021). Drugs can harm the fetus by: (1) crossing the placenta and acting directly on the fetus; (2) impairing the function of the placenta and restricting blood flow to the fetus; (3) causing the uterus to contract prematurely; and (4) lowering the woman's blood pressure, which can reduce blood flow to the fetus (Gunatilake and Patil 2023). The question is not whether to protect fetuses from risks, but how to best balance fetal protection with access to research participation for pregnant women (van der Graaf et al. 2018).

The Common Rule, subpart B, allows an IRB to approve research with pregnant women if it determines that the following conditions met:

(a) Where scientifically appropriate, preclinical studies, including studies on pregnant animals, and clinical studies, including studies on nonpregnant women, have been conducted and provide data for assessing potential risks to pregnant women and fetuses;
(b) The risk to the fetus is caused solely by interventions or procedures that hold out the prospect of direct benefit for the woman or the fetus; or, if there is no such prospect of benefit, the risk to the fetus is not greater than minimal and the purpose of the research is the development of important biomedical knowledge which cannot be obtained by any other means;
(c) Any risk is the least possible for achieving the objectives of the research;
(d) If the research holds out the prospect of direct benefit to the pregnant woman, the prospect of a direct benefit both to the pregnant woman and the fetus, or no prospect of benefit for the woman nor the fetus when risk to the fetus is not greater than minimal and the purpose of the research is the development of important biomedical knowledge that cannot be obtained by any other means,

her consent is obtained in accord with the informed consent provisions of subpart A of this part;

(e) If the research holds out the prospect of direct benefit solely to the fetus then the consent of the pregnant woman and the father is obtained in accord with the informed consent provisions of subpart A of this part, except that the father's consent need not be obtained if he is unable to consent because of unavailability, incompetence, or temporary incapacity or the pregnancy resulted from rape or incest.

(f) Each individual providing consent under paragraph (d) or (e) of this section is fully informed regarding the reasonably foreseeable impact of the research on the fetus or neonate;

(g) For children as defined in §46.402(a) who are pregnant, assent and permission are obtained in accord with the provisions of subpart D of this part;

(h) No inducements, monetary or otherwise, will be offered to terminate a pregnancy;

(i) Individuals engaged in the research will have no part in any decisions as to the timing, method, or procedures used to terminate a pregnancy; and

(j) Individuals engaged in the research will have no part in determining the viability of a neonate (45 CFR 46.204, emphasis added).

As one can see, Subpart B of the Common Rule allows pregnant women to enroll in research that is (1) minimal risk to the fetus and the knowledge cannot be obtained by other means; (2) research that is more than minimal risk but has the prospect of direct benefit for the woman or the fetus or both. Although Subpart B neither deters nor encourages inclusion of pregnant women of pregnant women in research, one might argue that it is somewhat restrictive because it does not allow pregnant women to enroll in non-beneficial research if the knowledge can be obtained by "other means," which might be interpreted as "can be obtained without enrolling pregnant women" (White et al. 2021). Subpart B also prohibits non-beneficial research that poses more than a minimal risk to the fetus.

Although the FDA does not have any special regulations for pregnant women in research, it has issued some guidance for industry (Food and Drug Administration 2018). The guidance discusses scientific and ethical considerations related to enrolling pregnant women in research and recommends that the requirements of Subpart B be satisfied for FDA regulated research.

The CIOMS guidelines are similar to Subpart B except they allow pregnant women to enroll in non-beneficial research that is a minor increase over minimal risk research when "the social value of the research for pregnant or breastfeeding women or their fetus or infant is compelling, and the research cannot be conducted in non-pregnant or non-breastfeeding women (Council for International Organizations of Medical Sciences 2016: 71)."

Subpart B does not preclude pregnant women from enrolling in Phase II and Phase III clinical trials because these studies may benefit the woman. However, an IRB must also decide whether the risks are reasonable in relation to the benefits (see Chap. 7) and IRBs may often decide that they are not. Evidence shows that pregnant women are usually not included in clinical trials due to the risks to the fetus, real or perceived (van der Graaf et al. 2018). For example, Phelan et al. (2016) examined 688 clinical trials of medications for Type 2 diabetes registered in ClinicalTrials.gov from 1995 to 2009 and found that 55% excluded pregnant women, even though many women who become pregnant also have Type 2 diabetes. Kons et al. (2022) examined COVID-19 clinical studies registered in ClinicalTrials.gov between May 1, 2020, to October 31, 2020, and found that 88/90 (97.8%) vaccine trials excluded pregnant women and 350/495 (70.7%) COVID treatment trials excluded pregnant women. A different study of registered trials by Taylor et al. (2021) found that 538/722 (75%) COVID treatment studies excluded pregnant women.

Although these data provide evidence that pregnant women are usually excluded from clinical trials, they do not tell us whether exclusion is justified to protect the fetus from harm. It may be the case that exclusion of pregnant women was entirely appropriate in most of the trials examined in these studies. However, it may also be the case that exclusion was not justified in many cases. Exclusion may simply be assumed as the default position for many researchers, sponsors, and IRBs.

Although excluding pregnant women from clinical research may help to protect the fetus, it undoubtedly has a negative impact on the health of pregnant women as a group (Grady and Denny 2008; Lyerly et al. 2009; Office of Research on Women's Health 2010). In response to this dilemma, some physicians and ethicists have argued that policies for enrolling pregnant women in research ought to be more inclusive. According to these authors, the burden of proof should be shifted from inclusion to exclusion; that it, pregnant women must be allowed to enroll in research unless the protocol includes a scientific or ethical reason justifies exclusion (Lyerly et al. 2008; Lyerly et al. 2011; Blehar et al. 2013; Little et al. 2024).

What do IRB policies have to say about the protection vs. inclusion dilemma for pregnant women? In the summer of 2024, I collaborated with two interns who were working with me at NIEHS, Irene Jonathan and Eliza Akers, and a professor of social medicine at the University of North Carolina, Anne Drapkin-Lyerly, to answer this question. The interns collected and coded data on 93 IRB policies regarding enrollment of pregnant women in research from the top-funded academic institutions in the US to determine how these policies handle the protection vs. inclusion dilemma.[9] We developed a set of questions for coding these policies based our knowledge of the background literature and an initial review of IRB policy documents. These questions were then refined based on discussions among members of the research team. The data for the study are shown in Table 8.2.

As one can see from Table 8.2, only 10.8% of the policies shift the burden from inclusion to exclusion and most of the policies (59.1%) basically reiterate Subpart B of Common Rule.

The question of burden of proof is often a key issue in law, ethics, and policy, because it frames standards of evidence, arguments, and procedures (Rothstein

[9]These policies were located by Jonathan et al. (2024).

Table 8.2 IRB policies for including pregnant women in research from 93 top-funded U.S. top research institutions

Policy requires that pregnant women be included unless a justification for exclusion is provided	10 (10.8%)
Policy essentially restates subpart B of the common rule	55 (59.1%)
Policy refers or describes pregnant women as vulnerable	11 (11.8%)
Policy is grouped with other vulnerable population policies	77 (83.7%)

et al. 1997). In criminal law, for example, the prosecution has the burden of proof to produce evidence that the defendant is guilty. If the prosecution fails to produce evidence that the defendant is guilty beyond reasonable doubt, the defendant is not convicted and does not have to face criminal punishments. Placing the burden of proof on the prosecution increases the probability that guilty people will go free, while placing it on the defense increase the probability that innocent people will be convicted. Society places the burden of proof on the prosecution because it values protecting the rights of the accused more than it values convicting people who are guilty. Thus, assignment of the burden in criminal law and other situations involves balancing competing values.

In the case of research with pregnant women, assignment of the burden of proof involves balancing fetal protection vs. access to research for pregnant women. On the one hand, placing the burden of proof on investigators who propose to include pregnant women in research is likely to increase exclusion of pregnant women from research because (1) investigators may not want to expend time and effort to gather evidence for inclusion and convince the IRB that pregnant women should be included; and (2) good evidence for inclusion may not be readily available. On the other hand, placing the burden of proof on investigators who propose to exclude pregnant women may decrease protection of fetuses and future children for largely the same reasons, i.e. inconvenience and lack of evidence for exclusion.

In my opinion, the best way to resolve this conflict is to adopt a more nuanced approach to inclusion and exclusion. To that end, I offer the following guidelines:

1. **Pregnant women should be allowed to participate (i.e., included) in non-beneficial research that poses only minimal risks to the fetus/future child**; it does not matter whether their participation is scientifically necessary.[10] As noted earlier, Subpart B and the CIOMS guidelines both say that pregnant women can be participate in non-beneficial research only if the scientific knowledge cannot be obtained by "other means," which might be interpreted as saying that the knowledge cannot be obtained without including pregnant women; that is, their participation is scientifically necessary.[11] This interpretation seems patently unethical to me because it treats pregnant women as second-class citizens. For example, a pregnant women would not be able to participate in a public opinion

[10]There are still important issues here about what constitutes "minimal risk" to the fetus (see Strong 2011; White et al. 2021).

[11]White et al. (2021) argue that the "other means" language needs to be clarified.

survey or biobanking survey involving a 50 ml blood draw if the knowledge can be obtained by "other means." When the research is minimal risk, pregnant women should be treated like other people and should have the same rights to participate in research.

2. **Pregnant women should be included in research that offers the prospect of direct benefit to the woman or the fetus unless exclusion is scientifically or ethically justified**; for example, there is evidence from animal or human studies that the risks to the fetus/future child are significant, or the research involves administration of a medication or invasive procedure for which the risks to the fetus/future child are unknown and a precautionary approach[12] is warranted. Although this approach places the burden of justifying exclusion on the investigators, I do not think it will result in under-protection of fetuses and future children because it the precautionary reasoning approach gives investigators some wide-ranging reasons for exclusion. The key point here is that exclusion from beneficial research must be justified and not assumed as part of the status quo.
3. **Pregnant women should not be included in non-beneficial research that poses more than minimal risk to the fetus/future child unless inclusion is scientifically or ethically justified**; for example, the knowledge is important for promoting the health of pregnant women or fetuses and cannot be obtained without participation of pregnant women. In this situation, the risk to the fetus should be no more than a minor increase over minimal. Because the risk to the fetus/future child are more than minimal and there are no direct benefits to the fetus or pregnant woman, the burden of proof should fall on inclusion.
4. **If investigators decide to exclude pregnant women from a study, they should require fertile females to take a pregnancy test just prior to enrollment, and the test may be repeated during the study, if necessary**. Women who become pregnant during the study would be withdrawn, unless there is a medical reason for allowing to continue to participate. Investigators should also ask fertile female participants to use an effective form of birth control while in the study and refrain from attempting to conceive a child for a limited period of time after the study ends to ensure that the body has eliminated the drug (Grady and Denny 2008).
5. **Since a father's toxic exposures may also impact the fetus and future child (Anderson et al.** 2014**), investigators should keep this in mind when designing clinical trials and informing men about potential risks to their children from exposures to radiation or drugs.** In some cases, birth control may be recommended for men who are participating in research to minimize these risks. This final recommendation is important for treating men and women fairly. Although the burdens of pregnancy almost entirely on women, men also need to be morally responsible for their contribution.

---

[12] See Resnik (2021).

8.7 Research Involving Adults with Impaired Decision-Making

The National Commission (1978) published a report recommending that the Common Rule include regulations for people who are "mentally infirm," but DHHS did not adopt these recommendations when it added special protections for pregnant women, fetuses, neonates, children, and prisoners in 1991 (Levine 1996). During the 1980s and 1990s, several bioethics commissions recommended that federal agencies should develop regulations for adults with impaired or uncertain DMC, but this advice was largely ignored (Yanos et al. 2009). In the absence of regulations, ethical guidance for enrolling people with impaired or uncertain DMC in research has been provided by several organizations, including the National Bioethics Advisory Commission (NBAC 1998), the National Institutes of Health (NIH 2009) and the Secretary's Advisory Committee for Human Research Protections (SACHRP 2009), and Council for the International Organizations of Medical Sciences (CIOMS 2016). And, as previously mentioned, DHHS announced regulations in May 2024 designed to prevent discrimination against disabled people in research, including people with cognitive disabilities (DHHS 2024: 40089).

In thinking about the ethics of research involving adults with impaired DMC, it is important to realize that this population is highly diverse (National Bioethics Advisory Commission 1998; Rosenstein and Miller 2008). Adults with impaired DMC may include:

- Adults with cognitive or developmental disabilities (such as Down's syndrome)
- Adults who are experiencing acute mental health symptoms (such as psychosis or severe depression)
- Adults with memory or reasoning deficits due to brain injuries
- Adults who are suffering from severe emotional problems (such as trauma)
- Adults with various forms of dementia
- Adults who are addicted to drugs or alcohol
- Adults who are temporarily unable to make decisions due to medical problems, such as shock, physical trauma, concussion, etc.

Because this group is so diverse, it is difficult to formulate guidelines or policies for addressing ethical issues related to their participation in research (Chen et al. 2002).

As we have discussed several times in this chapter, protection vs. inclusion is a recurring ethical dilemma in research involving vulnerable subjects. In the summer of 2024, a couple of interns who were working with me at NIEHS, David Appiah and Emily Nguyen, collected and coded data on 94 IRB policies regarding enrollment of adults with impaired DMC from the top-funded academic institutions in the US to determine how these policies handle the protection vs. inclusion dilemma. We developed a set of questions for coding these policies based on guidance provided by CIOMS, NBAC, NIH, and SACHRP and an initial review of IRB policy documents. These questions were then refined based on discussions among members of

Table 8.3 IRB policies for concerning inclusion of adults with impaired or uncertain DMC in research (Appiah et al. under submission)

	n = 94
The policy requires exclusion of people with impaired DMC inclusion if scientifically justified (E); requires unless exclusion is justified (I); leaves exclusion or inclusion decisions to the discretion of the investigator (D); or none of these options (N).	E = 41.5% (39) I = 5.3% (5) D = 3.2% (3) N = 50% (47)
The policy requires that inclusion is based on the risk to the participant	54.3% (51)
The policy provides guidance on assessing decision-making capacity	44.7% (42)
The policy provides guidance on using a legally authorized representative for decision-making	74.5% (70)
The policy provides guidance on consent or assent	77.7% (73)
The policy requires that the IRB include a member who is knowledgeable of the needs and concerns of people with impaired DMC	30.9% (29)

the research team. Mr. Appiah and Ms. Nguyen reviewed and coded the policies and resolved their disagreements. The data are reported in Table 8.3.

As reported in Table 8.2, 41.5% of the policies we analyzed **require exclusion** of people with impaired DMC unless inclusion is scientifically justified. Thus, the burden of proof falls on investigators to provide evidence that people with impaired DMC should be allowed to enroll in their study. For example, the University of North Carolina at Chapel Hill's policy states:

> Research involving subjects without the ability to provide consent or with impaired decision-making capacity **should only be conducted when the aims of the research cannot reasonably be achieved without their participation.** Participation of such subjects in research cannot be justified solely on their availability or the convenience for the investigator (University of North Carolina at Chapel Hill 2024; emphasis added).[13]

This policy is similar to the protectionistic philosophy expressed in the Belmont Report. A similar type of policy is found in the Helsinki Declaration, which states that:

> Medical research with individuals, groups, or communities in situations of particular vulnerability is **only justified if it is responsive to their health needs and priorities and the individual, group,** or community stands to benefit from the resulting knowledge, practices, or interventions. Researchers should only include those in situations of particular vulnerability when the research cannot be carried out in a less vulnerable group or community, or when excluding them would perpetuate or exacerbate their disparities (World Medical Association 2024, emphasis added).

One might argue, however, that these protectionistic policies unfairly exclude adults with impaired DMC from participating in research (DeCormier Plosky et al. 2022). Such policies would, for example, prevent adults with impaired DMC from participating in Phase III oncology trials because the aims of the research can be achieved

[13] Note this policy may violate Section 504 of the Rehabilitation Act by unnecessarily screening out people with cognitive disabilities (DHHS 2024: 40089).

without their participation. If I had an adult patient with Down's syndrome and advanced cancer, and their guardian and I determined it would be in their best medical interests to participate in an oncology clinical trial, we would be very frustrated with a policy that prevents them from enrolling in the study to protect them from harm. Or suppose that I am taking my adult son with Down's syndrome to a health fair and that I am also his legal guardian. There is a person at a booth recruiting people to be in a low-risk epidemiology/biobanking study that only involves a 50 ml blood sample and completion of a medical history form. Suppose that I volunteer for the study and my son also wants to, but the recruiter tells him he cannot be in the study because he cannot sign the consent form. I think my son would feel stigmatized and upset to learn that he cannot be in the study even though I can be. I would also feel angry and embarrassed.

The CIOMS guidelines prohibit the type of discrimination that one finds in protectionistic policies for potential participants with impaired DMC. According to the policy:

> Adults who are not capable of giving informed consent must be included in health-related research unless a good scientific reason justifies their exclusion (CIOMS 2016:61).

So, this policy is inclusionary because it places the burden of proof on investigators who plan to exclude adults with impaired DMC from their research. Only a few institutions in our study had inclusionary policies. The University of Vermont has such a policy:

> The IRB should be particularly cognizant of the special problems of research that involves the category of subjects who are vulnerable to coercion or undue influence, such as Children, Prisoners, Individuals with impaired decision-making capacity, Economically or educationally disadvantaged individuals...**Depending on the research, exclusion of any of the above populations might be construed as unfair and attempts should be made to include these populations**, with appropriate protections, if they are applicable to the research question (University of Vermont 2024, emphasis added).

Over half (54.3%) of the policies we analyzed had policies that base inclusion/exclusion on the level of risk to the research participants. One such policy was from Indiana University:

> Adult Individuals who lack consent capacity may be enrolled in research where the IRB finds:
>
> That the proposed research presents at least one of the following:
>
> - No greater than minimal risk to subjects
> - Greater probability of direct benefit to subjects than harm to subjects
> - Greater than minimal risk and no prospect of direct benefit to individual subjects, but is likely to yield generalizable knowledge about subjects' disorder or condition that is of vital importance for the understanding or amelioration of the disorder or condition (Indiana University 2024).

Risk-based policies may seem attractive to research institutions because (1) they are relatively simple and easy to follow and (2) they have some regulatory precedent insofar as they are similar to three categories of research that IRBs can approve for research with children under the Common Rule, Subpart D. However, risk-based policies may still unfairly exclude people from research participation because they are simplistic and lump adults with different abilities, impairments, and medical conditions into a single group (Rosenstein and Miller 2008; Dickert and Largent 2024). For example, suppose that a person with Parkinson's disease consents to participate in a long-term study of the aging process when they have full DMC. The goal is the study is to understand the aging process, not Parkinson's disease. The study includes more than minimal risk procedures, such as lumbar punctures. However, suppose that the person's cognitive abilities decline as the disease progresses. Should they, per Indiana University's policy, be withdrawn from the study because it is more than minimal risk? Should they be allowed to participate only in minimal risk procedures? Should they be allowed to participate in all procedures only if they have signed an advance directive when they had full DMC, which gives permission for research participation? Or consider the Down's syndrome case example mentioned above. Should that person not be allowed to participate in the epidemiology/biobanking study if it is a minor increase over minimal risk? Considerations like these led SACHRP (2009) to recommend that policies for adults with uncertain or impaired DMC should not be rigid but should be responsive to the diversity inherent in this population.

If a study is planning to enroll adults with impaired DMC, the protocol should describe procedures (1) assessing DMC of the potential participant; (2) identifying an LAR if the participant is unable to consent; and (3) obtaining assent from the participant. To identify an appropriate LAR, investigators must be familiar with local laws and customs, since these determine who can fulfill this role. Some potential LARs include: a guardian (if the subject has been adjudicated incompetent), a health care agent (if the subject has executed a health care power of attorney form), an immediate family member, or another family member or close friend (National Bioethics Advisory Commission 1998). When an LAR is used, adult subjects should provide their assent, if they are capable of doing so. The LAR may assist with decision-making or make the decision, depending on the subject's level of DMC (Chen et al. 2002).

In some cases, adults with declining mental abilities (e.g. patients with Alzheimer's or Parkinson's disease) may sign advance directives for future research participation when they still have good DMC (Stocking et al. 2006; Heinrichs 2021). Advance research directives (ARDs) are similar to other advance directives, such as living wills, in that they allow adults to express their preferences concerning important decisions before they lose their mental faculties. ARDs promote the person's autonomous decision-making by allowing them to decide whether to participate in research. When the person can no longer provide valid consent, LARs may consent on the adult's behalf in accordance with their previously expressed preferences,

One issue that may arise in assessing DMC is that investigators or members of the research team may have a conflict of interest that could bias their judgment. Since investigators are likely to be interested in enrolling subjects in a study to meet recruitment goals, this may bias their assessment in favor of affirming the subject's DMC, so that they can consent to research participation (National Bioethics

Advisory Commission 1998; Chen et al. 2002). To minimize the potential for bias, investigators could use a formal DMC assessment tool[14] or have an independent party assess DMC (Chen et al. 2002). Since the latter option may be inconvenient for investigators and research subjects and increase the costs of the study, it should be used only for more than minimal risk research (Chen et al. 2002).

Most commentators agree that enrolling adults with impaired DMC in minimal risk research does not raise any special ethical concerns because it does not expose them to significant risks without compensating benefits. Also, enrolling adults with impaired DMC in medically beneficial research does not raise special ethical concerns because it offers them benefits (such as treatment) which can offset risks. Most of the ethical debate has focused on enrolling adults with impaired DMC in more than minimal risk, non-beneficial research (Chen et al. 2002; Rosenstein and Miller 2008). The ethical issues with enrolling these adults in more than minimal risk, non-beneficial research are similar to those that arise with enrolling children in these types of studies, since the research subjects have not consented to participation.

In the Sect. 8.5, I argued that we can use socially accepted community service activities as a comparable standard for setting an upper bound for risk exposure in non-beneficial pediatric research. If we follow this line of reasoning, then we should look for a comparable community standard for enrolling adults with mental disabilities or diseases with impaired DMC. We do allow adults with impaired decision-making to participate in service activities that may expose them to more than minimal risks. For example, the American Red Cross allows people with disabilities, including cognitive disabilities, to donate blood (DeVault 2023). One might argue that adults with mental disabilities that impair decision-making can benefit psychologically from participating in research, because participation may enhance their sense of self-esteem by allowing them to contribute to society. While there are some good reasons for allowing adults with cognitive disabilities that impair DMC to participate in non beneficial, more than minimal risk research, there are also some good reasons to be cautious, because the potential for coercion, undue influence, and exploitation is great. If a cognitively disabled individual were to suffer a severe injury or die in a non-beneficial medical experiment, this could have a detrimental impact on the public's trust in research.

8.8 Research with Students and Employees

Before concluding this chapter, I will briefly discuss two other types of vulnerable subjects, students and employees. Students (including undergraduate and graduate students, interns, and postdoctoral trainees) often participate in research projects

[14] DMC assessment tools typically consist of lists of questions about the patient's ability to understand information and make decisions that clinicians can use to systematically assess their DMC. See Pennington et al. (2018).

conducted by their professors (including mentors or supervisors), such as surveys, interviews, behavioral experiments, or biospecimen collection or biomedical challenge studies (Resnik 2024). Student participation in research offers important benefits to students and their professors. Students can benefit from participation by learning more about how to conduct research and by feeling like there are making a contribution to the advancement of knowledge, and professors can benefit from having cooperative and conveniently-accessible research subjects. However, as illustrated by the Stanford Prison experiments (discussed in Chap. 2), research with students raises to key ethical concerns. First, students may feel pressured to participate in research to please their professors; and (2) student may be concerned that their professors (or research staff working with their professors) will learn private information about them. To minimize the potential for coercion, undue influence, and invasion of privacy, I recommend the following:

1. Whenever possible, professors should not directly recruit or enroll students into research projects; for example, recruitment could take place via advertisements or announcements and enrollment could be handled by research staff.
2. Steps should be taken to protect students' privacy, such as de-identifying data that can be accessed by professors or collecting and recording data anonymously.
3. If research participation is part of the student's grade, students must be offered an alternative to participation, such as writing a short essay, that takes roughly the same amount of time and effort and is worth the same number of grade-points.

Research involving employees also raises concerns regarding coercion, undue influence, and privacy. The Hwang case (see discussion in Chap. 2) illustrates some of the problems that can arise when subordinate employees participate in research. In this case, Hwang's lab assistants provided the human eggs used in his experiment. It likely that they faced pressure—real or perceived—to do this. Another case involving employee participation in research that is often discussed is Ellen's Roche's death in a human challenge study at Johns Hopkins University (JHU) in 2001 (Resnik 2016; see Chap. 2). However, it is unclear whether Roche's death had anything to do with her status as a JHU employee. OHRP said that Roche's death was due to the failure of the investigator thoroughly review the scientific literature on the risks of inhaling hexamethonium and the failure of the JHU IRB to carefully review the protocol (Crigger 2001).

Research with employees, like research with students, also raises concerns about coercion, undue influence, and privacy. Employees may feel significant pressure to participate in research. Indeed, some may fear that they will be fired, demoted, or sanctioned if they refuse to participate when asked. Employees may also be concerned that their employer will have access to private and sensitive information about them, such as information about their physical or mental health, and that their employer could use this information to take adverse actions against them, such as termination of employment. To protect employees who participate in research, I recommend the following:

1. Supervisors should not directly recruit or enroll subordinate employees into research projects; for example, recruitment could take place via advertisements or announcements and enrollment could be handled by non-supervisory research staff.
2. When recruitment and consent take place in the workplace, an independent party should observe the informed consent process to ensure that employees are not being pressured and that they have an opportunity to have their questions answered.
3. Whenever possible, private and sensitive information about employees should be collected and recorded anonymously, so that professors and staff will only have accessed to de-identified information.
4. Payment for research participation should be participation over and above what the employee would be otherwise entitled to.

8.9 Conclusion: Protection vs. Inclusion

In this chapter, we have seen that the conflict between protection vs. inclusion is an overarching theme in the ethics of research with vulnerable subjects (Friesen et al. 2023). *The Belmont Report* set the stage for this debate by emphasizing the importance of protecting vulnerable subjects from harm, coercion, undue influence, and exploitation (National Commission 1979). Federal agencies heeded this advice and adopted special protections for children, prisoners, and pregnant women/fetuses involved in research. However, in the 1980s, patient advocacy groups, politicians, physicians, and bioethicists argued that research policies were adversely affecting the health of protected populations, especially women and children (Mastroianni and Kahn 2001). Since then, government agencies and professional associations have taken steps to promote inclusion of different vulnerable groups in research, including women, racial and ethnic minorities, and disabled people.

One way of thinking about the protection vs. inclusion dilemma is to view it through the lens of justice (Kahn et al. 1998; Powers 1998; Mastroianni and Kahn 2001; Mackay and Saylor 2020). The principle of justice stated in the *Belmont Report* holds that the benefits and burdens of research should be distributed fairly (National Commission 1979). As noted above, federal research regulations require that the selection of research subjects be equitable, i.e. fair (45 CFR 46.111a3). One could argue that policies which exclude vulnerable populations from research are unfair because they prevent members of that group from obtaining the benefits of research (Kahn et al. 1998; Mackay 2016). Conversely, one could also argue that it would be unfair to impose risks on vulnerable research subjects in order to benefit other members of their groups (Mackay and Saylor 2020). Thus, the principle of justice could cut both ways, depending on whether one is concerned with unfairly denying benefits to a group or unfairly burdening research subjects with risks (Meltzer and Childress 2008).

How would the approaches to distributive justice discussed in Chap. 3—libertarianism, utilitarianism, and Rawlsian egalitarianism—address these issues? Libertarians would have favor policies that promote autonomy and procedural fairness, such as policies geared toward ensuring that informed consent is obtained from subject or their LAR. While libertarians would favor policies that protect people who cannot consent from harm and exploitation, they would oppose policies that restrict the choices of competent adults, such as policies that prevent pregnant women from participating in some types of research, because they would hold that the woman—and not a committee or institution—should have the right to decide whether the risk to the fetus is worth taking.

Utilitarians would support policies which are most likely to maximize benefits to society and minimize harms. A utilitarian could argue, for example, that pregnant women should be included in clinical trials, unless there is convincing scientific evidence supporting exclusion, because the benefits of inclusion for women's health are likely to far outweigh harms to the fetus and future children. However, a utilitarian would favor the opposite policy if the harms to the fetus and future children outweigh the benefits for women's health. For utilitarians, it all comes down to weighing the overall consequences of different policies.

Rawlsian egalitarians would favor policies that promote equality of opportunity and the well-being of the least advantaged members of society, such as socioeconomically disadvantaged people, racial or ethnic minorities, adults with impaired DMC and children. While Rawls intended his principles to apply to structure of social institutions and not to particular policy dilemmas occurring within those institutions, it seems clear that Rawls would favor policies that promote equality of opportunity related to participation in research (Mackay 2016). A Rawlsian could argue, for example, that research institutions should take steps to make that studies are accessible to who people who have disabilities or limited English proficiency (Resnik and Jones 2006).[15]

If we view the matter from the perspective of trust, there is very little debate about whether harms to vulnerable subjects can undermine the public's trust in research. As we saw in Chap. 2, some of the most egregious episodes in the history of research with human subjects, such as the Tuskegee Syphilis Study and the Nazi experiments on concentration camp prisoners—have involved vulnerable subjects. We also saw in Chap. 2, that most of the research regulations and ethical guidelines were developed, in large part, to restore the public's trust in research, especially guidelines governing research involving vulnerable subjects. However, members of the public may also lose confidence in investigators, institutions, sponsors, and the research enterprise if they feel that research is not addressing their needs and concerns (Mastroianni and Kahn 2001). Women's health advocates, for example, pressured the federal government not only to include women in clinical trials but also to fund more research on women's health (Dresser 2001). Likewise, HIV/AIDS

[15] A Rawlsian could argue that equality of opportunity requires that informed consent forms and other documents be translated into the person's native language and that researchers use interpreters to facilitate communication.

patients succeeded in convincing federal agencies to allocate more money toward research on HIV/AIDS (Dresser 2001). Thus, the trust ethos may favor protection or inclusion, depending on the expectations of the public.

Since the goal of fostering public trust does not provide simple solutions to the protection vs. inclusion dilemma, how should investigators, institutions, sponsors, and IRBs proceed? Perhaps that best way forward is to carefully consider the different factors in favor of inclusion or exclusion of vulnerable participants for each study, such as the risks of the research, the ability of the subject to provide informed consent, and the importance of the research for promoting the health of the population of which the group is a member. Nuanced approaches that allow for flexibility and contextualized decision-making may be preferable to rigid policies that require that some types of vulnerable subjects be automatically excluded from or included in research (Mastroianni and Kahn 2001; Meltzer and Childress 2008; Welch et al. 2015).

References

American Red Cross. 2024. *Requirements by donation type*. Available at https://www.redcrossblood.org/donate-blood/how-to-donate/eligibility-requirements.html. Accessed 20 Aug 2024.

Anderson, D., T.E. Schmid, and A. Baumgartner. 2014. Male-mediated developmental toxicity. *Asian Journal of Andrology* 16 (1): 81–88.

Atkins, A. 2018. *Famous misquotations: A civilization is measured by how it treats its weakest members*. Atkins Books Shelf, 21 February. Available at https://atkinsbookshelf.wordpress.com/2018/02/21/famous-misquotations-a-civilization-is-measured-by-how-it-treats-its-weakest-members/. Accessed 16 Aug 2024.

Bazzano, A.T., R. Mangione-Smith, M. Schonlau, M.J. Suttorp, and R.H. Brook. 2009. Off-label prescribing to children in the United States outpatient setting. *Academic Pediatrics* 9 (2): 81–88.

Blake, V. 2012. Minors' refusal of life-saving therapies. *Virtual Mentor* 14 (10): 792–796.

Blehar, M.C., C. Spong, C. Grady, S.F. Goldkind, L. Sahin, and J.A. Clayton. 2013. Enrolling pregnant women: Issues in clinical research. *Womens Health Issues* 23 (1): e39–e45.

Boldt, J. 2019. The concept of vulnerability in medical ethics and philosophy. *Philosophy, Ethics, and Humanity in Medicine* 14: 6.

Bonham, V.H., and J.D. Moreno. 2008. Research with captive populations: Prisoners, students, and soldiers. In *The Oxford handbook of clinical research ethics*, ed. E.J. Emanuel, C. Grady, R.A. Crouch, R.K. Lie, F.G. Miller, and D. Wendler, 461–474. New York: Oxford University Press.

Brunger, F., T. Russell, and D. Wall. 2024. Decolonizing research ethics: Best practices for the ethical conduct of research involving indigenous peoples. In *The Oxford handbook of research ethics*, ed. A.S. Iltis and D. Mackay, 581–600. New York, NY: Oxford University Press.

Camanni, G., O. Ciccone, A. Lepri, C. Tinarelli, M. Menna, A. Baglioni, N. Murgia, and S. Elisei. 2021. Intellectual disability and psychiatric disorders as exclusion criteria in randomized controlled trials (RCT). *Psychiatria Dunbina* 33 (Suppl 11): 36–39.

Caplan, A.L. 2007. Challenging teenagers' right to refuse treatment. *Virtual Mentor* 9 (1): 56–61.

Chen, D.T., F.G. Miller, and D.L. Rosenstein. 2002. Enrolling decisionally impaired adults in clinical research. *Medical Care* 40 (9 Suppl): V20–V29.

Christopher, P.P. 2024. Vulnerability in research with prisoners. In *The Oxford handbook of research ethics*, ed. A.S. Iltis and D. Mackay, 839–863. New York, NY: Oxford University Press.

Cincinnati Children's Hospital. 2024. *Leuprolide stimulation test*. Available at https://www.cincinnatichildrens.org/health/l/leuprolide. Accessed 19 Aug 2024.

Coleman, C.H. 2009. Vulnerability as a regulatory category in human subject research. *Journal of Law, Medicine & Ethics* 37 (1): 12–18.

Council for the International Organizations of Medical Sciences. 2016. *International ethical guidelines for health-related research involving humans*. https://cioms.ch/wp-content/uploads/2017/01/WEB-CIOMS-EthicalGuidelines.pdf Accessed 4 Aug 2024.

Crigger, B.J. 2001. What does it mean to "review" a protocol? Johns Hopkins & OHRP. *IRB* 23 (4): 13–15.

DeCormier Plosky, W., A. Ne'eman, B.C. Silverman, D.H. Strauss, L.P. Francis, M.A. Stein, and B.B. Bierer. 2022. Excluding people with disabilities from clinical research: Eligibility criteria lack clarity and justification: Study examines factors and practices that exclude people with disabilities from clinical research. *Health Affairs* 41 (10): 1423–1432.

Denny, C.C., and C. Grady. 2007. Clinical research with economically disadvantaged populations. *Journal of Medical Ethics* 33 (7): 382–385.

Department of Health and Human Services. 2024. Nondiscrimination on the basis of disability in programs or activities receiving federal financial assistance. *Federal Register* 89 (July 8): 40066–40195.

DeVault, N. 2023. *Call to action: People with disabilities can donate blood*. Ameridisability, 25 January. Available at https://www.ameridisability.com/call-to-action-people-with-disabilities-can-donate-blood/#:~:text=Sadly%2C%20the%20demand%20for%20blood,to%20roll%20up%20their%20sleeves. Accessed 21 Aug 2024.

Dickert, N.W., and E.A. Largent. 2024. Research in adults with diminished capacity. In *The Oxford handbook of research ethics*, ed. A. Iltis and D. Mackay, 699–714. New York, NY: Oxford University Press.

Dresser, R. 1992. Wanted. Single, white male for medical research. *Hastings Center Report* 22 (1): 24–29.

———. 2001. *When science offers salvation*. New York, NY: Oxford University Press.

DuBois, J.M., L. Beskow, J. Campbell, K. Dugosh, D. Festinger, S. Hartz, R. James, and C. Lidz. 2012. Restoring balance: A consensus statement on the protection of vulnerable research participants. *American Journal of Public Health* 102 (12): 2220–2225.

Food and Drug Administration. 2005. Chair's summary of the meeting of the pediatric ethics subcommittee ("407 panel"). Hilton Gaithersburg Hotel, Gaithersburg, November 15, 2005.

———. 2018. *Pregnant women: Scientific and ethical considerations for inclusion in clinical trials: Guidance for Industry (draft guidance)*. Available at https://www.fda.gov/media/112195/download?attachment. Accessed 27 Aug 2024.

Forsythe, N. 2022. *Navigating parental disputes over the medical care of minors*. Magmutual, December 2024. Available at https://www.magmutual.com/learning/article/navigating-parental-disputes-practice-setting/. Accessed 23 Aug 2024.

Friesen, P., L. Gelinas, A. Kirby, D.H. Strauss, and B.E. Bierer. 2023. IRBs and the protection-inclusion dilemma: Finding a balance. *American Journal of Bioethics* 23 (6): 75–88.

Gómez-Roig, M.D., R. Pascal, M.J. Cahuana, O. García-Algar, G. Sebastiani, V. Andreu-Fernández, L. Martínez, G. Rodríguez, I. Iglesia, O. Ortiz-Arrabal, M.D. Mesa, M.J. Cabero, L. Guerra, E. Llurba, C. Domínguez, M.J. Zanini, M. Foraster, E. Larqué, F. Cabañas, M. Lopez-Azorín, A. Pérez, B. Loureiro, C.R. Pallás-Alonso, D. Escuder-Vieco, and M. Vento. 2021. Environmental exposure during pregnancy: Influence on prenatal development and early life: A comprehensive review. *Fetal Diagnosis and Therapy* 48 (4): 245–257.

Goodin, R. 1986. *Protecting the vulnerable: A reanalysis of our social responsibilities*. Chicago: University of Chicago Press.

Gostin, L.O., C. Vanchieri, and A. Pope, eds. 2007. *Institute of medicine (US) committee on ethical considerations for revisions to DHHS regulations for protection of prisoners involved in research*, Ethical considerations for research involving prisoners. Washington, DC: National Academies Press.

Grady, C., and C. Denny. 2008. Research involving women. In *The Oxford handbook of clinical research ethics*, ed. E.J. Emanuel, C. Grady, R.A. Crouch, R.K. Lie, F.G. Miller, and D. Wendler, 407–422. New York, NY: Oxford University Press.
Gunatilake, R. and P.S. Avinash. 2023. *Safety of medications during pregnancy*. Merck Manual. Available at https://www.merckmanuals.com/home/women-s-health-issues/medication-and-substance-use-during-pregnancy/safety-of-medications-during-pregnancy. Accessed 28 Aug 2024.
Habitat for Humanity of the Charlotte Region. 2024. *Volunteer FAQs*. Available at https://www.habitatcltregion.org/volunteer/volunteer-faq/. Accessed 20 Aug.
Heinrichs, B. 2021. Advance research directives: Avoiding double standards. *BMC Medical Ethics* 22 (1): 137.
Herring, C., A. McManus, and A. Weeks. 2010. Off-label prescribing during pregnancy in the UK: An analysis of 18,000 prescriptions in Liverpool Women's hospital. *International Journal of Pharmacy Practice* 18 (4): 226–229.
Hicks, L. 2006. Research in public schools. In *Institutional review board: Management and function*, ed. E.A. Bankert and R. Amdur, 341–345. Boston: Jones and Bartlett.
Hornblum, A.M. 1997. They were cheap and available: Prisoners as research subjects in twentieth century America. *BMJ* 315: 1437–1441.
———. 1998. *Acres of skin: Human experiments at Holmesburg prison*, 1998. New York, NY: Routledge.
Hu, H., M.M. Téllez-Rojo, D. Bellinger, D. Smith, A.S. Ettinger, H. Lamadrid-Figueroa, J. Schwartz, L. Schnaas, A. Mercado-García, and M. Hernández-Avila. 2006. Fetal lead exposure at each stage of pregnancy as a predictor of infant mental development. *Environmental Health Perspectives* 114 (11): 1730–1735.
Iltis, A. 2007. Pediatric research posing a minor increase over minimal risk and no prospect of direct benefit: Challenging 45 CFR 46.406. *Accountability in Research* 14 (1): 19–34.
Indiana University. Adult individuals lacking consent capacity in research. 2024. Available at https://research.iu.edu/policies/human-subjects-irb/adult-individuals-lacking-consent-capacity.html#PolicyStatement. Accessed 21 Aug 2024.
Institute of Medicine. 2004. *Ethical conduct of clinical research involving children*. Washington, DC: National Academies Press.
Jonathan, I., E. Akers, M. Shi, and D.B. Resnik. 2024. Vulnerable research participant policies at U.S. academic institutions. *Journal of Empirical Research in Human Research Ethics*. Published online 18 October 2024. Available at https://journals.sagepub.com/doi/abs/10.1177/15562646241290093. Accessed 23 Oct 2024.
Kahn, J.P., A.C. Mastroianni, and J. Sugarman. 1998. Implementing justice in a changing research environment. In *Beyond consent: Seeking justice in research*, ed. J.P. Kahn, A.C. Mastroianni, and J. Sugarman, 166–174. New York, NY: Oxford University Press.
Kelley, A., A. Belcourt-Dittloff, C. Belcourt, and G. Belcourt. 2013. Research ethics and indigenous communities. *American Journal of Public Health* 103 (12): 2146–2152.
Kim, J.H., and A.R. Scialli. 2011. Thalidomide: The tragedy of birth defects and the effective treatment of disease. *Toxicological Sciences* 122 (1): 1–6.
King, P.A. 1998. Race, justice, and research. In *Beyond consent: Seeking justice in research*, ed. J.P. Kahn, A.C. Mastroianni, and J. Sugarman, 88–110. New York: Oxford University Press.
Kipnis, K. 2003. Seven vulnerabilities in the pediatric research subject. *Theoretical Medicine Bioethics* 24 (2): 107–120.
Klint, A.L., S.M. Leatherman, O. Taylor, P.A. Glassman, R.E. Ferguson, W.C. Cushman, and A. Ishani. 2013. Telephone informed consent in a pragmatic point-of-care clinical trial embedded in primary care. *Contemporary Clinical Trials* 131: 107239.
Kons, K.M., M.L. Wood, L.C. Peck, S.M. Hershberger, A.R. Kunselman, C. Stetter, R.S. Legro, and T.A. Deimling. 2022. Exclusion of reproductive-aged women in COVID-19 vaccination and clinical trials. *Womens Health Issues* 32 (6): 557–563.
Krause, H.D. 1995. *Family law*. St. Paul: West Publishing.

Levine, R.J. 1996. Research involving those institutionalized as mentally infirm: A consideration of their relevance in 1996. *IRB: Ethics & Human Research* 18 (5): 1–5.

Levine, C. 2004. The concept of vulnerability in disaster research. *Journal of Trauma and Stress* 17 (5): 395–402.

———. 2008. Research involving economically disadvantage participants. In *The Oxford handbook of clinical research ethics*, ed. E.J. Emanuel, C. Grady, R.A. Crouch, R.K. Lie, F.G. Miller, and D. Wendler, 431–436. New York: Oxford University Press.

Lewis, H. 2021. Why I'll keep saying 'pregnant women'. *The Atlantic*, 26 October. Available at https://www.theatlantic.com/ideas/archive/2021/10/pregnant-women-people-feminism-language/620468/. Accessed 20 Aug 2024.

Little, M.O., M.N. Wickremsinhe, and A. Drapkin Lyerly. 2024. Research in pregnancy: Frameworks for addressing a neglected and complex population. In *The Oxford handbook of clinical research ethics*, ed. A.S. Iltis and D. MacKay, 643–665. New York, NY: Oxford University Press.

Lo, B., and N. Garan. 2008. Research with ethnic and minority populations. In *The Oxford handbook of clinical research ethics*, ed. E.J. Emanuel, C. Grady, R.A. Crouch, R.K. Lie, F.G. Miller, and D. Wendler, 423–430. New York, NY: Oxford University Press.

Lyerly, A.D., M.O. Little, and R. Faden. 2008. The second wave: Toward responsible inclusion of pregnant women in research. *International Journal of Feminist Approaches Bioethics* 1 (2): 5–22.

Lyerly, A.D., L.M. Mitchell, E.M. Armstrong, L.H. Harris, R. Kukla, M. Kuppermann, and M.O. Little. 2009. Risk and the pregnant body. *Hastings Center Report* 39 (6): 34–42.

Lyerly, A.D., M.O. Little, and R. Faden. 2011. Reframing the framework: Toward fair inclusion of pregnant women as participants in research. *American Journal of Bioethics* 11 (5): 50–55.

MacKay, D. 2016. Fair subject selection in clinical research: Formal equality of opportunity. *Journal of Medical Ethics* 42: 672–677.

MacKay, D., and K.W. Saylor. 2020. Four faces of fair subject selection. *American Journal of Bioethics* 20 (2): 5–19.

Macklin, R. 2003. Bioethics, vulnerability, and protection. *Bioethics* 17 (5–6): 472–486.

Mastroianni, A.C., and L.M. Henry. 2024. Closing the gap: Improving women's health through research inclusion. In *The Oxford handbook of clinical research ethics*, ed. A.S. Iltis and D. MacKay, 601–619. New York, NY: Oxford University Press.

Mastroianni, A.C., and J.P. Kahn. 2001. Swinging on the pendulum. Shifting views of justice in human subjects research. *Hastings Center Report* 31 (3): 21–28.

Mayo Clinic. 2024. *Leuprolide risks*. Available at https://my.clevelandclinic.org/health/drugs/19844-leuprolide-depot-injection. Accessed 19 Aug 2024.

McDonald, K.E., A.E. Schwartz, and M. Sabatello. 2022. Eligibility criteria in NIH-funded clinical trials: Can adults with intellectual disability get in? *Disability Health Journal* 15 (4): 101368.

McManus, J., S.G. Mehta, A.R. McClinton, R.A. De Lorenzo, and T.W. Baskin. 2005. Informed consent and ethical issues in military medical research. *Academic Emergency Medicine* 12 (11): 1120–1126.

Meltzer, L.A., and J.F. Childress. 2008. What is fair participant selection? In *The Oxford handbook of clinical research ethics*, ed. E.J. Emanuel, C. Grady, R.A. Crouch, R.K. Lie, F.G. Miller, and D. Wendler, 377–385. New York, NY: Oxford University Press.

Menikoff, J. 2009. The vulnerability of the very sick. *Journal of Law, Medicine & Ethics* 37 (1): 51–58.

Miller, F.G., and A. Wertheimer. 2007. Facing up to paternalism in research ethics. *Hastings Center Report* 37 (3): 24–34.

Moskop, J.C. 1998. A moral analysis of military medicine. *Military Medicine* 163 (2): 76–79.

National Bioethics Advisory Commission. 1998. *Research involving people with mental disorders that may affect Decision making capacity*. Washington, DC: National Bioethics Advisory Commission.

———. 2001. *Ethical and policy issues in research with human participants*. Washington, DC: National Bioethics Advisory Commission.

National Commission for the Protection of Human Subjects of Biomedical or Behavioral Research. 1978. *Research involving those institutionalized as mentally infirm*. Washington, DC: Department of Health, Education and Welfare.

———. 1979. *The Belmont report: Ethical principles and guidelines for the protection of human subjects of research*. Washington, DC: Department of Health, Education, and Welfare.

National Institutes of Health. 2001. *NIH Policy and guidelines on the inclusion of women and minorities as subjects in clinical research*. Available at https://grants.nih.gov/grants/funding/women_min/guidelines_amended_10_2001.htm. Accessed 27 Aug 2024.

———. 2009. *Research involving individuals with questionable capacity to consent: Points to consider*. https://grants.nih.gov/grants/policy/questionablecapacity.htm. Accessed 4 Aug 2024.

———. 2024. *Best Pharmaceutical for Children Act*. Available at https://www.nichd.nih.gov/research/supported/bpca/about. Accessed 19 Aug 2024.

Nelson, D.K., D. Skinner, S. Guarda, S. Choudhury, J. Sideris, L. Barnum, K. Ten Haagen, Q. Heyward, and D.B. Bailey Jr. 2013. Obtaining consent from both parents for pediatric research: What does "reasonably available" mean? *Pediatrics* 131 (1): e223–e239.

Noggle, R. 2022. *The ethics of manipulation*. Stanford Encyclopedia of Philosophy. Available at https://plato.stanford.edu/archives/sum2022/entries/ethics-manipulation/. Accessed 16 Aug 2024.

North Carolina Department of Transportation. 2024. *Adopt at highway, get involved*. Available at https://www.ncdot.gov/initiatives-policies/environmental/adoptahighway/Pages/get-involved.aspx. Accessed 20 Aug 2024.

Office of Human Research Protections. 2016. *Use of electronic informed consent questions and answers*. Available at https://www.fda.gov/media/116850/download. Accessed 23 Aug 2024.

Office of Research on Women's Health. 2010. *Report: Enrolling pregnant women: Issues in CLINICAL Research*. Available at https://orwh.od.nih.gov/sites/orwh/files/docs/ORWH-EPW-Report-2010.pdf. Accessed 27 Aug 2024.

Offidani, C., F. Pomini, A. Caruso, S. Ferrazzani, M. Chiarotti, and A. Fiori. 1995. Fiori. Cocaine during pregnancy: A critical review of the literature. *Minerva Gynecology* 47 (9): 381–390.

Ouelette, A. 2024. People with disabilities in human subjects research: A history of exploitation, a problem of exclusion. In *The Oxford handbook of clinical research ethics*, ed. A.S. Iltis and D. MacKay, 666–682. New York, NY: Oxford University Press.

Packenham, J.P., R.T. Rosselli, S.K. Ramsey, H.A. Taylor, A. Fothergill, J. Slutsman, and A. Miller. 2017. Conducting science in disasters: Recommendations from the NIEHS working Group for Special IRB considerations in the review of disaster related research. *Environmental Health Perspectives* 125 (9): 094503.

Pennington, C., K. Davey, R. ter Meulen, E. Coulthard, and P.G. Kehoe. 2018. Tools for testing decision-making capacity in dementia. *Age and Ageing* 47 (6): 778–784.

Phelan, A.L., A.R. Kunselman, C.H. Chuang, N.T. Raja-Khan, and R.S. Legro. 2016. Exclusion of women of childbearing potential in clinical trials of type 2 diabetes medications: A review of protocol-based barriers to enrollment. *Diabetes Care* 39 (6): 1004–1009.

Popova, S., M.E. Charness, L. Burd, A. Crawford, H.E. Hoyme, R.A.S. Mukherjee, E.P. Riley, and E.J. Elliott. 2023. Fetal alcohol spectrum disorders. *Nature Reviews Disease Primers* 9 (1): 11.

Porro, C. 2022. Vulnerability. In *Handbook of bioethics*, ed. E. Di Nucci, J.Y. Lee, and I.A. Wagner, 149–156. Lanham: Rowman and Littlefield.

Powers, M. 1998. Theories of justice in the context of research. In *Beyond consent: Seeking justice in research*, ed. J.P. Kahn, A.C. Mastroianni, and J. Sugarman, 147–165. New York, NY: Oxford University Press.

Pratt, B. 2024. International research ethics. In *The Oxford handbook of research ethics*, ed. A.S. Iltis and D. Mackay, 796–822. New York, NY: Oxford University Press.

Radley, D.C., S.N. Finkelstein, and R.S. Stafford. 2006. Off-label prescribing among office-based physicians. *Archives of Internal Medicine* 166 (9): 1021–1026.

Rehman, W., L.M. Arfons, and H.M. Lazarus. 2011 Oct. 2011. The rise, fall and subsequent triumph of thalidomide: Lessons learned in drug development. *Therapeutic Advances in Hematology* 2 (5): 291–308.
Resnik, D.B. 2007. Are the new EPA regulations concerning intentional exposure studies involving children overprotective? *IRB* 29 (5): 15–19.
———. 2016. Employees as research participants: Ethical and policy issues. *IRB* 38 (4): 11–16.
———. 2021. *Precautionary reasoning in environmental and public health policy*. Cham: Springer.
———. 2024. Ethical issues in research involving students and employees. In *The Oxford handbook of clinical research ethics*, ed. A.S. Iltis and D. MacKay, 785–794. New York, NY: Oxford University Press.
Resnik, D.B., and C.W. Jones. 2006. Research subjects with limited English proficiency: Ethical and legal issues. *Accountability in Research* 13 (2): 157–177.
Rogers, W., and M.M. Lange. 2013. Rethinking the vulnerability of minority populations in research. *American Journal of Public Health* 103 (12): 2141–2146.
Rogers, W., C. Mackenzie, and S. Dodds. 2012. Why bioethics needs a concept of vulnerability. *International Journal of Feminist Approaches Bioethics* 5 (2): 11–38.
Rosenstein, D.L., and F.G. Miller. 2008. Research involving those at risk for impaired decision-making capacity. In *The Oxford handbook of clinical research ethics*, ed. E.J. Emanuel, C. Grady, R.A. Crouch, R.K. Lie, F.G. Miller, and D. Wendler, 437–445. New York, NY: Oxford University Press.
Ross, L.R. 2008. *Children in medical research: Access vs. protection*. New York, NY: Oxford University Press.
Rothstein, P.F., M.S. Raeder, and D. Crump. 1997. *Evidence*. 3rd ed. St. Paul: MN: West Publishing.
Sachs, G.A., and C.K. Cassel. 1990. Biomedical research involving older human subjects. *Law, Medicine, and Health Care* 18 (3): 234–243.
Schonfeld, T. 2013. The perils of protection: Vulnerability and women in clinical research. *Theoretical Medicine and Bioethics* 34 (3): 189–206.
Schroeder, D., and E. Gefenas. 2009. Vulnerability: Too vague and too broad? *Cambridge Quarterly of Healthcare Ethics* 18 (2): 113–121.
Scouting Wire. 2013. Scouts build character by building ramps. *Scouting Wire*, 20 August. Available at https://scoutingwire.org/scouts-build-character-by-building-ramps/. Accessed 20 Aug 2024.
Second Harvest. 2023. How old do you have to be to volunteer at the food bank. Available at https://localfoodbank.org/how-old-do-you-have-to-be-to-volunteer-at-the-food-bank/#:~:text=Food%20banks%20usually%20have%20a,bank%20to%20get%20the%20specifics. Accessed 20 Aug.
Secretary's Advisory Committee on Human Research Protections (SACHRP). 2009. Attachment: Recommendations Regarding Research Involving Individuals with Impaired Decision-making. https://www.hhs.gov/ohrp/sachrp-committee/recommendations/2009-july-15-letter-attachment/index.html.
Shea, A.K., and M. Steiner. 2008. Cigarette smoking during pregnancy. *Nicotine Tobacco Research* 10 (2): 267–278.
St John, B.M., E. Hickey, E. Kastern, C. Russell, T. Russell, A. Mathy, B. Peterson, D. Wigington, C. Pellien, A. Caudill, L. Hladik, and K.K. Ausderau. 2022. Opening the door to university health research: Recommendations for increasing accessibility for individuals with intellectual disability. *International Journal of Equity in Health* 21 (1): 130.
Stanford University. 2024. Disability language guide. Available at https://disability.stanford.edu/sites/g/files/sbiybj26391/files/media/file/disability-language-guide-stanford_1.pdf. Accessed 17 Aug 2024.
Stocking, C.B., G.W. Hougham, D.D. Danner, M.B. Patterson, P.J. Whitehouse, and G.A. Sachs. 2006. Speaking of research advance directives: Planning for future research participation. *Neurology* 66 (9): 1361–1366.

Strong, C. 2011. Minimal risk in research involving pregnant women and fetuses. *Journal of Law, Medicine, and Ethics* 39 (3): 529–538.

Taylor, M.M., L. Kobeissi, C. Kim, A. Amin, A.E. Thorson, N.B. Bellare, V. Brizuela, M. Bonet, E. Kara, S.S. Thwin, H. Kuganantham, M. Ali, O.T. Oladapo, and N. Broutet. 2021. Inclusion of pregnant women in COVID-19 treatment trials: A review and global call to action. *Lancet Global Health* 9 (3): e366–e371.

University of North Carolina at Chapel Hill. 2024. *Office of human research ethics SOP 1201: Vulnerable subjects in research*. Available at https://policies.unc.edu/TDClient/2833/Portal/KB/ArticleDet?ID=132228. Accessed 21 Aug 2024.

University of Vermont. 2024. *Subjects vulnerable to coercion or undue influence*. Available at https://legacy.drup2.uvm.edu/node/249877/#vul_II. Accessed 21 Aug 2024.

van der Graaf, R., I.S.E. van der Zande, H.M. den Ruijter, M.A. Oudijk, J.M. van Delden, K.O. Rengerink, and R.H.H. Groenwold. 2018. Fair inclusion of pregnant women in clinical trials: An integrated scientific and ethical approach. *Trials* 19: 78.

Webster's Dictionary of the English Language. 2024. *Vulnerable*. Available at https://www.merriam-webster.com/dictionary/vulnerable. Accessed 16 Aug 2024.CIOMS-EthicalGuidelines.pdf Accessed 4 Aug 2024.

Welch, M.J., R. Lally, J.E. Miller, S. Pittman, L. Brodsky, A.L. Caplan, G. Uhlenbrauck, D.M. Louzao, J.H. Fischer, and B. Wilfond. 2015. The ethics and regulatory landscape of including vulnerable populations in pragmatic clinical trials. *Clinical Trials* 12 (5): 503–510.

Wendler, D. 2010. *The ethics of pediatric research*. New York, NY: Oxford University Press.

———. 2012. A new justification for pediatric research without the potential for clinical benefit. *American Journal of Bioethics* 12 (1): 23–31.

Wendler, D., and E.J. Emanuel. 2005. What is a "minor" increase over minimal risk? *Journal of Pediatrics* 147 (5): 575–578.

Wendler, D., L. Belsky, K.M. Thompson, and E.J. Emanuel. 2005. Quantifying the federal minimal risk standard: Implications for pediatric research without a prospect of direct benefit. *Journal of the American Medical Association* 294 (7): 826–832.

Wenger, N.K., D.M. Lloyd-Jones, M.S.V. Elkind, G.C. Fonarow, J.J. Warner, H.M. Alger, S. Cheng, C. Kinzy, J.L. Hall, V.L. Roger, and American Heart Association. 2022. Call to action for cardiovascular disease in women: Epidemiology, awareness, access, and delivery of equitable health care: A presidential advisory from the American Heart Association. *Circulation* 145 (23): e1059–e1071.

White, A., C. Grady, M. Little, K. Sullivan, K. Clark, M. Ngwu, and A.D. Lyerly. 2021. IRB decision-making about minimal risk research with pregnant participants. *Ethics and Human Research* 43 (5): 2–17.

World Medical Association. 2024. *Declaration of Helsinki: Ethical principles for medical research involving human subjects (2024 revision)*. Available at https://www.wma.net/policies-post/wma-declaration-of-helsinki-ethical-principles-for-medical-research-involving-human-subjects/. Accessed 22 Oct 2024.

Yanos, P.T., B.S. Stanley, and C.S. Greene. 2009. Research risk for persons with psychiatric disorders: A decisional framework to meet the ethical challenge. *Psychiatric Services* 60 (3): 374–383.

Chapter 9
Scientific Integrity

Abstract This chapters explores a variety of topics related to integrity in research with human subjects including the importance of integrity in protecting human subjects; the definition of misconduct; questionable research practices; the prevalence of misconduct; misconduct vs. fraud; misconduct vs. error or negligence; procedures for reporting, investigating and adjudicating misconduct; famous cases of research misconduct; non-compliance in research; procedures for reporting, investigating, and correcting non-compliance; unanticipated problems in research involving increased risks of harm; serious or continuing non-compliance; serious adverse events; conflicts of interests; the definition of conflict of interests; empirical evidence concerning conflict of interest; why conflicts of interests are an ethical concern in science; the relationship between conflicts of interest, bias, and misconduct; famous cases of conflict of interest; policies for disclosing and managing conflicts of interest; and strategies for promoting integrity in research, such as support for education in responsible conduct of research; ethical leadership; development of policies for research ethics; auditing of data and research records; and mechanisms for providing consultation on ethics issues and protecting whistleblowers.

Keywords Ethics · Integrity · Trust · Misconduct · Non-compliance · Conflict of interest · Bias · Questionable research practices · Fraud · Law · Regulation

Chapters 5, 6, 7, and 8 of the book covered topics involving interactions between researchers and human subjects, such as informed consent, privacy, confidentiality, risk and benefits, and vulnerability. In this chapter we will switch gears a little bit and cover a topic involving interactions among scientists, research institutions, and the greater public: scientific integrity. Integrity comes from the two Greek words, *akeraios*, which means 'pure' or 'uncontaminated,' and *aorolokleros*, which means 'whole' or 'complete.' In modern English, integrity is equated with being honest, ethical, or honorable. **Scientific integrity** can be defined as (1) following the ethical, professional, and legal rules and guidelines (or standards) for the conduct of science; and (2) making morally wise decisions when the rules conflict, have multiple interpretations or provide insufficient guidance (Shamoo and Resnik 2022).

D. B. Resnik, *The Ethics of Research with Human Subjects*, The International Library of Bioethics 111, https://doi.org/10.1007/978-3-031-82757-0_9

Scientific integrity is a broad category that includes **research integrity** and **integrity in research with human subjects** as subsets (Miller et al. 1998; Office of Science and Technology Policy 2023). To illustrate the difference between these concepts, fabricating data is a violation of research integrity, whereas not giving disclosing a conflict of interest is a violation of scientific integrity, and preforming a research procedure without obtaining consent is a violation of integrity in research with human subjects. Some situations, such as falsifying data used to qualify a subject for enrollment in a study, may involve violations of three senses of integrity.

In this book, we have reviewed rules and guidelines pertaining to the conduct of research with human subjects, but not rules and guidelines for the conduct of science itself. Some of these include rules and guidelines for:

- Designing test and experiments, analyzing and interpreting data, evaluating hypotheses, models, and theories
- Recording, reporting, storing, sharing, publishing and data
- Assigning authorship credit
- Disclosing and managing conflicts of interest
- Working with collaborators, students, and trainees
- Reviewing manuscripts and grants
- Publishing research
- Protecting intellectual property
- Protecting the safety of laboratory personnel
- Handling hazardous materials
- Managing financial and other resources
- Investigating allegations of misconduct
- Protecting research subjects (animals and humans)

Since there is not enough space in this chapter to cover all these topics in depth (see Shamoo and Resnik 2022), I will focus on integrity issues which have an important bearing on the protection of human subjects. But first, I will explain why investigator integrity is essential to research with human subjects.

9.1 Integrity in Research with Human Subjects

Investigator behavior is probably the single most important factor in protecting the rights and welfare of human subjects (Miller et al. 1998). Regulations and guidelines are useless if investigators ignore them, and IRB review and oversight is ineffective if the investigator does not comply with the approved protocol. Investigators (and research staff members) have direct contact with human subjects and therefore have the greatest opportunity to either help or harm them. IRBs, institutions, and sponsors must therefore trust that investigators will follow the rules and act responsibly.

The importance of having standards governing investigator behavior (or conduct) is based on several key consideration. First, standards of conduct are important for achieving the aims of research, which include the advancement of knowledge and the improvement of the human condition. Research that produces biased, irreproducible, or fraudulent data does not help scientists uncover the truth about the world. Second, standards of conduct are important for promoting trust among scientists, which is essential for many different aspects of research, including working together in laboratories or research groups, collaborating with other researchers, reviewing articles for publication or grant proposals, and criticizing or building upon published research. Third, standards of conduct are important for fostering the public's trust in science (DeAngelis 2000; National Academies of Science, Engineering and Medicine 2017). People are not likely to fund or support research that they view as unethical or untrustworthy. And fourth, standards of conduct are important for holding scientists accountable to the public. People have a right to control how public money is spent and oversee research with important implications for society and human values (Shamoo and Resnik 2022). Table 9.1 lists some commonly accepted standards for the conduct of research.

The standards in Table 9.1 are guidelines for the conduct of scientific research that one finds expressed in various professional codes and official documents.[1] They are not absolute rules but *prima facie* ethical obligations that should be followed unless they conflict with other rules (Shamoo and Resnik 2022). For examples, as discussed in Chap. 6, openness sometimes conflicts with protection the confidentiality of human research subjects. When conflicts arise, one must make a well-reasoned decision based on the available information and options (see discussion of ethical decision-making in Chaps. 3 and 4).

9.2 Misconduct in Research

The most serious violations of ethical, legal, or professional standards of behavior in science are classified as misconduct (Shamoo and Resnik 2022). The U.S. government defines research misconduct as: **"fabrication, falsification, or plagiarism in proposing, performing, or reviewing research, or in reporting research results…misconduct does not include honest error or differences of opinion** (Office of Science and Technology Policy 2000:76262)." This definition, which is accepted by numerous federal agencies, including the NIH, NSF, and FDA, focuses on three types of misbehavior: fabrication, falsification, or plagiarism (FFP). Fabrication is "making up data or results and recording or reporting them," falsification is "manipulating research materials, equipment, or processes, or changing or omitting data or results such that the research is not accurately represented in the research record," and plagiarism is "the appropriation of another person's ideas, processes, results, or words without giving appropriate credit (Office of Science and

[1] See for example, National Academies of Science, Engineering and Technology (2009), World Conference on Research Integrity (2010), All European Academies (2023).

Table 9.1 Standards for the conduct of research. (Based on Shamoo and Resnik 2022)

Honesty	Do not fabricate, falsify, or deceptively manipulate data or images; do not plagiarize; communicate honestly with colleagues and the public.
Transparency	Disclose data, materials, methods, data, computer code, conflicts of interest, and other items needed to reproduce or evaluate your work.
Openness	Share data, materials, and methods with other researchers.
Objectivity	Use methods, procedures, experimental designs, and analytical frameworks that help to reduce or control bias; critically evaluate one's own work.
Reproducibility	Use methods, procedures, analytical frameworks and experimental designs that help to ensure that data and results are reproducible.
Freedom of expression	Do not interfere with the publication of research and expression of ideas; encourage open debate.
Fair sharing of credit	Assign authorship on publications and other works fairly; cite people who deserve to be cited; give credit where it is due but not where it is not due.
Respect for colleagues, students, trainees, and staff.	Treat colleagues, students, trainees, and staff with respect; do not harass, bully people, or discriminate against people.
Protection of human subjects	Protect the rights and welfare of human research subjects.
Protection of animal subjects	Protect the welfare of animal research subjects.
Social responsibility	Anticipate the consequences of your research and try to maximize the good consequences of your research and minimize the bad.
Professional responsibility	Strive to improve your expertise and skill within your field; educate and train the next generation of researchers; report illegal or unethical conduct in research.
Stewardship of resources	Take care of financial and material research resources; do not waste or misuse resources.
Safety	Maintain an environment that protects researchers from physical, chemical, biological, and psychosocial risks.
Regulatory compliance	Comply with laws, regulations, and institutional policies

Technology Policy 2000:76262)." For an act to qualify as misconduct, it must be committed "intentionally, knowingly, or recklessly (Office of Science and Technology Policy 2000:76262)."[2] Table 9.2 includes some examples of misconduct in research with human subjects.

While fabrication, falsification, and plagiarism are widely accepted as the core misbehaviors included within definition of misconduct, many countries and institutions go beyond FFP and include behaviors such as significant violations of human

[2] DHHS (2024) recently revised its research misconduct policies. The revised policies clarify and streamline administrative responsibilities and define key terms, such as intentional, knowing, and reckless. The difference between reckless behavior and merely negligent behavior is sometimes unclear. See Caron et al. (2023).

Table 9.2 Some cases of misconduct in research with human subjects

Scientist	Misconduct (admitted, proven, or alleged)
Roger Poisson	Admitted to fabricating and falsifying patient health records to allow people to qualify for a study (see Chap. 2).
Andrew Wakefield	Accused of fabricating and falsifying patient data to support the hypothesis that the MMR vaccine causes autism (see Chap. 2).
Woo Suk Hwang	Found to have fabricated images of cells in a paper on human therapeutic cloning (see Chap. 2).
Eric Poehlman	Admitted to fabricating or falsifying data on 15 federal grant applications and 17 papers (see Chap. 2).
Anil Potti	Found to have fabricated and falsified data for statistical method for selecting chemotherapy (see discussion in this chapter, below).
Joachim Boldt	Found to have fabricated and falsified data for his research purporting to show that colloids can increase blood volume during surgery (see Shamoo and Resnik 2022). So far, 220 of his papers have been retracted (Retraction Watch 2024). Fortunately, no participants were directly harmed from Boldt's misconduct. A review of the evidence related to using colloids to boost blood volume found that they are not more effective than other methods (Wise 2013).

or animal research regulations, failure to disclose a significant conflict of interest (COI), interfering with a misconduct investigation, and unethical authorship of publication practices (Resnik, Rasmussen et al. 2015a, Resnik, Neal et al. 2015b). Some have argued that harassment and bullying should be included in the definition of research misconduct, but there has not been widespread support for this idea (Resnik 2019a). During the late 1980s and 1990s, some federal agencies included "other serious deviations" from commonly accepted research practice in the definition of misconduct, but they decided to focus on FFP because they regarded this broader category as too vague and difficult to enforce (Shamoo and Resnik 2022).

Not all behaviors considered to be unethical in research rise to the level of misconduct. Questionable research practices (QRPs) are behavior that are regarded as unethical or ethically problematic but are not as serious as misconduct (Shamoo and Resnik 2022; Martinson et al. 2005; Sacco et al. 2019). Some QRPs include:

- Poor recordkeeping
- Honorary authorship (including someone as an author on a paper even though they did not make a significant contribution to the research)
- Ghost authorship (not including someone as an author on a paper even though they made a significant contribution)
- P-hacking and other questionable or deceptive use of statistics
- Over-stating the significance of one's research
- Refusing to share data, methods, or materials
- Poor supervision of the research group
- Poor mentoring of students or trainees
- Not disclosing a conflict of interest
- Violating the confidentiality of peer review
- Failing to comply with research regulations or institutional policies

There also behaviors that are very seriously wrong and qualify as misconduct, but they are not unique to research, so they are not considered research misconduct. Some types of "other misconduct" include:

- Theft of property
- Financial fraud or mismanagement of funds
- Harassment, bullying, assault and other threatening behavior
- Discrimination on the basis of gender, race, ethnicity, religion, sexual orientation, or disability.

The incidence of misconduct in research is difficult to accurately estimate but is thought to be rare in industrialized nations, such as the US, Europe, and Japan.[3] A survey of 3247 NIH-funded scientists published in 2005 found that 0.3% (3/1000) admitted to falsifying or cooking data in the last 3 years (Martinson et al. 2005), and a survey of 2212 NIH-funded scientists published in 2008 found that 3% per year observed someone else committing misconduct (Titus et al. 2008). Fanelli (2009) conducted a systematic review of 21 misconduct surveys and found that 1.97% of scientists had admitted to fabricating or falsifying data or committing plagiarism at least once in their careers and that 14.2% said they had observed someone else committing misconduct. It is worth noting that surveys that ask scientists to self-report their own misbehavior are likely to underestimate the incidence of misconduct because people may not want to admit to doing something illegal or unethical, and surveys that ask respondents to report misconducted they have observed in others may overestimate the incidence of misconduct because respondents who claim to have observed misconduct may not have enough evidence to determine whether it has occurred and some may observe the same act of misconduct, which may result in double-counting. Perhaps the best way to estimate the true rate of misconduct is to audit research. A study by Bik et al. published in 2016 did just this. The researchers examined 20,621 images from scientific papers and found that 3.8% were problematic and that half of these (about 1.9%) had evidence of deceptive manipulation (Bik et al. 2016).

It is not known whether the rate of misconduct is increasing, decreasing, or staying the same, although several trends indicate that it may be increasing. First, the rate of retractions has risen about ten-fold since the beginning of the century and most retractions are due to misconduct (Fang et al. 2012; Steen et al. 2013). However, the overall rate of retractions is still low (about 0.04% of published articles or 4/10,000) and the increase in retractions may be due to enhanced vigilance by journals and publishers to protect the integrity of the research record (Fanelli 2009; Steen et al. 2013; Brainard 2018). Second, entrepreneurs and unscrupulous scientists have developed different strategies for profiting unethically from publication, including predatory journals, paper mills, citation cartels, fake peer review, and commercialized authorship (Rivera and Teixeira da Silva 2021; Kojaku et al. 2021;

[3]There is evidence that the rate of misconduct may be much higher in China, India, and Africa (Okonta and Rossouw 2012; Dhingra and Mishra 2014; Liao et al. 2018).

Shamoo and Resnik 2022; Else 2023; Sanderson 2024).[4] Third, artificial intelligence (AI) tools, such as chatbots and image generators are making it easier to create fake data, images and papers without doing any research or actual writing (Májovský et al. 2023; Elali and Rachid 2023; Kim et al. 2024).

Research misconduct can harm human subjects in numerous different ways (George and Buyse 2015). First, fraudulent data can expose subjects to risks beyond what the IRB has approved or undermine safety-related aspects of the study. For example, if an investigator falsifies data related to patient safety, this could lead the DSMB, IRB, and sponsor to believe that the study is safer than it is and not take steps to protect human subjects, such as modifying or stopping the study. If an investigator falsifies data collected at enrollment to allow a subject to qualify for a study this would also expose subjects to risks that the IRB has not approved (see Poisson case, Chap. 2), and if an investigator falsifies patient data related to kidney or liver function, this could interfere with safe drug dosing. Second, fraudulent data can undermine the integrity and reproducibility of the study and therefore negate the benefits the research. As discussed in Chap. 7, exposure to risks in research is justified in relation to expected benefits. If there are not expected to be any benefits, then the risk exposure is unjustified.

An episode at Duke University demonstrates how misconduct can directly harm research subjects. In November 2015, the Office of Research Integrity (ORI), which oversees the integrity of NIH-funded research, found that former Duke University investigator Anil Potti had falsified data in one NIH grant application and 9 published papers. As part of the settlement, ORI required Potti to retract the papers and agree to supervision of his federally funded research for a period of 5 years. Earlier that year, Duke University had settled a lawsuit brought by eight participants in Potti's oncology clinical trial who claimed they were harmed by his fraudulent research (Shamoo and Resnik 2022).

Potti had been working with Joseph Nevins on cancer genomics in the early 2000s. They hypothesized that one could use a tumor's genomic signature to predict its response to chemotherapy. In 2006, they published a paper in *Nature Medicine* describing a predictive model that they had developed using computer programs and publicly available genomic data for the tumors (Potti et al. 2006). Shortly after the paper appeared in print, two statisticians, Keith Baggerly and Kevin Coombes, could not replicate the results and found errors in the model. They contacted Potti and Nevins about issues they had with the paper and published a critique in the same

[4] A predatory journal is an open access journal that promises authors a quick review and quick publication (for a fee) but has minimal or no standards for peer review. A citation cartel is a network of researchers who agree to cite each other to increase their research impact metrics. A paper mill journal in the business of publishing papers without any regard whatsoever for peer review. Fake peer review occurs when (a) author-suggested reviewers are fake and the authors use the associate email addresses to write favorable reviews for their own papers or (b) the reviewers are real, but the authors and reviewers agree to give each other favorable reviews. Commercialized authorship is selling and by authorship on a paper.

journal (Coombes et al. 2007). Potti and Nevins responded to these critiques and insisted that their model worked.

In 2008, Bradford Perez, a third year medical student at Duke University, who was working under Potti and Nevins, found problems with their predictive model and discovered that it had not been independently validated. Perez was worried that patients could be harmed because Potti and Nevins were preparing to begin clinical trials based on their research. Perez shared his concerns with university administrators, who encouraged him to work with Potti and Nevins to deal with these issues. They expressed the view that the problems he found amounted to a scientific disagreement rather than misconduct and warned him that he could damage his own career if he brought charges against his supervisors (Shamoo and Resnik 2022).

In 2009, Baggerly and Combes learned that clinical trials had begun at Duke University. They published another critique of Potti's research and expressed their concerns to the Duke IRB. The IRB suspended the clinical trials to review the research but then allowed them to start again. In July 2010, *The Cancer Letter* reported that Potti had falsely stated he was a Rhodes Scholar on his curriculum vitae (CV) and grant applications submitted to the NIH and American Cancer Society. Shortly thereafter, the IRB suspended Potti's clinical trials again and Duke University launched a misconduct investigation, which found that Potti had fabricated and falsified data. ORI accepted Duke's report and concluded its own review of the matter in 2015 (Shamoo and Resnik 2022).

Misbehavior that does not directly harm human subjects may nevertheless negatively impact society's trust in research if it leads to fraudulent, erroneous, or biased data or results that cause harm to society or increase the risk of harm. For example, Wakefield's study claiming to link the MMR vaccine to autism (see Chap. 2) probably contributed to the decline in MMR vaccination rates in the U.K. and other countries, and therefore may have increased infections of measles or other childhood diseases for which there is a vaccine. Merck's exclusion of some cardiovascular risk data from the VIGOR study may have harmed people who took Vioxx because they did not know some of the risks of the drug. If these risks had been fully disclosed to the public, fewer people may have taken the drug and, therefore, fewer may have been harmed. (See fuller discussion of this case below.) Although Poisson's alteration of his patient's medical records fortunately did not skew the overall results of the clinical trial, it easily could have. Conflicts of interest (discussed below) are a significant concern in research because they may increase the risk of bias or misbehavior (Shamoo and Resnik 2022).

Misconduct investigations are legal proceedings in the U.S. and most other countries. ORI has developed policies and procedures that institutions are required to follow when they perform misconduct investigations involving NIH-funded research, but most institutions apply these to all research, regardless of the funding source (Shamoo and Resnik 2022). Research institutions are required to have a research integrity official (RIO) who receives formal allegations of misconduct. If the RIO determines that an allegation meets the definition of misconduct and is credible, they will appoint a committee to conduct an inquiry to determine whether there is enough evidence to support an investigation. If the inquiry committee concludes that an investigation is warranted, the RIO will appoint another committee to conduct an investigation. Both of these committees can review research records and other materials and interview witnesses. The investigation committee is tasked with deciding whether the evidence shows that the accused party (i.e. the respondent) has committed misconduct. A finding of misconduct must be supported by a preponderance of evidence, i.e. probability >50% that the person committed misconduct (Office of Science and Technology Policy 2000). The RIO will act on the investigation committee's findings and recommend a punishment if misconduct is found. Sanctions may include termination of the respondent's employment, demotion, increased supervision of the respondent's work during a probationary period, education in RCR, or a letter of reprimand.

If the research is federally funded, the institution will report its findings to the appropriate agency. The ORI reviews findings for NIH-funded research. The federal agency may accept the institution's findings, ask for additional evidence or clarification, or conduct its own investigation. If the agency reaches a finding of misconduct, it will usually require the investigator to forego federal research funding for a period of time (sometimes indefinitely) and to correct the publication record (if appropriate) (Shamoo and Resnik 2022). Although researchers rarely face criminal or civil penalties for misconduct, Eric Poehlman served a year and a day in a federal prison and was required to pay the government $180,000 (see Chap. 2).

Institutions are required to follow rules for legal due process when conducting misconduct inquiries and investigations. Respondents are permitted to obtain legal counsel, interview witnesses, and review the evidence. They can also appeal misconduct findings to a federal court. To protect the reputations of the parties, institutions are required to maintain confidentiality during misconduct proceedings. However, sometimes deliberations or findings are unlawfully leaked to the press. ORI publishes cases on its website once it has made official findings of misconduct (Shamoo and Resnik 2022).

Research misconduct inquires and investigations are stressful, time-consuming, and labor-intensive proceedings. Although federal regulations and institutional policies set deadlines for different decisions made in these proceedings, the institution may extend a deadline for valid reasons, such as the respondent needs more time. Appeals by the respondent can also add to the length of the process. For example, an allegation brough in 1986 against Thereza Imanishi-Kari, a researcher at the Whitehead Institute associated with the Massachusetts Institute of Technology and Tufts University, was not resolved until 1996, when a DHHS appeals panel found that the evidence presented against her was not sufficient to prove her guilt (Shamoo and Resnik 2022).

Scientific journals have adopted policies for dealing with misconduct allegations. However, their legal authority to investigate allegations of misconduct is limited. Editors will usually refer such matters to the authors' institutions for resolution. While an investigation is ongoing, the editors may publish an expression of concern, pending the outcome. If there is an official finding of misconduct, editors will usually retract the article, with or without the permission of the authors. Editors may face difficult ethical dilemmas when dealing with alleged misconduct. On the one hand, they have a duty to the journal's reputation and the integrity of the research record. On the other hand, they have a duty to not falsify accuse innocent people of wrongdoing. They may also be concerned about being sued for libel by authors. The Committee on Publication Ethics (2024) has developed some guidelines for editors and journals for handling misconduct cases.

When misconduct involves research with human subjects, the IRB, the misconduct committee, RIO, and other institutional officials will coordinate their actions and share information to protect human research participants. For example, the IRB may decide to stop the study or halt enrollment while the misconduct proceedings are taking place. The IRB can (and should) take action to protect human participants as soon as it is notified about the allegations. In Potti case, the Duke University IRB did temporarily stop the study when it learned about his fraudulent CV, but the IRB started it again when it determined that this issue did not affect the research.

9.3 Noncompliance

Not complying with human research regulations or policies is one the of QRPs mentioned above. The Common Rule requires that IRBs have written procedures for reporting of "unanticipated problems involving risks to subjects or others or any serious or continuing noncompliance (45 CFR 46.108a4)." Serious or continuing noncompliance could also qualify as research misconduct under federal policies if it involves FFP. As noted above, many academic institutions and some countries classify serious noncompliance with human research regulations as a form of misconduct.

The NIH's intramural human research protection program defines noncompliance as:

> Failure of an investigator to follow the applicable laws, regulations, or institutional policies governing the protection of human subjects in research, or the requirements or determinations of the Institutional Review Board (IRB), whether the failure is intentional or not (National Institutes of Health 2024a)

Continuing non-compliance is defined as:

> [A] pattern of recurring non-compliance that either has resulted, or, if continued, may result in harm to subjects or otherwise materially compromise the rights, welfare and/or safety of subjects, affect the scientific integrity of the study or validity of the results. The pattern may comprise repetition of the same non-compliant action(s), or different non-compliant events. Such non-compliance may be unintentional (e.g. due to lack of understanding, knowledge, or commitment), or intentional (e.g. due to deliberate choice to ignore or compromise the requirements of any applicable regulation, organizational policy, or determination of the IRB) (National Institutes of Health 2024a, b)

Serious non-compliance is defined as:

> Non-compliance, whether intentional or not, that results in harm or otherwise materially compromises the rights, welfare and/or safety of the subject. Non-compliance that materially affects the scientific integrity or validity of the research may be considered serious non-compliance, even if it does not result in direct harm to research subjects (National Institutes of Health 2024a)

As one can see from these definitions, there are different types of non-compliance, ranging from minor, unintended protocol deviations to serious and deliberate transgressions of the rules that may harm participants or threaten the integrity of the study data. See Table 9.2 for examples of non-compliance.

Anyone involved in research (i.e. investigators, staff, subjects) may report non-compliance. Investigators often report their own noncompliance; for example, when they discover through a self-audit or a report from a staff member. External audits (routine or for-cause) may also uncover noncompliance. The IRB receives reports of noncompliance and investigates them accordingly. Some types of non-serious, non-continuing, non-compliance, such as minor protocol deviations, may be collected together and reported at continuing review, depending on the IRB's policy.

Investigations of noncompliance in research with human subjects are administrative proceedings which are usually not nearly as legalistic as misconduct investigations. The IRB and leaders of the institute (such as the vice president for research or clinical director) must work together to investigate allegations on non-compliance and impose sanctions. If needed, the IRB may obtain assistance from an *ad hoc* committee, administrative staff, or special expert. If the noncompliance involves an allegation of research misconduct, then it will also be referred to the RIO for review, as discussed above. If the IRB makes a finding of noncompliance, it may recommend further action, such as terminating or suspending the study; amending the study; requiring additional education or training for the investigator or staff; requiring the investigator to develop policies and procedures to prevent noncompliance from occurring again; or requiring the investigator to inform subjects about the noncompliance if it impacts them. An investigator may ask an IRB to reconsider its findings or disciplinary actions (Table 9.3).

9.4 Conflict of Interest

Conflicts of interest (COIs) are important ethical concern in research with human subjects because they often lead to misconduct, QRPs, or non-compliance. While having a COI is not, itself, unethical, it can be a risk factor for unethical behavior because it can bias one's judgment and decision-making. **A COI, for a person, can be defined as a situation or relationship that is likely to bias the person's judgment or decision-making and increase the risk that they will violate their ethical, legal or professional obligations** (Kafaee et al. 2022; Shamoo and Resnik 2022). COIs can be financial, personal, professional, or political (see Table 9.4). Because COIs can bias judgment and decision-making, they can undermine the integrity of research and the public's trust in science (Thompson 1993; DeAngelis

Table 9.3 Some examples of non-compliance

Not serious or continuing[a]	Serious or continuing
Collecting data at the wrong time because the subject was late for their appointment	Enrolling subjects in a study without IRB approval or after IRB approval has expired
Using an outdated copy of the consent form when are were no significant differences between the old and new versions	Multiple instances of using an outdated copy of the consent form
Not collecting data at a required time point because the subject did not show up for their appointment	Not reporting protocol deviations, non-compliance, unanticipated problems, or serious adverse events in a timely fashion, as required by institutional policy
Drawing 50 ml of blood when the protocol only calls for 15 ml (assuming no impact on participant safety)	Violating consent requirements, such as: Performing procedures without consent; not following correct procedures for obtaining consent from an LAR or obtaining consent from a person limited English proficiency; or posting advertisements that have not been approved by the IRB
Not performing a required lab test but failure to do so has no impact on the subject's safety	Giving the wrong medication or wrong dose to a participant
Performing procedures out of correct sequence with no impact on subject's safety	Violating the participant's confidentiality
The subject misses a follow-up visit or does the visit 2 days late	Fabricating or falsifying data
Accidental destruction or loss of research records or biospecimens with no impact on participant safety or data integrity	Performance of a procedure by an unqualified researcher
A problem with confidentiality that does not result in disclosure of private information; for example, a letter asking for update contact information is delivered to the wrong person	Enrolling a participant who does not meet the study's inclusion criteria

[a]Many of the examples in this table are minor protocol deviations, i.e., deviations from the approved IRB protocol that do not affect the rights or welfare of the subject or compromise the integrity of the data.

2000; Shamoo and Resnik 2022). Several cases involving misconduct or questionable behavior we have discussed in this book have involved COIs that have negatively impacted the investigator's behavior and the public's trust in science (see Table 9.5).

Institutions may also have COIs if the organization or its leaders have financial or other interests which are likely to impact institutional decision-making (Shamoo and Resnik 2022). For example, if a private company is giving a major gift to a university or sponsoring millions of dollars of research on campus, institutional leaders may make decisions that favor the company and compromise the university's academic mission or scientific integrity. Commentators on the Gelsinger case

Table 9.4 Some types of COIs

Type of interest	Examples
Financial	Ownership of stock in a company that supports one's research; receiving money for lecturing, consulting, expert witnessing related to one's research; owning intellectual property, such as patents or receipt of royalties; receipt of funding to attend conferences
Personal	Reviewing a grant proposal submitted by a collaborator or former advisor or student
Professional	Reviewing a paper submitted by a directly competing lab or research group; having one's professional career closely aligned with a particular hypothesis (or theory) such that one's future employment, income, and prestige hinges on the hypothesis being accepted by the scientific community
Political	Involvement, such as board membership or consulting arrangements, with groups engaged in political activity, such as lobbying, campaigning, and petitioning

Table 9.5 Some cases of conflicts of interest in clinical research

Researcher(s)	Interest(s)	Misbehavior (suspected or confirmed)
James Wilson	Significant financial connections to Genovo, a company he founded; ownership of stock in Genovo; patents on gene therapy methods	Not fully informing Jesse Gelsinger of risks of a phase I gene therapy trial and his financial interests (see Chap. 2)
Andrew Wakefield	Funding from a law firm that was suing vaccine manufacturers; relationship with the anti-vaccine community	Fabricating and falsifying data in study linking the MMR vaccine to autism (see Chap. 2)
Anil Potti	Patents on methods for selecting chemotherapy based on the genomic signatures of tumors	Fabricating and falsifying data (see discussion in this chapter)
David Golde	Patents on a cell line developed from John Moore's removed spleen	Not telling John Moore about these patents and other financial interests as part informed consent for medical treatment (see Chap. 2)
VIGOR investigators	Funding from Merck; ownership of stock in Merck; consulting agreements with Merck	Suppressing data on the cardiovascular risks of Vioxx, a drug manufactured by Merck (see discussion in this chapter)
Albert Kligman	Ownership of stock, patents	Not disclosing funding from Johnson and Johnson and colluding with the company to deprive his university of profits (see discussion in this chapter)
Scheffer Tseng	Ownership of stock in Spectra	Not disclosing a significant conflict of interest (see discussion in this chapter)
Trey Sunderland	Consulting fees	Sunderland earned $285,000 in consulting fees from Pfizer without getting approval from NIH supervisors and shared human biospecimens with the company without IRB approval. In 2005, he was indicted for violating federal COI disclosure laws (Lenzer 2006)

(discussed in Chap. 2) were concerned that the University of Pennsylvania had financial interests related to the research which could have impacted its oversight of the gene therapy experiment.

A COI is different from a conflict of obligation or duty. For example, a physician who is enrolling patients in a placebo-controlled clinical trial in which an effective therapy already exists has a conflict between his duties as a physician and his duties as a researcher, but not a COI, if he has no significant interests at stake (see discussion of this conflict in Chap. 7). A conflict of obligation could become a COI if the investigator's interests favor one obligation over another.

A COI is also different from a conflict of commitment, which occurs when investigators have different professional commitments competing for their time and effort. For example, a university professor who spends a significant amount of time consulting for industry might not have enough time to meet her teaching, advising, or research obligations. Conflicts of commitment can usually be handled by proper time management. Most universities have policies concerning the amount of time and effort that one may devote to outside activities, such as private consulting (Shamoo and Resnik 2022).

COIs, especially financial ones, have increased in research since the 1980s, due to the growing commercialization of science and the complex web of interactions between academic institutions and private industry (Krimsky 2003; Resnik 2007; Michaels 2008). Two key pieces of legislation passed in the early 1980s, the Bayh-Dole Act and the Technology Transfer Act, allowed scientists to patent inventions developed using federal funds and to transfer those inventions to universities. Congress passed these laws to encourage the transfer of technology from the public to the private sector and to spur economic development.

In the 1980s, pharmaceutical, biotechnology, and information technology companies began investing more money in research and development (R &D) in their own laboratories and on academic campuses. At the same time, government support of non-military scientific research stagnated, due to budgetary constraints related to economic downturns and increased spending on defense and entitlement programs. R & D funding in the U.S. shifted from about 50% industry and 50% non-industry (government, university, and private philanthropy) in the 1970s to about 65% industry by the beginning of the twenty-first century (Shamoo and Resnik 2022).

In response to dwindling government support, universities have been seeking R & D contracts with private companies and accepting large corporate gifts, which they have used to develop their research infrastructure. Today, many universities have buildings, laboratories, or endowed professorships named after industry donors and some have corporate campuses (Krimsky 2003). Universities also now generate income from intellectual property developed by faculty. Universities have established technology transfer offices to help scientists apply for patents and to manage the institution's intellectual property, including copyrights. A university scientist who patents an invention is normally required to transfer the patent to the institution in exchange for a share of the royalties (usually around 50%). Universities also provide faculty with legal and financial support needed to form start-up companies to commercialize products and services (Shamoo and Resnik 2022). Faculty members began to rely more on private funding to support their research and they accepted money from industry for consulting, writing, or lecturing. Faculty members and universities also acquired stock or equity in private companies that sponsor R &D. Some universities formed private foundations to manage these investments and dole out venture capital for faculty companies (Krimsky 2003).

Financial relationships between clinical researchers and drug companies began to attract the attention of the media, the public, and politicians in the mid-1980s (Shamoo and Resnik 2022). For example, in 1985, Tseng published an article showing that an ointment containing Vitamin A is safe and effective at treating dry-eye, but he did not disclose that the FDA had granted him exclusive rights to market the product or that he held 530,000 shares in Spectra, a drug company which was planning to purchase those rights for $310,000. The price of Spectra's stock soared after the paper appeared in print and Tseng soon sold his shares. Subsequent research refuted Tseng's findings (Resnik 2007). In 1988, the *Journal of the American Medical Association* published an article and an editorial claiming that Retin-A, a product made by Johnson and Johnson, could reverse the effects of aging on the skin. After the article was published, sales of Retin-A increased 500%. Later, a reporter discovered that University of Pennsylvania researcher Albert Kligman, the author of the editorial and one of the authors of the research article, had not disclosed that he had received financial compensations from Johnson and Johnson. It was subsequently determined that the study had exaggerated the safety of Retin-A and was not well-controlled for bias because of the absence of a double-blind design. In 1990, the University of Pennsylvania sued Kligman for colluding with Johnson and Johnson to deprive the university of profits from Retin-A and violating its patent disclosure policy (Marshall 1990, 1992).

9.4.1 Evidence Concerning the Impact of Conflicts of Interest

There is a growing body of evidence on the association between financial interests, industry funding, and research outcomes (Hampson et al. 2008). Numerous studies have demonstrated the existence of a funding effect in science: studies sponsored by the manufacturer of a product are more likely to favor the product than studies sponsored by an independent organization, such as a government agency (see Bekelman et al. 2003; Sismondo 2008a; Lundh et al. 2018 for a review of the literature). See Table 9.6.

There are several plausible explanations for the association between financial interests (or other) and research outcomes. First, funding decisions made by research sponsors can have a direct impact on outcomes (Resnik 2007). For example, drug companies are required to conduct preliminary studies pertaining to a drug's chemistry, pharmacology, and toxicity before proceeding to clinical trials. Thus, when a company decides to fund a clinical trial, it has already screened hundreds or even thousands of compounds and has some data suggesting the drug in intends to test

Table 9.6 Some studies on the association between financial interests and research outcomes

Stelfox et al. (1998) examined articles published between March 1995 and September 1996 on the controversy over using calcium channel agonists to treat hypertension and found that 96% of authors who supported calcium channel antagonists had relationships to manufacturers compared to 37% who were critical and had no relationships
Friedberg et al. (1999) examined 44 articles published between 1988 and 1998 on the cost-effectiveness of drugs used in oncology and found that 95% of articles with industry funding reported positive results for cancer treatments as opposed to 62% without industry funding
Als-Nielsen et al. (2003) analyzed a random sample of Cochrane review articles and found that the experimental drug was recommended as the treatment of choice in 16% of clinical trials funded by nonprofit organizations, 30% of trials not reporting funding, 35% of trials funded by both nonprofit and for-profit organizations, and 51% of trials funded by for-profit organizations
Friedman and Richter (2004) reviewed 398 original articles published in the New England Journal of Medicine and the Journal of the American Medical Association in 2001 and found a strong association between one of more authors having a financial COI (such as stock or fees for consulting or speaking) and positive findings, e.g. evidence that a treatment is safe and effective. Articles that disclosed a financial COI were 2.3 times more likely to report positive findings than those that did not disclose a financial COI
Lesser et al. (2007) reviewed 2006 nutrition research articles published between January 1, 1999 and December 31, 2003 and found that studies funded by industry were significantly less likely to report unfavorable findings related to a product in question as compared to those without industry funded (0–37%)
Friedman and Friedman (2016) found a strong association between financial COIs and research outcomes in 373 original articles published in occupational and environmental health journals in 2012. Articles disclosing financial COIs were 4.31 times more likely to report that the authors had found no evidence that a chemical or agent is dangerous

may be safe and effective. Additionally, companies may decide to withdraw funding from studies that are not yielding favorable results.

Second, companies may decide not to publish data or results that are contrary to their financial interests. Although data suppression is widely regarded as unethical, companies are not legally required to publish their research and can treat data and results as trade secrets (Shamoo and Resnik 2022). For example, if a company sponsors a clinical trial comparing its drug to competing drugs and the trial does not show that its drug is better than the competitors, it could decide to suppress the results of this study. It would still need to report the data to the FDA, but the FDA would treat the data as confidential.

A salient example of this phenomenon occurred in 1995, when Boots Pharmaceuticals forced University of San Francisco pharmacologist Betty Dong to withdraw a paper from publication in the *Journal of the American Medical Association*. Boots had sponsored Dong's study, which compared its hypothyroidism drug Synthroid (levothyroxine) to generic alternatives. The study found that Synthroid was not superior to the generic drugs and that the U.S. government could save millions of dollars per year by switching Medicaid and Medicare patients with hypothyroidism from Synthroid to a cheaper generic drug. Dong had signed a contract with Boots giving the company the right to prevent her from publishing the research without the company's written approval. Dong withdrew her paper after the company threatened to sue her for violating the agreement. Fortunately, the company eventually decided to allow Dong to publish the study in the New England Journal of Medicine, but only after it had published an article favorable to Synthroid (Resnik 2007).

Sometimes a company's decision to suppress data can place human subjects at risk. For example, Nancy Olivieri and her collaborators at the University of Toronto (UT) and Toronto General Hospital (TGH) published a paper in the *New England Journal of Medicine* in 1995 which found that deferiprone is effective at treating the thalassemia and has few side effects. The drug's manufacturer, Apotex, sponsored their research. Olivieri soon noticed that many of her patients had increasing iron levels that placed them at risk of heart failure or death. In May 1996, she decided to inform TGH's research ethics board (REB) about this problem, but the company told her not to do this. After discussing the issue with representatives from Apotex, Olivieri decided to submit a report to REB. Within 3 days, the Apotex terminated the research and withdrew all supplies of the drug from the hospital's pharmacy. The company also threatened to sue Olivieri if she told her patients, regulatory agencies, or researchers about her concerns with the drug. After receiving additional threatening letters from the company, Olivieri withdrew a presentation on the drug she had submitted to a scientific meeting. In 1998, Apotex, which was in the middle of negotiation large donations to UT and TGH, pressured these institutions to take action against Olivieri, which they did. The institutions tried to discredit Olivieri and her research and took steps to dismiss her from her position. Fortunately, some of Olivieri's colleagues intervened on her behalf prevented her from being dismissed (Olivieri 2003).

Data suppression can also have adverse effects on public health, as illustrated by Merck suppression of data concerning its drug Vioxx (rofecoxib). In 1999, the FDA approved Vioxx for the treatment of arthritis and pain. Merck claimed that one of the benefits of the drug is that it produces fewer adverse gastrointestinal (GI) effects than other non-steroidal, anti-inflammatory drugs commonly used to treat arthritis and pain. Vioxx generated an estimated $2.5 billion in annual sales. In 2001, researchers published the results of a Merck-sponsored clinical trial, named VIGOR, in the *New England Journal of Medicine*. VIGOR showed that Vioxx produces fewer GI side effects than other drugs although it increases the risk of cardiovascular disease. The investigators, most of whom had financial ties to Merck, claimed that the difference in cardiovascular disease risks was due to the protective effects of the other drugs. However, the VIGOR study did not include all of the cardiovascular risk data collected by the researchers. If it had, it would have shown that Voixx patients had five times the risk of having a heart attack or stroke as compared to patients taking other medications. In 2002, the FDA notified Merck that it had misrepresented Vioxx's safety profile and required the company to include a black-box warning on the drug label. In 2002, another study of Vioxx was stopped early to protect patients receiving the drug from cardiovascular risks. In 2004, Merck withdrew Vioxx from the market due to safety and liability concerns. Thousands of patients who had a heart attack or stroke while taking Vioxx have sued the company (Shamoo and Resnik 2022).

The same year that Merck withdrew Vioxx from the market, a systematic review found that several selective serotonin reuptake inhibitors (SSRIs), including Paxil (paroxetine), Zoloft (sertraline), and Prozac (fluoxetine) increase the risks of suicide when prescribed to children and adolescents. Previous published studies had not shown that these drugs increase the risk of suicide in children or adolescents. However, the systematic review included data that had been reported to the Committee on Safety in Medicines (a U.K. agency similar to the FDA) but had not been previously published. Parents and psychiatrists were understandably furious when they found out that companies had suppressed this data, and New York Attorney General Eliot Spitzer sued the manufacturer of Paxil, Glaxo Smith Klein, for fraud (Shamoo and Resnik 2022).

The issue of whether data suppression constitutes research misconduct or fraud is complex. Data suppression is ethically problematic because it is highly deceptive and could be construed as a type of data falsification because it may involve excluding data for an inappropriate reason. In some cases, it may be appropriate to exclude data outliers from the analysis if they are due to experimental or human error or are more than two standard deviations from the mean (Shamoo and Resnik 2022). However, excluding data without providing a sound scientific justification, as appears to have occurred in the VIGOR study, may constitute data falsification. Not publishing all the data from different studies is different from excluding data from a particular study because the data are from different datasets and do not necessarily address the same hypothesis or research question. Moreover, there may be some good scientific reasons, such as poor experimental design or inadequate statistical power, for not publishing data from a particular study. Selective publication of studies may constitute a pattern of deceptive behavior even if it is not data falsification. To adhere to principles of honesty and transparency in research, investigators who exclude data should explain why certain data points were not included (Shamoo and Resnik 2022).

A third explanation of the association between financial (or other) interests and research outcomes is that researchers or sponsors may sometimes deliberately bias results to promote their financial interests. For example, Poehlman (see discussion Chap. 2) admitted that financial pressures influenced his behavior. Potti (see earlier discussion in this chapter) was probably also impacted by the money and the prestige he expected to gain from his research. Most of the evidence concerning research misconduct in the U.S. has come from surveys of government-funded researchers or official findings of misconduct published by the NIH or NSF, and we know very little about misconduct cases, patterns, or trends in privately funded research (Shamoo and Resnik 2022). However, it is likely that the incidence of misconduct in privately-funded research is not very different from the incidence in publicly-funded research.

A fourth explanation of the association between financial (and other) interests and research outcomes is that interests may exert subconscious influences on researchers' judgment (Kafaee et al. 2022; Resnik 2023). Scientific research involves many different decisions that require the exercise of judgment, such as choosing the study design, sample size, interventions, procedures, measurements, and statistical methods. For many of these choices, there may be more than one option that researchers would regard as legitimate. Scientists who have financial interests related to a study may be unaware that they are making choices that tend to bias the study in a particular direction (Resnik 2007). Studies have shown that financial interests can exert powerful, subconscious influences over human behavior. Even small gifts, such as pens and free lunches, can influence our decision-making at subconscious level (Katz et al. 2003).

9.4.2 Dealing with Conflicts of Interest

There are several strategies for dealing with conflicts of interest in research. The most basic strategy is to disclose COIs to interested parties, such as institutions, IRBs, research participants, journals, granting agencies, etc. (Institute of Medicine 2009; Shamoo and Resnik 2022). Disclosure embodies the virtues of honesty, openness, and transparency and allows interested parties to decide how to respond to the COI. Readers of journals articles may use COI disclosures to assess the credibility of the research (Resnik and Elliott 2013). A scientist who learns that the authors of a paper have significant financial relationships to a company that sponsored the research may decide to give the paper additional scrutiny to ascertain the validity of the data, methods, and approach. Human research subjects may consider COI disclosures relevant to their decision to enroll in a study (Resnik 2004). Jesse Gelsinger might not have participated in the study that led to his death if he had known the extent of the investigator's and university's financial COIs. Studies have found that most research participants or potential participants would like to receive information concerning financial interests and sources of funding related to the research (Kim et al. 2004; Grady et al. 2006, Coors et al. 2015).

Disclosure can also promote trust by assuring interested parties that significant interests are out in the open. Failing to disclose COIs can destroy trust, since parties that discover undisclosed interests may feel that the person with the COI has been trying to deceive them (Shamoo and Resnik 2022). While some have argued that disclosing COIs to human subjects may make them suspicious of investigators' motivations and undermine their trust in research (Thompson 1993), studies indicate that disclosure can have the opposite effect if done properly (Levinson et al. 2005; Coors et al. 2015). COIs disclosures should be expressed in non-technical language that is understandable to the subjects (Resnik 2004).

Granting agencies, such as the NIH and NSF, require researchers to disclose significant financial interests to the agency and their institution. The NIH, for example, requires funded extramural investigators or those seeking funding to disclose financial interests greater than \$5000 to the agency and institution (National Institutes of Health 2024b). The NIH and NSF both require institutions to establish committees to receive and manage COI disclosures (Shamoo and Resnik 2022). As noted above, the NIH also requires intramural researchers to disclose COIs and has oversight mechanisms for managing them. Most scientific journals also require authors to disclose financial or other interests related to submitted manuscripts (Cooper et al. 2006; Resnik et al. 2017). IRBs also require investigators to disclose COIs to the committee and to human subjects (Shamoo and Resnik 2022).

While disclosure is an important strategy for dealing for dealing with COIs, it may not be sufficient to promote integrity and trustworthiness in many cases because it does not deter unethical behavior and may not assure the public that enough has been done to address the COI (Elliott 2008). Strategies that involve more than disclosure include avoiding the conflict and managing it (Shamoo and Resnik 2022). Avoidance involves taking steps to prevent the conflict from occurring, such as prohibiting the person from engaging in the activity or relationship that creates the conflict or restructuring the situation that gives rise to the conflict. For example, the NIH and NSF prohibit individuals from reviewing research proposals from current or former (within the last 3 years) students, collaborators, or advisors (Shamoo and Resnik 2022). As mentioned in Chap. 3, the federal regulations forbid IRB members reviewing studies in which they have COIs. IRBs may allow a member with a COI to share information about the study with the committee but prohibit them from taking part in discussion or voting (45 CFR.107d). Journal editors may ask peer reviewers not to evaluate manuscripts if they have COIs (Shamoo and Resnik 2022).

COIs can be a key concern when institutions are handling investigations of misconduct, non-compliance, or other ethical or legal problems. To ensure that investigations and inquiries are conducted impartially, people involved in them should not have COIs, such as personal or professional relationships with the person(s) under investigation. However, COIs can be difficult to avoid, especially when the people involved may know each other personally and professionally. Because institutions may have a great deal at stake in these cases, such as reputations or funding, these cases may involve institutional COIs. Institutional officials may be hesitant to bring charges of misconduct against well-established, prestigious investigators because they do not want to lose grant dollars or face public backlash (Shamoo and Resnik 2022).

Managing the conflict of interest may be the best option when avoidance is not feasible or it likely to have adverse consequences for science or society. A general strategy that institutions can use to manage financial COIs is to carefully review contracts with private research sponsors to ensure that they do not include any clauses that could undermine scientific integrity (Resnik 2007). For example, contracts should not unreasonably restrict the researcher's rights to publish or share their data, findings, materials, or methods. Sponsors should be able to review data and findings prior to publication, but they should not be able to suppress data or findings. Contracts should also not restrict researchers' rights to develop or modify research and should include provisions for fair sharing of intellectual property rights among the researchers, institutions, and the sponsor. Another important strategy for managing COIs is to share information among different offices on campus, such as the COI committee,[5] the IRB. Sharing of information is important for making appropriate decisions regarding the conflict, such as deciding whether is human subjects should be informed about it or deciding whether a human subject protocol should be approved.

A strategy that may be useful in some cases is to establish a committee at the institution to help manage the conflict. For example, suppose that an investigator has patented a tool to be used in cardiac catheterizations and she is planning to conduct a clinical trial of the device at an academic medical center. The investigator has also formed a startup company to develop, test, manufacture, and market the device. The investigator also owns stock in the company. The investigator transferred the patent to the institution but will receive a share of royalties from it. The startup company will license the device from the institution. The institution also owns stock in the company. In this situation, there are numerous individual and institutional COIs that could affect the integrity and trustworthiness of the research. The COI committee reviewing this situation could mandate that the investigator and institution avoid the conflict by requiring the investigator to conduct the clinical trial at another institution or by requiring the investigator and institution to sell their stock in the company. However, this proposal could undermine product development and research. The investigator might be the best person to conduct and manage the study, and moving the study to another institution might compromise the quality of the research. If the investigator and institution sell their stock in the company, the company may suffer significant losses and go bankrupt, because it needs the capital investment. The best way to handle this situation might be to allow the study to take place on campus but appoint a committee to provide additional oversight of the research to ensure that human subjects are protected, that the research is well-designed and properly implemented, and that data and results are published in a timely fashion (Shamoo and Resnik 2022).

Institutional COIs can be difficult to manage because there may not be individuals or groups within the organization with the authority or expertise to make fair and unbiased decisions concerning the institution's financial other interests (Resnik 2019b). The Board of Trustees or similar body is responsible for ensuring that the institution achieves its academic, social, and economic goals and complies with ethical and professional standards. The Board could establish a committee to review and oversee COIs pertaining to the institution and its leaders. Other steps that institutions can take to manage their own COIs include separating research oversight from financial decision-making. For example, IRBs and compliance officers should be able to make decisions which are not influenced by the interests of the institution or its leaders. Institutions can also create foundations to hold institutional investments and intellectual property rights (Resnik 2019b).

[5] A COI committee is a committee that reviews COI disclosures and decides how COIs should be managed.

To decide whether avoidance or management is the best way to handle a COI, one should consider the following factors (Institute of Medicine 2009; Shamoo and Resnik 2022):

- **The strength of the conflict**: how likely is it that the conflict will lead to bias, noncompliance, or other ethical or legal problems in research? How much money is at stake? What types of relationships, financial or otherwise, are involved?
- **The difficulty of managing the conflict**: does the institution have the resources to manage the conflict? Can the institution create special committees or other mechanisms to manage the conflict?
- **The consequences of avoiding the conflict**: what are the likely consequences of avoiding the conflict for human subjects, the investigator, institution, science, and society? Will avoiding the conflict prevent important research or product development from taking place? Will not avoiding the conflict undermine public trust?

Looking beyond how institutions or organizations deal with COIs, there are larger questions about how society should deal with them. Some scholars have proposed radical ways of addressing relationships between private companies and academic institutions. James Robert Brown (2000, 2002) has argued that the best way to minimize bias, corruption, secrecy, and other ethical problems related to financial interests in research is to sever the relationship between academic institutions and private corporations: Academic research should be funded by government grants, not private money. Brown also argues that biomedical patenting should cease so that investigators will share data more freely and drugs, biologics, other medical products and technologies will be accessible to public at low cost. Brown's solution to the problem of financial interests is to socialize academic research.[6]

Brown's proposal is unrealistic and counterproductive. First, private industry supports about 65% of R & D and governments are not likely to have the funds pay for research at current levels if one follows Brown's proposal (Shamoo and Resnik 2022). Brown's proposal would radically reduce the amount of research conducted on academic campuses and increased the amount conducted in private laboratories or corporate facilities. Since companies protect research under trade secrecy laws, shifting research from academic campuses to private venues is not likely to increase openness, data sharing, or publication in science. Indeed, it would probably have the opposite effect. Moreover, purging private research funding from academic institutions is not likely to eliminate conflicts of interest, bias, or other problems, since ethically vexing situations can still arise when scientists compete for government funds (Resnik 2007).

Second, foregoing biomedical patenting would have many adverse consequences for the development of drugs, biologics, medical devices, and other biotechnological inventions. Since the 1700s, the overarching rationale for the patent law has been to stimulate science and the practical arts (Resnik 2003; Miller and Davis 2011).

---

[6] Radder (2019) defends a socialistic view that it less radical than Brown's.

Prior to the advent of the patent system, scientists and companies relied on secrecy to protect their intellectual property. Governments started granting patents to encourage inventors to share their trade secrets with the public. In exchange for this disclosure, government allowed inventors to monopolize their work for a limited time. Under U.S. law, a patent grants the patent owner the right to exclude others from using, making, or commercializing his or her invention for 20 years from the date of the application (Miller and Davis 2011). The patent holder must submit an application to the patent office describing the invention and how to make and use it. Once the patent is awarded, the application becomes public (Miller and Davis 2011). Patenting provides economic incentives not only for inventors but also for investors. Private companies invest in R & D in order to profit from patented products or methods. Indeed, pharmaceutical companies derive most of the profits from new drugs when they are still under patent. While patented drugs are more expensive than drugs not protected by patents, once a patent expires prices drop precipitously, since other companies can manufacture and sell the drug, which introduces competition to the market (Resnik 2003). Although patenting encourages secrecy while scientists are working on patentable research, publication and data sharing can take place when the patent is awarded. Overall, the benefits of patenting for biomedical science, biomedicine, and society outweigh the harms (Resnik 2003). Third, academic researchers and institutions benefit from collaborations with industry. In addition to receiving money to support research, academic scientists and institutions also benefit from exchanging information, knowledge, and ideas with their colleagues in industry (Resnik 2007; Shamoo and Resnik 2022). They have the opportunity to work with companies who are developing cutting-edge products, methods, and technologies. Students can benefit from learning about real-world problems, internships with private companies, and employment opportunities in the private sector after graduation. Society can also benefit from collaborations between academic and industry that lead to the development of useful products, services, and technologies.

Others have made more modest proposals for dealing with relationships between academic institutions and private companies (Moses 2001; Morin et al. 2002; Krimsky 2003; Resnik 2007; DeAngelis and Fontanarosa 2008; Angell 2008; Radder 2019). Some of these include:

- Increasing government sponsorship of clinical research to provide a source of data independent from industry. For example, the government could sponsor more studies designed to compare different treatments available on the market so physicians and patients will have information concerning the comparative benefits, risks, and costs of these treatments.
- Properly enforcing and interpreting patent laws to restrict the scope of patents so that inventors (and the companies or institutions that own their patents) cannot control more intellectual property than is warranted. For example, a patent on a gene therapy method should apply only to that method and not broadly to all genetic medicine.

- Requiring registration of clinical trials in a public database, such as ClinicalTrials.gov, prior to initiation. Registration involves providing some basic information about the trial, such as the aims and objectives, procedures, interventions, target population, institutions, sponsors, and outcomes (when the trial is finished). The FDA and most journals require registration of clinical trials (Department of Health and Human Services 2016; International Committee of Medical Journal Editors 2024).
- Strengthening authorship criteria used by journals to clearly prohibit ghost authorship of scientific papers and prevent private companies from hiding their connections to the research project (DeAngelis and Fontanarosa 2008, see discussion Chap. 2).
- Enforcing COI rules. Researchers that fail to disclose COIs to the appropriate parties should be punished appropriately.

9.5 Conclusion: Promoting Scientific Integrity in Research with Human Subjects

In this chapter I have examined several scientific integrity issues which have a direct bearing on the protection of human research subjects. Because investigators are not perfect, misbehavior, bias, and honest mistakes are likely to occur in research involving human subjects. While it is important to report and investigate ethics problems, prevention is even more important. Prevention can head-off problems before they occur and avoid the damage to human subjects, investigators, institutions, and the scientific enterprise that often results from significant violations of ethical or legal standards in research. Forms of prevention include policies, processes, actions, and commitments that help to foster a culture of integrity within the research institution (National Academies of Science, Engineering, and Medicine 2017; Shamoo and Resnik 2022):

- **Leadership**. Institutional leaders should affirm their commitment to ethics through their words and deeds and stress the importance of ethics throughout the organization, from top to bottom. They should stress the importance of following ethical and legal standards for the conduct of research in their communications with investigators, students, trainees, staff, and administrators. Leaders who set a bad example through their own questionable behavior or indifference to research integrity may encourage corruption, misconduct, non-compliance, and other ethical and legal problems.
- **Policy development**: Institutions should develop policies that address scientific integrity issues. Policies should define expected standards of behavior and establish procedures for reporting, investigating, and adjudicating misconduct, non-compliance, and other significant problems in research. Institutions should inform investigators, student, trainees, and research staff about their policies and keep them abreast of changes. They should also publicize their policies on websites and other venues.

- **Education**. Institutions should educate investigators, research staff, students, trainees, and administrators about scientific integrity topics, such as misconduct, compliance, data management, conflicts of interest, etc. Education and training should take place when an individual takes a position at an institution or matriculates, as well as at regular intervals subsequently (e.g., period updates). Researchers who work with human subjects should have special instruction in the relevant regulations, policies, and guidelines.
- **Mentoring**. Investigators should mentor their students and trainees in scientific ethics. Mentoring involves teaching by example and through informal discussions. For example, a clinical investigator could mentor a medical student by showing her how to keep research records or explaining to her how the IRB submission process works. Junior faculty members can also benefit from mentoring from senior faculty.
- **Auditing**. Institutions should regularly audit research records, including data, case reports, and informed consent documents. Auditing can discourage misconduct and noncompliance and catch problems at the early stages, before they become worse.
- **Administrative support**. Institutions should provide adequate administrative support for research ethics and compliance activities, including staff members who work with the IRB as well as officials who can consult with researchers, students, and trainees on ethical or legal issues. Many institutions have an upper-level administrator whose job is to oversee ethics and compliance activities.
- **Accreditation**. The Association for Accreditation of Human Research Protections Programs (AAHRPP) accredits institutions which conduct, support, or review research with human subjects. AAHRPP has established standards pertaining to research oversight, policy development, education, administrative support that applicants must meet to receive accreditation. AAHRPP has accredited hundreds of organizations since 2001 (Association for Accreditation of Human Research Protections Programs 2024). Accreditation can promote compliance with ethics policies and guidelines and signal the institution's commitment to integrity in research with human subjects.
- **Protection for whistleblowers**. Institutional policies should protect people who report violations of ethical or legal rules from retaliation, harassment, or intimidation and should encourage people to report misbehavior. Policies should also provide funding and other forms of support for students or trainees who report their supervisors, in case they lose their funding and need to find a new supervisor or need to transfer to another laboratory or academic institution to continue their research. To encourage whistleblowers to report allegations of misconduct or noncompliance, many institutions permit anonymous reports. However, it may not be possible to guarantee anonymity if the whistleblower is asked to testify as a witness.

This concludes my discussion of integrity in research with human subjects. In the next, and final, chapter of this book I will summarize my conclusions and arguments and consider some proposals for reforming the oversight system.

References

All European Academies. 2023. *The European code of conduct for research integrity, revised edition 2023*. Available at https://allea.org/code-of-conduct/. Accessed 3 Sept 2024.

Als-Nielsen, B., W. Chen, C. Gluud, and L.L. Kjaergard. 2003. Association of funding and conclusions in randomized drug trials: A reflection of treatment effect or adverse events? *Journal of the American Medical Association* 290 (7): 921–928.

Angell, M. 2008. Industry-sponsored clinical research: A broken system. *Journal of the American Medical Association* 300 (9): 1069–1071.

Association for the Accreditation of Human Research Protection Programs. 2024. Available at https://www.aahrpp.org/. Accessed 12 Sept 2024.

Bekelman, J.E., Y. Li, and C.P. Gross. 2003. Scope and impact of financial conflicts of interest in biomedical research: A systematic review. *Journal of the American Medical Association* 289 (4): 454–465.

Bik, E.M., A. Casadevall, and F.C. Fang. 2016. The prevalence of inappropriate image duplication in biomedical research publications. *MBio* 7 (3): e00809–e00816.

Brainard, J. 2018. Rethinking retractions. *Science* 362 (6413): 390–393.

Brown, J.R. 2000. Essays on science and society: Privatizing the university—The new tragedy of the commons. *Science* 290 (5497): 1701–1702.

———. 2002. Funding, objectivity and the socialization of medical research. *Science and Engineering Ethics* 8 (3): 295–308.

Caron, M. M., S.B. Dohan, M. Barnes, and B.E. Bierer. 2023. Defining "recklessness" in research misconduct proceedings. *Accountability in Research*, published online 5 September 2023. Available at https://doi.org/10.1080/08989621.2023.2256650. Accessed 3 Sept 2024.

Committee on Publication Ethics. 2024. Allegations of misconduct. Available at https://publicationethics.org/misconduct. Accessed 2 Sept 2024.

Coombes, K.R., J. Wang, and K.A. Baggerly. 2007. Microarrays: Retracing steps. *Nature Medicine* 13 (11): 1276–1277.

Cooper, R.J., M. Gupta, M.S. Wilkes, and J.R. Hoffman. 2006. Conflict of interest disclosure policies and practices in peer-reviewed biomedical journals. *Journal of General Internal Medicine* 21 (12): 1248–1252.

Coors, M.E., K.M. Raymond, S.K. McWilliams, C.J. Hopfer, and S.K. Mikulich-Gilbertson. 2015. What adolescents enrolled in genomic addiction research want to know about conflicts of interest. *Drug and Alcohol Dependence* 147: 272–275.

DeAngelis, C.D. 2000. Conflict of interest and the public trust. *Journal of the American Medical Association* 284 (17): 2237–2238.

DeAngelis, C.D., and P.B. Fontanarosa. 2008. Impugning the integrity of medical science: The adverse effects of industry influence. *Journal of the American Medical Association* 299 (15): 1833–1835.

Department of Health and Human Services. 2016. Clinical trials registration and results information submission. *Federal Register* 81 (183): 64982–65157.

———. 2024. Public Health Service policies on research misconduct. 42 CFR 93.

Dhingra, D., and D. Mishra. 2014. Publication misconduct among medical professionals in India. *Indian Journal of Medical Ethics* 11 (2): 104–117.

Elali, F.R., and L.N. Rachid. 2023. AI-generated research paper fabrication and plagiarism in the scientific community. *Patterns* 4 (3): 10076.

Elliott, K.C. 2008. Scientific judgment and the limits of conflict-of-interest policies. *Accountability in Research* 15 (1): 1–29.

Else, H. 2023. Multimillion dollar trade in paper authorships alarms publishers. *Nature* 613: 617–618.

Fanelli, D. 2009. How many scientists fabricate and falsify research? A systematic review and meta-analysis of survey data. *PLoS One* 4 (5): e5738.

Fang, F.C., R.G. Steen, and A. Casadevall. 2012. Misconduct accounts for the majority of retracted scientific publications. *Proceedings of the National Academy of Sciences USA* 109 (42): 17028–17033.
Friedberg, M., B. Saffran, T.J. Stinson, W. Nelson, and C.L. Bennett. 1999. Evaluation of conflict of interest in economic analyses of new drugs used in oncology. *Journal of the American Medical Association* 282 (15): 1453–1457.
Friedman, L.S., and M. Friedman. 2016. Financial conflicts of interest and study results in environmental and occupational health research. *Journal of Occupational and Environmental Medicine* 58 (3): 238–247.
Friedman, L.S., and E.D. Richter. 2004. Relationship between conflicts of interest and research results. *Journal of General Internal Medicine* 19 (1): 51–46.
George, S.L., and D. Buyse. 2015. Data fraud in clinical trials. *Clinical Investigation* 5 (2): 161–173.
Grady, C., E. Horstmann, J.S. Sussman, and S.C. Hull. 2006. The limits of disclosure: What research subjects want to know about investigator financial interests. *Journal of Law, Medicine & Ethics* 34 (3): 592–599.
Hampson, L.A., J.E. Bekelman, and C.P. Gross. 2008. Empirical data on conflict of interest. In *The Oxford handbook of clinical research ethics*, ed. E.J. Emanuel, C. Grady, R.A. Crouch, R.K. Lie, F.G. Miller, and D. Wendler, 767–779. New York: Oxford University Press.
Institute of Medicine. 2009. *Conflict of interest in medical research, education, and practice*. Washington, DC: National Academies Press.
International Committee of Medical Journal Editors. 2024. *Clinical trials*. Available at https://www.icmje.org/recommendations/browse/publishing-and-editorial-issues/clinical-trial-registration.html. Accessed 12 Sept 2024.
Kafaee, M., M.T. Kheirkhah, R. Balali, and S. Gharibzadeh. 2022. Conflict of interest as a cognitive bias. *Accountability in Research* 29 (6): 379–396.
Katz, D., A.L. Caplan, and J.F. Merz. 2003. All gifts large and small: Toward an understanding of the ethics of pharmaceutical industry gift-giving. *American Journal of Bioethics* 3 (3): 39–46.
Kim, S.Y., R.W. Millard, P. Nisbet, C. Cox, and E.D. Caine. 2004. Potential research participants' views regarding researcher and institutional financial conflicts of interest. *Journal of Medical Ethics* 30 (1): 73–79.
Kim, J.H., R.S. Um, J.W. Lee, and O. Ajilore. 2024. Generative AI can fabricate advanced scientific visualizations: ethical implications and strategic mitigation framework. *AI and Ethics*, March 2024. Available at https://doi.org/10.1007/s43681-024-00439-0. Accessed 9 Sept 2024.
Kojaku, S., G. Livan, and N. Masuda. 2021. Detecting anomalous citation groups in journal networks. *Science Reports* 11: 14524.
Krimsky, S. 2003. *Science in the private interest—Has the lure of profits corrupted biomedical research?* Lanham: Rowman and Littlefield.
Lenzer, J. 2006. Researcher received undisclosed payments of 300,000 dollars from Pfizer. *BMJ* 333 (7581): 1237.
Lesser, L.I., C.B. Ebbeling, M. Goozner, D. Wypij, and D.S. Ludwig. 2007. Relationship between funding source and conclusion among nutrition-related scientific articles. *PLoS Medicine* 4 (1): e5.
Levinson, W., A. Kao, A.M. Kuby, and R.A. Thisted. 2005. The effect of physician disclosure of financial incentives on trust. *Archives of Internal Medicine* 165 (6): 625–630.
Liao, Q.J., Y.Y. Zhang, Y.C. Fan, M.H. Zheng, Y. Bai, G.D. Eslick, X.X. He, S.B. Zhang, H.H. Xia, and H. He. 2018. Perceptions of Chinese biomedical researchers towards academic misconduct: A comparison between 2015 and 2010. *Science and Engineering Ethics* 24 (2): 629–645.
Lundh, A., J. Lexchin, B. Mintzes, J.B. Schroll, and L. Bero. 2018. Sponsorship and research outcome: Systematic review with meta-analysis. *Intensive Care Medicine* 44 (10): 1603–1612.
Májovský, M., M. Černý, M. Kasal, M. Komarc, and D. Netuka. 2023. Artificial intelligence can generate fraudulent but authentic-looking scientific medical articles: Pandora's box has been opened. *Journal of Medicine and Internet Research* 25: e46924.

Marshall, E. 1990. Penn charges Retin-A inventor with a conflict. *Science* 247: 1028–1029.
———. 1992. Intellectual property. A new wrinkle in Retin-A dispute. *Science* 256 (5057): 607–608.
Martinson, B., M. Anderson, and R. De Vries. 2005. Scientists behaving badly. *Nature* 435: 737–738.
Michaels, D. 2008. *Doubt is their product: How industry's assault on science threatens your health*. New York, NY: Oxford University Press.
Miller, A., and M. Davis. 2011. *Intellectual property*. 4th ed. St. Paul: West Publishing.
Miller, F.G., D.L. Rosenstein, and E.G. DeRenzo. 1998. Professional integrity in clinical research. *Journal of the American Medical Association* 280 (16): 1449–1454.
Morin, K., H. Rakatansky, F.A. Riddick Jr., L.J. Morse, J.M. O'Bannon, M.S. Goldrich, P. Ray, M. Weiss, R.M. Sade, and M.A. Spillman. 2002. Managing conflicts of interest in the conduct of clinical trials. *Journal of the American Medical Association* 287 (1): 78–84.
Moses, H., and J.B. Martin. 2001. Academic relationships with industry: A new model for biomedical research. *Journal of the American Medical Association* 285 (7): 933–935.
National Academies of Science, Engineering, and Medicine. 2009. *On being a scientist*. 3rd ed. Washington, DC: National Academies Press.
———. 2017. *Fostering integrity in research*. Washington, DC: National Academies Press.
National Institutes of Health. 2024a. *Policy manual, non-compliance*. Available at https://policymanual.nih.gov/3014-001#N. Accessed 3 Sept 2024.
———. 2024b. *Financial conflict of interest*. Available at https://grants.nih.gov/grants/policy/coi/index.htm. Accessed 11 Sept 2024.
Office of Science. 2023. *Scientific integrity policy*. Available at https://www.whitehouse.gov/wp-content/uploads/2023/06/OSTP-SCIENTIFIC-INTEGRITY-POLICY.pdf. Accessed 12 Sept 2024.
Office of Science and Technology Policy. 2000. Federal policy on research misconduct. *Federal Register* 65 (235): 76260–76,264.
Okonta, P., and T. Rossouw. 2012. Prevalence of scientific misconduct among a group of researchers in Nigeria. *Developing World Bioethics* 13 (3): 149–157.
Olivieri, N.F. 2003. Patients' health or company profits? The commercialisation of academic research. *Science and Engineering Ethics* 9 (1): 29–41.
Potti, A., H.K. Dressman, A. Bild, R.F. Riedel, G. Chan, R. Sayer, J. Cragun, H. Cottrill, M.J. Kelley, R. Petersen, D. Harpole, J. Marks, A. Berchuck, G.S. Ginsburg, P. Febbo, J. Lancaster, and J.R. Nevins. 2006. Genomic signatures to guide the use of chemotherapeutics. *Nature Medicine* 12 (11): 1294–1300.
Radder, H. 2019. *From commodification to common good: Reconstructing science, technology, and society*. Pittsburgh: University of Pittsburgh Press.
Resnik, D.B. 2003. *Owning the genome: A moral analysis of DNA patenting*. Albany: SUNY Press.
———. 2004. Disclosing conflicts of interest to research subjects: An ethical and legal analysis. *Accountability in Research* 11 (2): 141–159.
———. 2007. *The price of truth: How money affects the norms of science*. New York, NY: Oxford University Press.
———. 2019a. Is it time to revise the definition of research misconduct? *Accountability in Research* 26 (2): 123–137.
———. 2019b. Institutional conflicts of interest in academic research. *Science and Engineering Ethics* 25 (6): 1661–1669.
———. 2023. Disclosing and managing non-financial conflicts of interest in scientific publications. *Research Ethics* 19 (2): 121–138.
Resnik, D.B., and K.C. Elliott. 2013. Taking financial relationships into account when assessing research. *Accountability in Research* 20 (3): 184–205.
Resnik, D.B., L.M. Rasmussen, and G.E. Kissling. 2015a. An international study of research misconduct policies. *Accountability in Research* 22 (5): 249–266.

Resnik, D.B., T. Neal, A. Raymond, and G.E. Kissling. 2015b. Research misconduct definitions adopted by U.S. research institutions. *Accountability in Research* 22 (1): 14–21.

Resnik, D.B., B. Konecny, and G.E. Kissling. 2017. Conflict of interest and funding disclosure policies of environmental, occupational, and public health journals. *Journal of Occupational and Environmental Medicine* 59 (1): 28–33.

Retraction Watch. 2024. *Leaderboard.* Available at https://retractionwatch.com/the-retraction-watch-leaderboard/. Accessed 3 Sept 2024.

Rivera, H., and J.A. Teixeira da Silva. 2021. Retractions, fake peer reviews, and paper mills. *Journal of Korean Medical Science* 36 (24): e165.

Sacco, D.F., M. Brown, and S.V. Bruton. 2019. Grounds for ambiguity: Justifiable bases for engaging in questionable research practices. *Science and Engineering Ethics* 25 (5): 1321–1337.

Sanderson, K. 2024. High profile effort will tackle paper mills. *Nature* 626: 17–18.

Shamoo, A.E., and D.B. Resnik. 2022. *Responsible conduct of research.* 4th ed. New York, NY: Oxford University Press.

Sismondo, S. 2008a. Pharmaceutical company funding and its consequences: A qualitative systematic review. *Contemporary Clinical Trials* 29 (2): 109–113.

Steen, R.G., A. Casadevall, and F.C. Fang. 2013. Why has the number of scientific retractions increased? *PLoS One* 8 (7): e68397.

Stelfox, H.T., G. Chua, K. O'Rourke, and A.S. Detsky. 1998. Conflict of interest in the debate over calcium-channel antagonists. *New England Journal of Medicine* 338 (2): 101–106.

Thompson, D.F. 1993. Understanding financial conflicts of interest. *New England Journal of Medicine* 329 (8): 573–576.

Titus, S.L., J.A. Well, and L.J. Rhoades. 2008. Repairing research integrity. *Nature* 453: 980–982.

Wise, J. 2013. The great pretender. *BMJ* 346: 16–17.

World Conference on Research Integrity. 2010. *Singapore statement.* Available at http://www.singaporestatement.org/statement.html. Accessed 3 Sept 2024.

Chapter 10
Conclusion

Abstract This chapter summarizes the arguments and conclusions developed in the book and discusses proposed reforms to the system of overseeing research with human subjects including mandating that sponsors of institutions provide fair compensation to subjects for research-related injuries; creating a national registry for healthy volunteers to protect them from harm and promote data integrity; harmonizing regulations from different federal agencies, including DHHS, FDA, and EPA; closing loopholes in the federal regulations; simplifying and streamlining the process for adopting reliance agreements between institutions that are collaborating on a research project; developing federal regulations or policies for participation of adults with impaired decision-making in research; providing additional logistical, financial, and educational support for IRBs; enhancing support for local participation on IRBs; enhancing public and community engagement with human subjects research; promoting evidence-based IRB decisions-making in research with subjects; and controlling IRB mission-creep. The chapter also discusses some areas for future research, such as the ethics of testing emerging technologies on human beings, such as human gene editing (germline and somatic); nanomedicine; stem cell therapy and regenerative medicine, robotics, bionics, neural implants, wearable devices, and machine-human interfaces; microbiome interventions, and artificial intelligence.

Keywords Ethics · Trust · Regulation · Law · Policy · Reform proposals · Future research

10.1 Summary of the Book

In this book, I have described and defended what I call a trust-based account of the ethics of research with human subjects. My approach is not radically different from standard views of the ethics of research with human subjects, except that it places

D. B. Resnik, *The Ethics of Research with Human Subjects*, The International Library of Bioethics 111, https://doi.org/10.1007/978-3-031-82757-0_10

greater emphasis on trust than other accounts. I view trust as part of the moral fiber of enterprise of ethical research with human subjects and as a principle that complements other principles, including respect for human dignity, non-maleficence, beneficence, and justice. The idea that trust is essential for ethical research with human subjects is hardly new. Indeed, hundreds of articles, books, and reports on the ethics of research with human subjects mention the importance of trust. What is new about my book is that it provides an in-depth analysis of trust in research with human participants by explaining what it is and why it is important, and showing how it can guide decision-making and policy formation.

In Chap. 1, I described some high-profile cases of research with human subjects that have generated considerable discussion and argued that they illustrate how moral predicaments involving research with human beings often boil down to a conflict between protecting individual rights/welfare and benefitting the public by advancing human knowledge. I argued that while most people agree that both values are important, ethical disagreements can arise with respect to priority-setting. I also suggested that a moral framework that includes fostering trust as a key component can help us set priorities when values conflict.

In Chap. 2, I set the stage for my trust-based approach by reviewing the history of research with human subjects and arguing that existing regulations and guidelines have arisen, for the most part, in response to abuses of the rights and welfare of human subjects and the need to restore and fortify the public's trust in research. I also discussed the role of regulation and public oversight in research with human subjects and reflected on the dilemma of under-protection vs. overprotection. I argued that this question involves the same type of value-conflict (i.e., individual vs. society) that was discussed in the opening chapter, because protecting human subjects from harm and exploitation often hinders the advancement of research.

In Chap. 3, I examined some prominent moral theories from the Western philosophical tradition to ascertain whether they provide a satisfactory approach to making decisions involving conflicting values. I argued that none of them perform this task perfectly although each one offers an important perspective on our moral experience. I defended an approach to moral decision-making based on four widely accepted principles of bioethics: respect for human dignity, non-maleficence, beneficence, and justice. I argued that these principles should be used in decision-making involving ethical dilemmas in research with human subjects because they capture key insights from different moral theories and are easy to apply to real-world cases.

In Chap. 4, I explicated my trust-based approach in greater depth. I reflected on the nature and importance of trust, distinguished between trust and verification, showed how trust plays a crucial role in research with human subjects, and discussed some strategies for promoting trust among the various stakeholders in involved in research, i.e. subjects, investigators, institutions, communities, sponsors, IRBs, and the public. I argued for a conceptually distinct principle of trust and showed how it complements the four other principles in decision-making. Although difficult questions may still arise when trust conflicts with other principles, the principle can augment ethical decision-making in research with human subjects.

In Chaps. 5, 6, 7, 8, and 9, I applied my trust-based approach to a multitude of controversial topics in the ethics and regulation of research with human subjects, including:

- Obtaining and documenting informed consent
- Determining whether a person has the capacity to consent
- Conducting research without consent
- Using general (or blanket) consent forms
- Using opt-out consent procedures
- Deceiving research subjects
- Paying people for research participation
- Safeguarding privacy and confidentiality
- Sharing data and biospecimens
- Assessing and balancing risks and benefits
- Minimizing risks
- Defining minimal risk
- Choosing the appropriate clinical trial design (e.g., whether to use a placebo control group)
- Conducting challenge studies on healthy volunteers
- Upper limits for risks to healthy volunteers
- Dealing with risks to third parties, communities, and society
- Sharing individualized results with research participants
- Sharing research benefits with subjects, communities, and populations
- Providing ancillary care to subjects and post-trial access to medications
- Enrolling and protecting vulnerable subjects in research
- Issues with special populations, such children, pregnant people, and adults with impaired decision-making capacity
- Promoting research integrity
- Investigating and reporting misconduct and non-compliance
- Disclosing and managing conflicts of interest in research

In my discussion of these topics, I explored connections between trust and other key concepts, such as consent, privacy, risk management, integrity, regulatory oversight, and public and community engagement, and I argued that reflecting on the importance of trust can help us address ethical issues and dilemmas that arise in research.

In the remainder of this book, I will discuss some proposals for reforming the system of ethical and regulatory oversight of research with human subjects and topics that merit further study.

10.2 Reform Proposals

Compensate Subjects for Research-Related Injuries As discussed in Chap. 7.18, since the 1970s, numerous commentators and organizations have recommended that institutions and/or sponsors be legally required to compensate participants for research-related injuries. Although many European countries and some African countries have laws that require compensation for research-related injuries, the U.S. does not. Many ethicists and researchers have argued that U.S. federal regulations should be changed to include language mandating that institutions or sponsors provide compensation for research-related injuries. Some have proposed no-fault compensation systems similar to the National Vaccine Injury Compensation Program. Without such a change to the regulations, compensation, if it occurs at all, will be haphazard and distributed unfairly (see Pike 2012).

Create a National Registry for Healthy Volunteers Several commentators have argued that the U.S. should have a national registry of healthy volunteer participants in clinical research. Some European countries have such registries, but the U.S. does not. Research has shown that many healthy volunteers view research participation as a key source of income and may enroll in dozens of Phase I trials each year (Fisher 2020). Most Phase I studies exclude subjects who have participated in another Phase I trial within a particular period, typically 30 days. Exclusion of these subjects serves to protect them from harm from drug interactions and to minimize confounding factors that could undermine the integrity of the study data. Registries help to prevent multiple or overlapping enrollment without the investigator's knowledge or consent. (Resnik and Koski 2011).

Harmonize U.S. Federal Regulations As we saw in Chap. 2, there are differences among federal agencies with respect to protecting research subjects. Some key differences include:

- While 21 federal agencies have adopted Subpart A, only DHHS and FDA have adopted Subpart D, and only DHHS has adopted Subparts B and C;
- The FDA includes special rules for waiving informed consent for emergency medical research, but the Common Rule does not include such provisions;
- The EPA regulations include special protections for children and pregnant and nursing women which go beyond the protections found in the Common Rule's protections for vulnerable subjects in Subparts B and D;
- The Common Rule and FDA regulations have different definitions of research involving human subjects.

The lack of harmonization among different agency rules can create confusion among investigators and IRBs as they attempt to apply the regulations to studies covered by conflicting requirements. For example, if the EPA and the NIH collaborate on a study involving children, investigators would need to decide how to resolve potential conflicts between EPA and NIH rules concerning risks. Harmonization of federal regulations could improve the efficiency and consistency of human research review and oversight (Emanuel et al. 2004). While many commentators have argued that harmonization is important, it has proven to be difficult to achieve, due to political and institutional resistance. Some additional steps toward harmonization that could be taken without too much effort include:

- Requiring that all agencies adopt common definitions of 'minimal risk,' 'research,' 'human subject,' and other important terms;
- Requiring that all agencies adopt Subparts B, C, and D of the Common Rule;
- Requiring that all agencies adopt parts of Subpart A that cover IRB composition, functions, and responsibilities.

Close Loopholes in U.S. Federal Regulations In the U.S., not all research with human subjects is covered by the federal regulations. The federal research regulations cover human subjects research funded by federal agencies or submitted to the FDA or EPA for regulatory decision-making. However, the regulations do not cover privately funded research that is not submitted to the FDA or EPA. For example, Facebook's study (discussed in Chap. 1) would not have been covered by any research regulations if it had not included collaborators from academic institutions. Another important loophole is research on people who are newly dead or brain-dead (Wicclair and DeVita 2004; Walker et al. 2014). Since the federal regulations define a human subject as a "living individual" (45 CFR 46.102e) testing drugs or medical devices on a brain-dead body that is kept alive artificially is not covered by these rules. Other types of research not covered by the federal regulations include privately funded marketing studies, public opinion surveys, and non-FDA regulated consumer product testing. The Advisory Committee on the Human Radiation Experiments (1995) and the National Bioethics Advisory Commission (2001) recommended that the federal government close these loopholes and develop a national system for oversight of human subjects research. Others have made similar recommendations (Shamoo and Schwartz 2008). U.S. states, California and Maryland, have developed their own statutes to regulate human research not covered by federal law. Closing the gaps in the existing regulations would help to ensure that common ethical standards apply to all human subjects, regardless of the source of funding, state of residence, or the purpose of the research. It could also help to promote the public's trust in the research enterprise by providing oversight for studies not covered by the current regulatory framework. However, extending the federal research regulations would increase the government's regulatory reach and impose significant costs and burdens on researchers and businesses (Schneider 2015). A reasonable compromise might be to extend the regulations to apply only to more than minimal risk research not currently covered by the regulations and research on newly dead or brain-dead people. This type of extension would close important loopholes but without affecting a great of research done by private companies, such as opinion polls and consumer products testing and surveys.

Simplify and Streamline the Process for Adopting Reliance Agreements One of the major reforms to the Common Rule enacted in 2017 was the requirement that there be one IRB of record for U.S.-based multisite research, with some exceptions (45 CFR 46.114). The single-IRB mandate was adopted to prevent duplicative review of the same study and reduce unnecessary administrative burdens on investigators and IRBs (Emanuel and Menikoff 2011). Evidence indicates that the single-IRB mandate has not clearly reduced overall administrative burdens but has shifted them from investigators and IRBs members to IRB staff and legal counsel, who are responsible for negotiating and implementing reliance agreements, which are used when IRB is used for multisite research (Resnik et al. 2018; Lidz et al. 2018). Streamlining and simplifying this process by, for example, using common, OHRP-approved templates and forms and creating IRB consortia, could help to reduce these burdens.

Include Special Protections for Adults with Impaired Decision-Making in the Federal Regulations As discussed in Chap. 8, numerous commentators have recommended that the federal regulations should include additional protections for adults with impaired decision-making capacity (National Bioethics Advisory Commission 1998; Moreno et al. 1998; Yanos et al. 2009). Including special protections for adults with impaired DMC is important to ensure that institutions strike a reasonable balance between protecting adults with impaired DMC from risks and including them in research. In the absence of federal regulations, institutions may adopt policies that lead to over-protection or under-protection.

Develop and Refine Policies for Managing COIs in Research Review and Oversight (Klitzman 2011; Green and Rosenfeld 2023) As noted in Chap. 9, financial and non-financial COIs can threaten the integrity and trustworthiness of research. COIs also occur in the review and oversight of research with human subjects. Although the federal regulations prohibit IRB members from voting on studies for which they have a COI (45 CFR 46.107d), they do not define COIs. Additionally, institutional COIs may impact IRB deliberations. For example, private, for-profit IRBs have interests in generating revenue that conflict with their obligations to review protocols carefully and thoroughly. A private IRB may lose income if does not approve a study, requires numerous modifications for approval, or takes a long time to approve it. At universities, hospitals, or academic medical centers, institutional COIs may occur when a powerful person (such as a president, vice president, or department head) is an investigator on a protocol submitted to the IRB or when a human research study (such as a clinical trial) generates a great deal of income for the institution. Development and refinement of policies for handling these types of COIs can play an important role in ensuring the integrity and trustworthiness of research with human subjects.

Enhance Support for Local Participation in IRB Review and Oversight As mentioned in Chap. 2, IRBs are required to have at least one member who is a not a scientist and at least one member who is not affiliated with the institution (i.e., a local or community member). Often, the same person fills both roles (Klitzman 2012). Community members play an important role in review and oversight of research with human subjects because they can provide the board with information about the community's perspective on issues related to the consent process, consent documents, recruitment methods, survey instruments, and the reasonableness (or acceptability) of risks. And, as discussed in Chap. 4, local representation on the IRB is important for fostering trust between research institution and the community and holding the institution accountable to people in the community. However, because many IRBs have difficulty recruiting and retaining community members, very often there is only one community member attending the meeting (Klitzman 2012: Nicholls et al. 2023). Research indicates that while most community members feel they play an important role in research oversight, but that they also sometimes feel intimidated, and many would like to have better training for the work they do (Sengupta and Lo 2003; Anderson 2006; Nicholls et al. 2023). Increasing the proportion of community members on IRBs and improving educational support for the work they do is important for ensuring the ethical review of research studies (Nicholls et al. 2023). Furthermore, one of the critiques of the single-IRB mandate is that it eliminates local representation on the board and therefore may undermine meaningful consideration of the local context (e.g., community values, traditions, and knowledge) in research review (Klitzman et al. 2017; Guetterman et al. 2024). Therefore, taking steps to ensure that the local context is adequately incorporated into IRB decision-making is also important for enhancing support for the community's participation in research oversight.

Enhance Community and Public Engagement As discussed in Chap. 4, community and public engagement activities and processes, such as seminars, conferences, and lectures open to the public; townhall meetings; public forums; participation in local health fairs; collaboration with local clinics, hospital, and religious or civic organizations; community advisory boards, and local representation on IRBs are crucial building and maintaining good relationships among researchers, institutions, local communities, and the public. While many academic institutions are starting to incorporate community and public engagement activities into their human research protection programs (see example, Washington University 2022), this should become standard practice. However, it is unclear whether additional regulations are needed to achieve this goal, since making it a part of AAHRPP accreditation standards may be sufficient.

Enhance Support for Evidence-Based IRB Decision-Making As discussed in Chap. 7, IRB members often make decisions concerning the reasonableness of risks, the level of risk, the potential for benefits, and other important ethical issues based on intuitions and emotions and not empirical evidence. Although intuitions and emotions and feelings play an important role ethical decision-making, they can also lead to biased and inconsistent decision-making. Good decision-making in ethics incorporates logical reasoning, credible evidence, and sound intuition (Resnik 2017). Institutions should take steps to support evidence-based decision-making by IRBs, such as providing members with education and training to enhance their ethical and scientific knowledge related to matters under consideration and developing databases of past IRB decisions that can serve as a precedent for future decisions (Anderson and Dubois 2012; Resnik 2017; Seykora et al. 2021; Smith and Anderson 2022).

Guard Against IRB Mission Creep A frequently voiced criticism of IRBs is that they have taken on responsibilities that stray far away from the original mission, such as evaluating research design; assessing risks to third parties, communities, and society; and reviewing student projects and other activities that should not be considered research (Gunsalus et al. 2006; Schrag 2010; Schneider 2015; Waltz et al. 2023). Investigators and IRB members both complain about mission creep. Investigators are concerned that IRBs are interfering with their research and burdening them with unnecessary paperwork and IRB members are concerned about their increased workloads and involvement in making decisions about matters that are beyond their expertise. Some of the revisions to the Common Rule adopted in 2017 (and effective in 2018), such as clarifying the definition of research, expanding exemptions, and eliminating continuing review requirements for some types of low-risk studies, were designed to reduce administrative burdens and counteract some of the IRB's mission creep. However, some degree of mission creep may be unavoidable because the IRB is often the only organization at an institution that is able to address novel and emerging ethical and legal issues involving research with human subjects. Take, for example, risks to communities or populations. As discussed in Chap. 7, some types of research may produce information that could exacerbate stigma or discrimination against communities or populations. The Common Rule instruct the IRB to focus its risk assessment on the risks to the human subjects not on "possible long-range effects of applying knowledge gained in the research (45 CFR 46.111a2)." But if the IRB does not address these risks, who will? Investigators and institutions may fail to address these risks because of the have interests (professional or financial) in conducting the research and don't want to deal with anything that could interfere with it. The chief function of the IRB is to be an independent, disinterested body that can deal with these and other ethical issues in a fair and reasonable way (Emanuel et al. 2000). By emphasizing the importance of the IRB's role in providing independent oversight of research, I do not mean to imply that mission creep is acceptable, however, since it does raises serious issues. Institutions, sponsors, government agencies and other stakeholders in the research enterprise should continue to look for creative ways of limiting the IRB's responsibilities. For example, scientific review committees (discussed in Chap. 7) can relieve the IRB of some of its responsibilities for overseeing research design, and COI committees (discussed Chap. 8) can relieve the IRB of some of its responsibilities concerning COI review.

10.3 Topics for Future Research

Research on Trust in Human Subjects Research As discussed in Chap. 4, because trust is an essential part of the ethics of research with human subjects, more research is needed on:

- The psychological, social, cultural, and economic factors that can positively or negatively impact trust among participants, researchers, institutions, communities, and the public;
- How policies, procedures, and personnel affect trust;
- The relationship between trust, integrity, verification, ethics, compliance, consent, privacy, vulnerability, and other key concepts in human research ethics;
- Strategies for building, maintaining, and restoring trust.

Research That Supports the Quality of IRB Decision-Making More research is needed that supports that quality of IRB decision-making, including:

- Research on the risks of different tests and procedures;
- Research on psychological, social, economic, institutional, and factors that can positively or negatively impact informed consent;
- Research on privacy and confidentiality protections;
- Research on the return of individualized research results;
- Research on benefit-sharing in research;
- Research on community-engaged research;
- Research on the IRB decision-making process, such as the cognitive, emotive, and psychosocial aspects of IRB deliberations and the role of empirical evidence and intuition/emotion in decision-making (Anderson and Dubois 2012; Resnik 2021).

Risks and Benefits of Enrolling Pregnant Women in Clinical Research As discussed in Chap. 8, it is important to include pregnant women in clinical research to promote women's health. Although there is some disagreement about how inclusive of pregnant women clinical research should be, there is little debate about the importance of basing inclusion/exclusion decisions on sound scientific evidence concerning the risks and benefits of different drugs, tests, and interventions for the fetus and the woman. Unfortunately, this evidence is often lacking (Sewell et al. 2022). More scientific research is needed, therefore, to inform IRB decision-making and strike a reasonable balance between protecting the fetus and promoting women's health. However, research on pregnant women may involve a "Catch-22" conundrum[1] if one holds that (a) it is unethical to do experiments on pregnant women until one

[1] The phase "Catch 22," which comes from Jospeh Heller's novel of the same name, refers to a situation in which it is impossible to do something (A), because being able to do (A) depends on being able to do something else (B) and (B) depends on (A), so you are stuck in circle you can't break out of.

knows the risks of the experiments, but (b) to know these risks one must first perform the experiments (Bastian 2017). One can avoid this paradoxical situation by obtaining evidence about risks that does not involve enrolling pregnant women in experiments, such as research involving experiments on animals or non-pregnant humans, or non-experimental (i.e., observational) research on pregnant women.

Research on the Ethics of Testing Emerging Technologies on Human Beings including such technologies as:

- Human gene editing (somatic and germline)
- Nanomedicine
- Stem cell therapy and regenerative medicine
- Robotics, bionics, neural implants, wearable devices, and machine-human interfaces
- Microbiome interventions
- Artificial intelligence (see also below)

Research on the Ethics of Using Artificial Intelligence in Human Subjects Research Artificial intelligence (AI) is a rapidly evolving technology that is significantly transforming science and society (Resnik and Hosseini 2024). AI is being used in data analysis, writing, diagnostic imaging, imaging generation, music, art manufacturing, transportation, investing, advertising, education, business decision-making, and the list goes on. Using AI in human subjects research raises important issues that have not been discussed in-depth in the book (or hardly anywhere else for that matter; see Secretary's Advisory Committee on Human Research Protections. 2022; Resnik et al. 2024; Office of Human Research Protections 2024). Some of these include:

- Should human subjects be informed that their data may be uploaded into an AI system and possibly used to train the system? As noted in Chap. 5, must consent forms include check boxes for subjects to permit researchers to store and use their data and share it with other researchers for a variety of purposes, but it not clear whether checking the box covers uploading data into an AI system, especially AI system may not have the type of privacy and confidentiality protections found in other data processing systems.
- How can privacy and confidentiality be protected when data is uploaded into an AI system? What should human subjects be told about these protections?
- Is it acceptable for IRB members or support staff to use AI to assist with the review process? What sorts of precautions should be taken if this done?
- Do investigators have an obligation to minimize biases or errors resulting from the use of AI in research? How can they do this? Should issues relating to bias and error be addressed in research protocols and IRB applications? Should the IRB be concerned with the potential for bias when it reviews a research study using AI to analyze data?

- Should AI tools/systems (such as chat bots) be used to assist the informed consent process? Does AI-assisted consent undermine human-to-human consent and trust?

This concludes my book on the ethics and regulation of research with human subjects. I hope that it will be useful to readers and prompt further discussion, inquiry, and reflection.

References

Advisory Committee on Human Radiation Experiments. 1995. *Final report*. Washington, DC: U.S. Government Printing Office.

Anderson, E.E. 2006. A qualitative study of non-affiliated, non-scientist institutional review board members. *Accountability in Research* 13 (2): 135–155.

Anderson, E.E., and J.M. DuBois. 2012. IRB decision-making with imperfect knowledge: A framework for evidence-based research ethics review. *Journal of Law, Medicine & Ethics* 40 (4): 951–969.

Bastian, H. 2017. Catch-22, clinical trial edition: Protecting women and children. Absolutely Maybe, 30 October. Available at: https://absolutelymaybe.plos.org/2017/10/30/catch-22-clinical-trial-edition-protecting-women-and-children/. Accessed 18 Sept 2024.

Emanuel, E.J., and J. Menikoff. 2011. Reforming the regulations governing research with human subjects. *New England Journal of Medicine* 365 (12): 1145–1150.

Emanuel, E.J., D. Wendler, and C. Grady. 2000. What makes clinical research ethical? *Journal of the American Medical Association* 283 (20): 2701–2711.

Emanuel, E.J., A. Wood, A. Fleischman, A. Bowen, K.A. Getz, C. Grady, C. Levine, D.E. Hammerschmidt, R. Faden, L. Eckenwiler, C.T. Muse, and J. Sugarman. 2004. Oversight of human participants research: Identifying problems to evaluate reform proposals. *Annals of Internal Medicine* 141 (4): 282–291.

Fisher, J.A. 2020. *Adverse events: Race, inequality, and the testing of new pharmaceuticals*. New York: New York University Press.

Green, J.M., and S. Rosenfeld. 2023. Conflicts of interest in institutional review boards are a threat to ethical research. *Nature Medicine* 29 (11): 2701–2703.

Guetterman, T.C., A. Haggins, S. Montas, J. Black, D. Harney, M.D. Fetters, R. Silbergleit, and N.W. Dickert. 2024. Institutional review boards' assessment of local context: A mixed methods study. *Ethics and Human Research* 46 (1): 2–13.

Gunsalus, C.K., E.M. Bruner, N.C. Burbules, L. Dash, M. Finkin, J.P. Goldberg, W.T. Greenough, G.A. Miller, and M.G. Pratt. 2006. Mission creep in the IRB world. *Science* 312 (5779): 1441.

Klitzman, R. 2011. "Members of the same club": Challenges and decisions faced by US IRBs in identifying and managing conflicts of interest. *PLoS One* 6 (7): e22796.

———. 2012. Institutional review board community members: Who are they, what do they do, and whom do they represent? *Academic Medicine* 87 (7): 975–981.

Klitzman, R.L., E. Pivovarova, and C.W. Lidz. 2017. Single IRBs in multisite trials: Questions posed by the new NIH policy. *Journal of the American Medical Association* 317 (20): 2061–2062.

Lidz, C.W., E. Pivovarova, P. Appelbaum, D.F. Stiles, A. Murray, and R.L. Klitzman. 2018. Reliance agreements and single IRB review of multisite research: Concerns of IRB members and staff. *AJOB Empirical Bioethics* 9 (3): 164–172.

Moreno, J., A.L. Caplan, P.R. Wolpe, and the Members of the Project on Informed Consent, Human Research Ethics Group. 1998. Updating protections for human subjects involved in research. *Journal of the American Medical Association* 280 (22): 1951–1958.

National Bioethics Advisory Commission. 1998. *Research involving people with mental disorders that may affect decisionmaking capacity*. Washington, DC: National Bioethics Advisory Commission.

———. 2001. *Ethical and policy issues in research involving human participants*. Washington, DC: National Bioethics Advisory Commission.

Nicholls, S.G., H.A. Taylor, R. James, E.E. Anderson, P. Friesen, T. Schonfeld, and E.I. Summers. 2023. A cross sectional survey of recruitment practices, supports, and perceived roles for unaffiliated and non-scientist members of IRBs. *AJOB Empirical Bioethics* 14 (3): 174–184.

Office of Human Research Protections. 2024. Workshop: The evolving landscape of human research with AI—putting ethics into practice. Available at: https://www.hhs.gov/ohrp/education-and-outreach/exploratory-workshop/2024-workshop/index.html. Accessed 19 Sept 2024.

Pike, E.R. 2012. Recovering from research: A no-fault proposal to compensate injured research participants. *American Journal of Law and Medicine* 38 (1): 7–62.

Resnik, D.B. 2017. The role of intuition in risk/benefit decision-making in human subjects research. *Accountability in Research* 24 (1): 1–29.

———. 2021. Standards of evidence for institutional review board decision-making. *Accountability in Research* 28 (7): 428–455.

Resnik, D.B. and M. Hosseini. 2024. The ethics of using artificial intelligence in scientific research: New guidance needed for a new tool. *AI Ethics*, 27 May. Available at: https://doi.org/10.1007/s43681-024-00493-8. Accessed 19 Sept 2024.

Resnik, D.B., and G. Koski. 2011. A national registry for healthy volunteers in phase 1 clinical trials. *Journal of the American Medical Association* 305: 1236–1237.

Resnik, D.B., E.M. Smith, and M. Shi. 2018. How U.S. research institutions are responding to the single Institutional Review Board mandate. *Accountability in Research* 25 (6): 340–349.

Resnik, D.B., M. Hosseini, and N. Cobb. 2024. Artificial intelligence and the ethics of clinical research. *Bioethics Today*: 29. Available at: https://bioethicstoday.org/blog/artificial-intelligence-and-the-ethics-of-clinical-research/. Accessed 18 Sept 2024.

Schneider, C.E. 2015. *The Censor's hand: The misregulation of human-subject research*. Cambridge: MIT Press.

Schrag, Z.M. 2010. *Ethical imperialism: Institutional review boards and the social sciences, 1965–2009*. Baltimore: Johns Hopkins University Press.

Secretary's Advisory Committee on Human Research Protections. 2022. IRB considerations on the use of artificial intelligence in human subjects research. Available at: https://www.hhs.gov/ohrp/sachrp-committee/recommendations/irb-considerations-use-artificial-intelligence-human-subjects-research/index.html. Accessed 18 Sept 2024.

Sengupta, S., and B. Lo. 2003. The roles and experiences of nonaffiliated and non-scientist members of institutional review boards. *Academic Medicine* 78 (2): 212–228.

Sewell, C.A., S.M. Sheehan, M.S. Gill, L.M. Henry, C. Bucci-Rechtweg, C. Gyamfi-Bannerman, A.D. Lyerly, L.C. McKinney, K.P. Hatfield, G.R. Baer, L. Sahin, and C.P. Nguyen. 2022. Scientific, ethical, and legal considerations for the inclusion of pregnant people in clinical trials. *American Journal of Obstetrics and Gynecology* 227 (6): 805–811.

Seykora, A., C. Coleman, S.J. Rosenfeld, B.E. Bierer, and H.F. Lynch. 2021. Steps toward a system of IRB precedent: Piloting approaches to summarizing IRB decisions for future use. *Ethics Human Research* 43 (6): 2–18.

Shamoo, A.E., and J. Schwartz. 2008. Universal and uniform protections of human subjects in research. *American Journal of Bioethics* 8 (11): 3–5.

Smith, E., and E.E. Anderson. 2022. Reimagining IRB review to incorporate a clear and convincing standard of evidence. *Accountability in Research* 29 (1): 55–62.

Walker, R.L., E.T. Juengst, W. Whipple, and A.M. Davis. 2014. Genomic research with the newly dead: A crossroads for ethics and policy. *Journal of Law, Medicine, and Ethics* 42 (2): 220–231.

Waltz, M., J.A. Fisher, and R.L. Walker. 2023. Mission creep or mission lapse? Scientific review in research oversight. *AJOB Empirical Bioethics* 14 (1): 38–49.

Washington University. 2022. Community outreach. Available at: https://hrpo.wustl.edu/about-us/community-outreach/. Accessed 17 Sept 2024.

Wicclair, M.R., and M. DeVita. 2004. Oversight of research involving the dead. *Kennedy Institute of Ethics Journal* 14 (2): 143–164.

Yanos, P.T., B.S. Stanley, and C.S. Greene. 2009. Research risk for persons with psychiatric disorders: A decisional framework to meet the ethical challenge. *Psychiatric Services* 60 (3): 374–383.

Appendixes

Appendix A: Types of Studies Involving Human Subjects

There are many different types of studies involving human subjects conducted by investigators from a wide variety of disciplines using diverse methods (Nesbit 2004; Gallin and Ognibene 2017). See Table 1.

A fundamental distinctions concerning the types of human subjects research is the distinction between **observational** and **experimental** research. In observational research, the investigators do not control or manipulate the conditions of the study; in experimental research, they do.

The purposes of controlling the conditions of a study are to minimize bias and confounding factors and enhance rigor and reproducibility. For example, to study whether a new drug is effective at treating COVID-19, an investigator at a hospital could collect data on COVID-19 patients in their facility who receive the new drug as treatment for COVID-19 and compare them to patients who receive other forms of treatment. This would be an observational study, with no attempt to manipulate the conditions. However, this study is susceptible to various forms of bias, such as bias related to the selection of patients that receive the drug. If the patients who receive the drug are healthier than those who do not, this could affect the outcome of treatment. The placebo effect might also impact the results if patients' belief that they are receiving an effective treatment affects their response to the drug. To address these and other issues, an investigator could perform an experiment in which they randomly assign COVID-19 patient to receive the new drug, other forms of treatment, or perhaps no treatment (i.e., a placebo). The investigator could also blind the study by preventing the subjects and the researchers from knowing who is receiving or the new drug as opposed to other forms or treatment or a placebo.

Observational studies usually, but not always, impose fewer risks on subjects than experiments. In the example above, the observational study of COVID-19

D. B. Resnik, *The Ethics of Research with Human Subjects*, The International Library of Bioethics 111, https://doi.org/10.1007/978-3-031-82757-0

Table 1 Types of research and examples

Observational	Experimental
Prospective cohort study: following a group of people over time and comparing them to a control group For example, investigators collected data on a group of infants who were fed breast milk, over a several-year period, to determine whether breast feeding is associated with protection against infections requiring hospitalization (Christensen et al. 2020).	**Clinical trials** **Phase I: small studies (n = 50 or so), usually performed on healthy subjects, to collect data on the safety of an intervention and establish safe dosing levels.** **Phase II: larger studies (n = about 300), performed on patients with a disease, to test the effectiveness and safety of an intervention.** **Phase III: larger studies (n = 1000 to 3000), performed on patients with a disease, to test the effectiveness and safety of an intervention.** **Phase IV: studies conducted after a product has been approved to collect additional safety and efficacy data.** For example, investigators compared the safety and efficacy administering a short course of zidovudine to prevent mother-child transmission of HIV as compared to placebo-control groups (Dabis et al. 1999).
Retrospective case-control study: comparing a group people with a disease (cases) to group of people who are similar except they do not have the disease. For example, investigators compared patients with amyotrophic lateral sclerosis (ALS) to a matched control group to determine with ALS is associated with genetic and environmental factors (Savettieri et al. 1991).	**Challenge studies, i.e., challenging human physiology, metabolism, or immunity to obtain an experimental result.** For example, investigators exposed subjects to diesel exhaust to understand how it affects brain function (Gawryluk et al. 2023)
Cross-sectional study: collecting data on one a group at one time point. For example, researchers collected data on patients who were treated for self-harm at a hospital to identify factors associated with self-harm (Breet et al. 2018).	**Socio-behavioral experiments: experiments that manipulate the human subject to learn more about human behavior and psychology.** For example, an investigator conducted experiments to determine the degree to which subjects are willing to follow order given by an authority figure (Milgram 1974).
Surveys, interviews, questionnaires, focus groups: Studies that gather data on human oral or written responses to questions. For example, researchers surveyed adolescents about the effects experiencing or witnessing domestic violence (Lepistö et al. 2010).	**Public health experiments: experiments that test the effectiveness of a public health intervention.** Investigators conducted experiments to determine whether mask-wearing reduces transmission of respiratory diseases (Brooks and Butler 2021)

(continued)

Table 1 (continued)

Observational	Experimental
Ethnography: studies of human culture in natural settings, such as works sites, sporting events, or religious services. Researchers spent several months observing and talking to emergency department employees to better understand the culture of the department (Person et al. 2013)	**Quality improvement experiments: experiments that test the effectiveness of a intervention designed to improve the quality of a clinical process or service.** For example, investigators conducted an experimental to determine whether using a checklist reduces IV-catheter infections (Pronovost et al. 2006).
Secondary analysis of data or samples: Studies that analyze data or samples collected for another purpose. Researchers analyzed data collected as part of large cohort of subjects who provided information for a health and nutrition study to understand the relationship between alcohol and drug abuse and deafness or being hard of hearing (Anderson et al. 2018).	**Educational Experiments: experiments that test the effectiveness of an education intervention, such as a pedagogical method or technique.** Researchers conducted an experiment to assess the effectiveness of ethics education on ethical knowledge and reasoning (May and Luth 2013).

would not be exposing subjects to risk that they would not have otherwise been exposed to, since the investigators are just collecting data on COVID-19 patients in their hospital. In the experimental study, subjects would be exposed to risks that would not have been exposed if they were not in the study, such as the risk of receiving the new treatment or no treatment. Some observational studies, such as surveys, impose almost no risks on subjects.

Appendix B: Human Subjects Research Decision Tree Under the Common Rule (45 CFR 46)

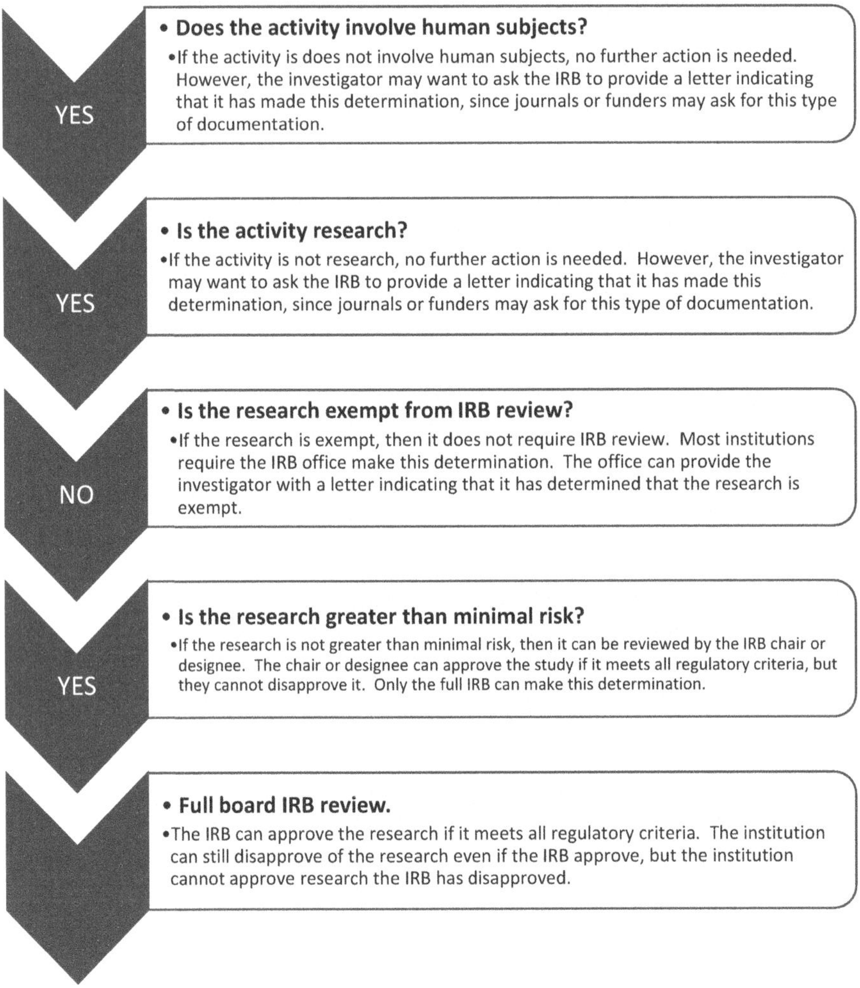

Definitions

Human Subject

45 CFR 46.102

(e)(1) Human subject means a living individual about whom an investigator (whether professional or student) conducting research:

(i) Obtains information or biospecimens through intervention or interaction with the individual, and uses, studies, or analyzes the information or biospecimens; or

(ii) Obtains, uses, studies, analyzes, or generates identifiable private information or identifiable biospecimens.

(2) Intervention includes both physical procedures by which information or biospecimens are gathered (e.g., venipuncture) and manipulations of the subject or the subject's environment that are performed for research purposes.
(3) Interaction includes communication or interpersonal contact between investigator and subject.
(4) Private information includes information about behavior that occurs in a context in which an individual can reasonably expect that no observation or recording is taking place, and information that has been provided for specific purposes by an individual and that the individual can reasonably expect will not be made public (e.g., a medical record).
(5) Identifiable private information is private information for which the identity of the subject is or may readily be ascertained by the investigator or associated with the information.
(6) An identifiable biospecimen is a biospecimen for which the identity of the subject is or may readily be ascertained by the investigator or associated with the biospecimen.

Research
45 CFR 46.102

(l) Research means a systematic investigation, including research development, testing, and evaluation, designed to develop or contribute to generalizable knowledge. Activities that meet this definition constitute research for purposes of this policy, whether or not they are conducted or supported under a program that is considered research for other purposes. For example, some demonstration and service programs may include research activities. For purposes of this part, the following activities are deemed not to be research:

1. Scholarly and journalistic activities (e.g., oral history, journalism, biography, literary criticism, legal research, and historical scholarship), including the collection and use of information, that focus directly on the specific individuals about whom the information is collected.
2. Public health surveillance activities, including the collection and testing of information or biospecimens, conducted, supported, requested, ordered, required, or authorized by a public health authority. Such activities are limited to those necessary to allow a public health authority to identify, monitor, assess, or investigate potential public health signals, onsets of disease outbreaks, or conditions of public health importance (including trends, signals, risk factors, patterns in diseases, or increases in injuries from using consumer products). Such activities include those associated with providing timely situ-

ational awareness and priority setting during the course of an event or crisis that threatens public health (including natural or man-made disasters).

3. Collection and analysis of information, biospecimens, or records by or for a criminal justice agency for activities authorized by law or court order solely for criminal justice or criminal investigative purposes.
4. Authorized operational activities (as determined by each agency) in support of intelligence, homeland security, defense, or other national security missions.

Minimal Risk
45 CFR 46.102

(j) Minimal risk means that the probability and magnitude of harm or discomfort anticipated in the research are not greater in and of themselves than those ordinarily encountered in daily life or during the performance of routine physical or psychological examinations or tests.

Exempt Categories
45 CFR 46.104

(d) Except as described in paragraph (a) of this section, the following categories of human subjects research are exempt from this policy:

(1) Research, conducted in established or commonly accepted educational settings, that specifically involves normal educational practices that are not likely to adversely impact students' opportunity to learn required educational content or the assessment of educators who provide instruction. This includes most research on regular and special education instructional strategies, and research on the effectiveness of or the comparison among instructional techniques, curricula, or classroom management methods.
(2) Research that only includes interactions involving educational tests (cognitive, diagnostic, aptitude, achievement), survey procedures, interview procedures, or observation of public behavior (including visual or auditory recording) if at least one of the following criteria is met:

(i) The information obtained is recorded by the investigator in such a manner that the identity of the human subjects cannot readily be ascertained, directly or through identifiers linked to the subjects;
(ii) Any disclosure of the human subjects' responses outside the research would not reasonably place the subjects at risk of criminal or civil liability or be damaging to the subjects' financial standing, employability, educational advancement, or reputation; or
(iii) The information obtained is recorded by the investigator in such a manner that the identity of the human subjects can readily be ascertained, directly or through identifiers linked to the subjects, and an IRB conducts a limited IRB review to make the determination required by §46.111(a)(7).

(3) (i) Research involving benign behavioral interventions in conjunction with the collection of information from an adult subject through verbal or written responses (including data entry) or audiovisual recording if the subject prospectively agrees to the intervention and information collection and at least one of the following criteria is met:

(A) The information obtained is recorded by the investigator in such a manner that the identity of the human subjects cannot readily be ascertained, directly or through identifiers linked to the subjects;
(B) Any disclosure of the human subjects' responses outside the research would not reasonably place the subjects at risk of criminal or civil liability or be damaging to the subjects' financial standing, employability, educational advancement, or reputation; or
(C) The information obtained is recorded by the investigator in such a manner that the identity of the human subjects can readily be ascertained, directly or through identifiers linked to the subjects, and an IRB conducts a limited IRB review to make the determination required by §46.111(a)(7).

(ii) For the purpose of this provision, benign behavioral interventions are brief in duration, harmless, painless, not physically invasive, not likely to have a significant adverse lasting impact on the subjects, and the investigator has no reason to think the subjects will find the interventions offensive or embarrassing. Provided all such criteria are met, examples of such benign behavioral interventions would include having the subjects play an online game, having them solve puzzles under various noise conditions, or having them decide how to allocate a nominal amount of received cash between themselves and someone else.
(iii) If the research involves deceiving the subjects regarding the nature or purposes of the research, this exemption is not applicable unless the subject authorizes the deception through a prospective agreement to participate in research in circumstances in which the subject is informed that he or she will be unaware of or misled regarding the nature or purposes of the research.
(4) Secondary research for which consent is not required: Secondary research uses of identifiable private information or identifiable biospecimens, if at least one of the following criteria is met:

(i) The identifiable private information or identifiable biospecimens are publicly available;
(ii) Information, which may include information about biospecimens, is recorded by the investigator in such a manner that the identity of the human subjects cannot readily be ascertained directly or through identifiers linked to the subjects, the investigator does not contact the subjects, and the investigator will not re-identify subjects;
(iii) The research involves only information collection and analysis involving the investigator's use of identifiable health information when that use is regulated under 45 CFR parts 160 and 164, subparts A and E, for the

purposes of "health care operations" or "research" as those terms are defined at 45 CFR 164.501 or for "public health activities and purposes" as described under 45 CFR 164.512(b); or
(iv) The research is conducted by, or on behalf of, a Federal department or agency using government-generated or government-collected information obtained for nonresearch activities, if the research generates identifiable private information that is or will be maintained on information technology that is subject to and in compliance with sect. 208(b) of the E-Government Act of 2002, 44 U.S.C. 3501 note, if all of the identifiable private information collected, used, or generated as part of the activity will be maintained in systems of records subject to the Privacy Act of 1974, 5 U.S.C. 552a, and, if applicable, the information used in the research was collected subject to the Paperwork Reduction Act of 1995, 44 U.S.C. 3501 et seq.

(5) Research and demonstration projects that are conducted or supported by a Federal department or agency, or otherwise subject to the approval of department or agency heads (or the approval of the heads of bureaus or other subordinate agencies that have been delegated authority to conduct the research and demonstration projects), and that are designed to study, evaluate, improve, or otherwise examine public benefit or service programs, including procedures for obtaining benefits or services under those programs, possible changes in or alternatives to those programs or procedures, or
possible changes in methods or levels of payment for benefits or services under those programs. Such projects include, but are not limited to, internal studies by Federal employees, and studies under contracts or consulting arrangements, cooperative agreements, or grants. Exempt projects also include waivers of otherwise mandatory requirements using authorities such as sections 1115 and 1115A of the Social Security Act, as amended.

(i) Each Federal department or agency conducting or supporting the research and demonstration projects must establish, on a publicly accessible Federal Web site or in such other manner as the department or agency head may determine, a list of the research and demonstration projects that the Federal department or agency conducts or supports under this provision. The research or demonstration project must be published on this list prior to commencing the research involving human subjects.
(ii) [Reserved]

(6) Taste and food quality evaluation and consumer acceptance studies:

(i) If wholesome foods without additives are consumed, or
(ii) If a food is consumed that contains a food ingredient at or below the level and for a use found to be safe, or agricultural chemical or environmental contaminant at or below the level found to be safe, by the Food and Drug Administration or approved by the Environmental Protection Agency or the Food Safety and Inspection Service of the U.S. Department of Agriculture.

(7) Storage or maintenance for secondary research for which broad consent is required: Storage or maintenance of identifiable private information or identifiable biospecimens for potential secondary research use if an IRB conducts a limited IRB review and makes the determinations required by §46.111(a)(8).
(8) Secondary research for which broad consent is required: Research involving the use of identifiable private information or identifiable biospecimens for secondary research use, if the following criteria are met:

(i) Broad consent for the storage, maintenance, and secondary research use of the identifiable private information or identifiable biospecimens was obtained in accordance with §46.116(a)(1) through (4), (a)(6), and (d);
(ii) Documentation of informed consent or waiver of documentation of consent was obtained in accordance with §46.117;
(iii) An IRB conducts a limited IRB review and makes the determination required by §46.111(a)(7) and makes the determination that the research to be conducted is within the scope of the broad consent referenced in paragraph (d)(8)(i) of this section; and
(iv) The investigator does not include returning individual research results to subjects as part of the study plan. This provision does not prevent an investigator from abiding by any legal requirements to return individual research results.

IRB Approval Criteria
45 CFR 46.111

(a) In order to approve research covered by this policy the IRB shall determine that all of the following requirements are satisfied:

(1) Risks to subjects are minimized:

(i) By using procedures that are consistent with sound research design and that do not unnecessarily expose subjects to risk, and
(ii) Whenever appropriate, by using procedures already being performed on the subjects for diagnostic or treatment purposes.

(2) Risks to subjects are reasonable in relation to anticipated benefits, if any, to subjects, and the importance of the knowledge that may reasonably be expected to result. In evaluating risks and benefits, the IRB should consider only those risks and benefits that may result from the research (as distinguished from risks and benefits of therapies subjects would receive even if not participating in the research). The IRB should not consider possible long-range effects of applying knowledge gained in the research (e.g., the possible effects of the research on public policy) as among those research risks that fall within the purview of its responsibility.
(3) Selection of subjects is equitable. In making this assessment the IRB should take into account the purposes of the research and the setting in which the research will be conducted. The IRB should be particularly cognizant of

the special problems of research that involves a category of subjects who are vulnerable to coercion or undue influence, such as children, prisoners, individuals with impaired decision-making capacity, or economically or educationally disadvantaged persons.

(4) Informed consent will be sought from each prospective subject or the subject's legally authorized representative, in accordance with, and to the extent required by, §46.116.

(5) Informed consent will be appropriately documented or appropriately waived in accordance with §46.117.

(6) When appropriate, the research plan makes adequate provision for monitoring the data collected to ensure the safety of subjects.

(7) When appropriate, there are adequate provisions to protect the privacy of subjects and to maintain the confidentiality of data.

(i) The Secretary of HHS will, after consultation with the Office of Management and Budget's privacy office and other Federal departments and agencies that have adopted this policy, issue guidance to assist IRBs in assessing what provisions are adequate to protect the privacy of subjects and to maintain the confidentiality of data.

(ii) [Reserved.]

(8) For purposes of conducting the limited IRB review required by §46.104(d)(7)), the IRB need not make the determinations at paragraphs (a)(1) through (7) of this section, and shall make the following determinations:

(i) Broad consent for storage, maintenance, and secondary research use of identifiable private information or identifiable biospecimens is obtained in accordance with the requirements of §46.116(a)(1)–(4), (a)(6), and (d);

(ii) Broad consent is appropriately documented or waiver of documentation is appropriate, in accordance with §46.117; and

(iii) If there is a change made for research purposes in the way the identifiable private information or identifiable biospecimens are stored or maintained, there are adequate provisions to protect the privacy of subjects and to maintain the confidentiality of data.

(b) When some or all of the subjects are likely to be vulnerable to coercion or undue influence, such as children, prisoners, individuals with impaired decision-making capacity, or economically or educationally disadvantaged persons, additional safeguards have been included in the study to protect the rights and welfare of these subjects.

Appendix C: IRB Worksheet

Research design and implementation	Is the research well-designed and rigorous? Are the aims and objectives clearly defined? Are the methods, procedures, tests, and interventions appropriate and valid? Is the study statistically well-designed? Are the personnel well-qualified? Have the investigators done an adequate review of the prior literature to provide a context for the study? Are there sufficient institutional and financial resources to conduct and complete the study? Has the study undergone prior scientific review? Are there clearly defined plans to publish and share the data and results? Are there any conflicts of interest that could impact the conduct of the study?
Risks	What are the risks of the study to the subjects and identifiable individuals or communities, including medical, psychosocial, cultural, legal, and financial risks? Risks may include physical or emotional harm, injury, pain, suffering, discrimination, or bias. Are the risks not greater than minimal or greater than minimal? What are the risks of particular tests, interventions, or procedures? What are the cumulative risks? Are there any societal risks that need to be addressed?
Risk minimization	What steps are being taken to minimize risks? Can additional measures be taken to minimize risks that do not compromise design and implementation of the study? Is there adequate monitoring of data and safety? Does the study require clinical or other monitoring of individual safety? Follow-up after the study is completed? Are there adequate institution resources (such as access to medical care) if subjects are significantly harmed? Will subjects be compensated or treated for research injuries?
Benefits	Does the study offer important benefits to science, participants, affected communities or populations, or society? Are there measures that can be taken to enhance benefits (such as providing some form of ancillary care or clinically useful information) that do not compromise the research design and implementation?
Risk/benefits	Are the risks of the study justified (or reasonable or proportional) in relation to the potential benefits? Is the distribution of risks and benefits fair?
Consent process	Will informed consent be sought? Are there are any circumstances or conditions, such as those related recruitment, enrollment, or financial incentives, that could undermine the validity of informed consent? Will the participants be provided with the information that one would need to make a well-reasoned decision to participate in the study? Is there additional information that a reasonable person would want to know? Are the investigators seeking a waiver of consent requirements? Is a waiver justified? Are the subjects capable of providing consent or will legally authorized representative provide consent? How will consent capability be assessed? Are interpreters or representatives needed to assist with the consent process?

(continued)

Consent documentation	Will consent be documented? How will it be documented? Does the consent form (and other documents) need to be translated into other languages? Is the consent form clearly written in language understandable to the subjects? Does the consent form include technical terms or complex sentences that need to be rewritten? Are the investigators asking for a waiver of documentation? Is a waiver justified?
Privacy/ confidentiality	Are there adequate protections in place to protect the privacy of the subjects and the confidentiality and security of their data and biospecimens?
Equitable subject selection	Is the subject selection fair and equitable? Are the inclusion and exclusion criteria for the study scientifically and ethically appropriate? Are some categories of subjects unfairly excluded?
Vulnerable subjects	Will the study include subjects who are likely to be vulnerable to exploitation, manipulation, or harm? What steps are being taken to protect these subjects? Are additional steps needed?

References

Anderson, M.L., B.H. Chang, and N. Kini. 2018. Alcohol and drug use among deaf and hard-of-hearing individuals: A secondary analysis of NHANES 2013–2014. *Substance Abuse* 39 (3): 390–397.

Breet, E., J. Bantjes, and I. Lewis. 2018. Substance use and self-harm: A cross-sectional study of the prevalence, correlates and patterns of medical service utilisation among patients admitted to a South African hospital. *BMC Health Services Research* 18 (1): 157.

Brooks, J.T., and J.C. Butler. 2021. Effectiveness of mask wearing to control community spread of SARS-CoV-2. *Journal of the American Medical Association* 325 (10): 998–999.

Christensen, N., S. Bruun, J. Søndergaard, H.T. Christesen, N. Fisker, G. Zachariassen, P.T. Sangild, and S. Husby. 2020. Breastfeeding and infections in early childhood: A cohort study. *Pediatrics* 146 (5): e20191892.

Dabis, F., P. Msellati, N. Meda, C. Welffens-Ekra, B. You, O. Manigart, V. Leroy, A. Simonon, M. Cartoux, P. Combe, A. Ouangré, R. Ramon, O. Ky-Zerbo, C. Montcho, R. Salamon, C. Rouzioux, P. Van de Perre, and L. Mandelbrot. 1999. 6-month efficacy, tolerance, and acceptability of a short regimen of oral zidovudine to reduce vertical transmission of HIV in breastfed children in Côte d'Ivoire and Burkina Faso: A double-blind placebo-controlled multicentre trial. DITRAME Study Group. DIminution de la Transmission Mère-Enfant. *Lancet* 353 (9155): 786 792.

Gallin, J.I., and F.P. Ognibene. 2017. *Principles and practice of clinical research*. 3rd ed. Amsterdam: Elsevier.

Gawryluk, J.R., D.J. Palombo, J. Curran, A. Parker, and C. Carlsten. 2023. Brief diesel exhaust exposure acutely impairs functional brain connectivity in humans: A randomized controlled crossover study. *Environmental Health* 22 (1): 7.

Lepistö, S., P. Astedt-Kurki, K. Joronen, T. Luukkaala, and E. Paavilainen. 2010. Adolescents' experiences of coping with domestic violence. *Journal of Advances in Nursing* 66 (6): 1232–1245.

May, D.R., and M.T. Luth. 2013. The effectiveness of ethics education: A quasi-experimental field study. *Science and Engineering Ethics* 19 (2): 545–568.

Milgram, S. 1974. Obedience to authority. New York: Harper and Rowe.

Nesbit, L.A., ed. 2004. *Clinical research: What it is and how it works*. Sudbury: Jones and Bartlett.

Person, J., L. Spiva, and P. Hart. 2013. The culture of an emergency department: An ethnographic study. *International Emergency Nursing* 21 (4): 222–227.

D. B. Resnik, *The Ethics of Research with Human Subjects*, The International Library of Bioethics 111, https://doi.org/10.1007/978-3-031-82757-0

Pronovost, P., D. Needham, S. Berenholtz, D. Sinopoli, H. Chu, S. Cosgrove, B. Sexton, R. Hyzy, R. Welsh, G. Roth, J. Bander, J. Kepros, and C. Goeschel. 2006. An intervention to decrease catheter-related bloodstream infections in the ICU. *New England Journal of Medicine* 355 (26): 2725–2732.

Savettieri, G., G. Salemi, A. Arcara, M. Cassata, M.G. Castiglione, and B. Fierro. 1991. A case-control study of amyotrophic lateral sclerosis. *Neuroepidemiology* 10 (5–6): 242–245.

Index

D. B. Resnik, *The Ethics of Research with Human Subjects*, The International Library of Bioethics 111, https://doi.org/10.1007/978-3-031-82757-0

The manufacturer's authorised representative in the EU is Springer
Nature Customer Service Centre GmbH, Europaplatz 3, 69115 Heidelberg,
Germany. If you have any concerns regarding our products, please
contact ProductSafety@springernature.com

Printed and bound by CPI Group (UK) Ltd, Croydon, CR0 4YY
01/10/2026
02225950-0014